DEVELOPMENTS IN IONIC POLYMERS—2

CONTENTS OF VOLUME 1

DEVELOPMENTS IN IONIC POLYMERS—2

Edited by

ALAN D. WILSON

and

HAVARD J. PROSSER

Laboratory of the Government Chemist, Department of Trade and Industry, London, UK

ELSEVIER APPLIED SCIENCE PUBLISHERS
LONDON and NEW YORK

ELSEVIER APPLIED SCIENCE PUBLISHERS LTD
Crown House, Linton Road, Barking, Essex IG11 8JU, England

Sole Distributor in the USA and Canada
ELSEVIER SCIENCE PUBLISHING CO., INC.
52 Vanderbilt Avenue, New York, NY 10017, USA

WITH 31 TABLES AND 100 ILLUSTRATIONS

© ELSEVIER APPLIED SCIENCE PUBLISHERS LTD 1986
© IMPERIAL CHEMICAL INDUSTRIES PLC 1986—Chapter 7
Softcover reprint of the hardcover 1st edition 1986

British Library Cataloguing in Publication Data

Developments in ionic polymers.
2
1. Addition polymerization—Periodicals
547.8'4   QD281.P6

The Library of Congress has cataloged this serial publication as
follows:

Developments in ionic polymers.—1– —London; New York:
  Applied Science Publishers, c1983–
    v.: ill.; 23 cm.—(The Developments series)
  ISSN 0264-7982 = Developments in ionic polymers.

  1. Polymers and polymerization—Collected works.   I. Series
  QD380.D478   547.8'4—dc19   84-645186

ISBN-13: 978-94-010-8360-7   e-ISBN-13: 978-94-009-4187-8
DOI: 10.1007/978-94-009-4187-8
The selection and presentation of material and the opinions expressed in this publication
are the sole responsibility of the authors concerned.

PREFACE

Ionic polymers, like elephants, are easier to recognise than to define. Several methods of classification have been attempted but none is wholly satisfactory because of the extreme diversity of ionic polymers, which range from the organic, water-soluble polyelectrolytes, through hydrogels and ionomer carboxylate rubbers, to the almost infusible inorganic silicate minerals. For this reason, a general classification is not only difficult, but has minimal utility.

However, there are some characteristics of these materials that should be highlighted. The role of counterions is the significant one. These ions, either singly or as clusters, take part in the formation of ionic bonds which have a varying structural role. Often they act as crosslinks, but in the halato-polymers the ionic bonds form an integral part of the polymer backbone itself. Conversely, in polymers containing covalent crosslinks, such as the ion-exchange resins, the counterions have virtually no structural role to play, since they dwell in cage-like structures without affecting the crosslinking, and are readily exchanged. They are, perhaps, best described as ion-containing polymers rather than structural ionic polymers.

Another crucial factor is the role of water in ionic polymers. The presence of ionic bonds means that there is a tendency for these materials to interact with water. Where the ionic polymer contains a high proportion of ionic units, it acts as a hydrogel and may be highly soluble. Such interactions with water decrease sharply as the ionic content is reduced, though even then water can act as a plasticiser.

This book is the second volume in the *Developments in Ionic Polymers* series, the basis of which may be regarded as Holliday's *Ionic Polymers,* published in 1975. Since then certain areas have remained static and require no further treatment, while others have advanced dramatically. The two volumes in the Developments Series are together concerned totally with the latter areas.

This volume commences with a very comprehensive review of preparative methods for ionic polymers, both anionic and cationic (Chapter 1). Chapters following describe various ionic polymer types in terms of constitution and properties.

Hybrid silicate–organic structures are recognised as being potentially of great importance—indeed, one hypothesis of the origin of life assigns that essential role to them—but they have tended to be neglected as ionic polymers until now. This situation is remedied in Chapter 2.

The review of carboxylated rubbers which appeared in *Ionic Polymers* is now updated in Chapter 3.

The majority of ionic polymers consist of an anionic polymer and a small cationic counterion. However, not all ionic polymers conform to this pattern. Ionenes, conversely, consist of a small anion and a polymeric cation. They have tended to be neglected in the past but recently they have excited much interest. They are the subject of Chapter 4, which is perhaps one of the most comprehensive reviews to have appeared in recent years.

Another class of ionic polymers comprises the so-called polyelectrolyte complexes, where both the cation and the anion are polymeric, and encompasses some very useful advanced materials; one may cite, for example, electrically conducting polymers which are at the forefront of developments in high technology. Since polyelectrolyte complexes are important biopolymers of the living cell, interpolymer complexes prepared in the laboratory have been used as models to aid elucidation of complex biopolymer functions in their native environment in the living cell. A comprehensive review of their properties is presented in Chapter 5.

The applications of ionic polymers have not been neglected in this volume. One of their most important applications is as membranes. Certain members of this class are characterised by a high chemical resistance and find particular application in production of materials, involving corrosive intermediates, by electrolytic techniques. This topic is fully covered in Chapter 6.

Polyelectrolytes have found increasing use in medicine, pharmacy and biology in the last decade or so. Some possess intrinsic biological activity while others which are inert are used as carriers for the sustained release of drugs. The subject of biomedical applications is treated in Chapter 7.

Another interesting field is the formation of ionic polymers by electrodeposition, a technique which is based on their fundamental characteristic. This technology finds application in the electrodeposition of coating systems and is described in Chapter 8.

Finally, it must be emphasised that the volume is not meant to provide an encyclopaedic coverage of the subject, nor does it even aim to cover all the latest developments. Rather, the object has been to present a selection of the most interesting contemporary developments. As such, it is essentially complementary to Volume 1 of the series and together the two volumes represent an updating of Holliday's earlier work on ionic polymers. For topics that have remained essentially unchanged, recourse should be made to Holliday's earlier work.

A. D. WILSON
H. J. PROSSER

CONTENTS

LIST OF CONTRIBUTORS

KOJI ABE

Department of Functional Polymer Sciences, Shinshu University, Ueda 386, Japan.

FRITZ BECK

Fachbereich 6—Elektrochemie, Universität Duisburg, (Gesamthochschule), Lotharstrasse 65, D-4100 Duisburg, Federal Republic of Germany.

PAULINE J. BROOKMAN

Laboratory of the Government Chemist, Department of Trade and Industry, Cornwall House, Waterloo Road, London SE1 8XY, UK.

F. G. HUTCHINSON

Imperial Chemical Industries plc, Pharmaceuticals Division, Mereside, Alderley Park, Macclesfield, Cheshire SK10 4TG, UK.

GERHARD LAGALY

Institut für anorganische Chemie der Universität Kiel, Olshausen-strasse 40, D-2300 Kiel, Federal Republic of Germany.

JOHN W. NICHOLSON

Laboratory of the Government Chemist, Department of Trade and Industry, Cornwall House, Waterloo Road, London SE1 8XY, UK.

M. PINERI

Commissariat à l'Energie Atomique, Institut de Recherches Fondamentales, Centre d'Etudes Nucléaires de Grenoble, Avenue des Martyrs, 85X 38041, Grenoble Cedex, France.

The late R. A. M. THOMSON

Department of Chemical and Physical Sciences, The Polytechnic, Queensgate, Huddersfield, West Yorkshire HD1 3DH, UK.

EISHUN TSUCHIDA

Department of Polymer Chemistry, Waseda University, Tokyo 160, Japan.

TETSUO TSUTSUI

Graduate School of Engineering Sciences, Kyushu University, 39, Kasuga-shi, Fukuoka 816, Japan.

Chapter 1

PREPARATION OF IONIC POLYMERS

The late R. A. M. THOMSON

Department of Chemical and Physical Sciences, The Polytechnic, Huddersfield, UK

1 INTRODUCTION

Ionic polymers have been defined[1] as organic or inorganic materials which contain both ionic and covalent bonds in their chain or network structure.

Methods of classification of such materials are generally based on the degree of ionic character and the degree of crosslinking of the polymer.[2,3] Increase in the former is associated with increase in electrolytic behaviour, although the counterion is of great importance in determining the nature of the material. Increase in extent of crosslinking is associated with reduction in solubility and in swelling character.

The present chapter will be restricted to a discussion of organic polymers displaying little, if any crosslinking, but a wide range of ionic character resulting from the presence of a range of numbers of charged or potentially charged sites per repeat unit. In most cases, the polymers under discussion are those which consist of water-soluble chains with charged groups pendent to the backbone, and associated with small counterions. Such materials are normally termed polyelectrolytes, and although they are of considerable commercial importance, relatively few comprehensive reviews of their formation have appeared in the literature.[4–15]

Organic polyelectrolytes may be subdivided into anionics, cationics and ampholytics, according to the sign of the charges carried, and each

1

class may be further subdivided according to the variability of the charge concentration with change in environment, particular pH.

The usual routes to such materials are:

(i) polymerization or copolymerization of charged monomers with suitable charged or uncharged comonomers either by addition or step-reaction,

(ii) suitable modification of uncharged preformed or other functionalized polymer or copolymer.

Route (i) may prove impracticable because of inability of the monomers to undergo suitable polymerization by virtue of charge or reactivity factors, or alternatively, the inability to copolymerize in the correct proportions. If the method is practical, however, it will normally yield consistent, well-defined products. Problems associated with polymer modification, route (ii), include difficulties in ensuring good mixing, temperature control and constant reaction conditions, particularly if the reacting system is viscous or heterogeneous—a common situation. Differences in the nature of reactant and product polymers in solubility, may pose problems and side reactions, such as chain degradation and crosslinking, may deleteriously affect the product. Neighbouring-group effects may also exert a significant influence.

The main types of polyelectrolyte discussed in this chapter will be:

(i) anionics, e.g. ions derived from carboxylic ($-COOH$), sulphonic ($-SO_3H$), phosphate ($-O \cdot P(O)(OH)_2$) and phosphonate ($-PH(O)(OH)$) groups;

(ii) cationics, e.g. protonated amines ($-\overset{+}{N}H_3X^-$), $-\overset{+}{N}RH_2X^-$ or $-\overset{+}{N}R_2HX^-$), quaternary ammonium compounds ($-\overset{+}{N}R_3X^-$), sulphonium ($-\overset{+}{S}R_3X^-$) and phosphonium salts, ($-\overset{+}{P}R_3X^-$), the latter two being relatively unimportant;

(iii) ampholytics, e.g. polyelectrolytes containing each of the above types of group.

In the above list, carboxylic and amino groups behave as 'weak' electrolytes; the remainder are effectively completely ionized at all pH values and behave as 'strong' electrolytes. In general, the former class displays complicated ionization behaviour. The polymer has a greater pK_a value than that of the corresponding monomer and it increases with increase in degree of ionization as explained later, i.e. with increase in pH in the case of the polyacids and decrease in pH in the case of the polybases. It also increases with the decrease in ionic strength of the system and the increase in base counterion radius.[16]

Polymerization and post-polymerization modification techniques for the formation of these classes of polymer will be discussed.

2 ANIONIC POLYELECTROLYTES

2.1 Carboxylic-based Polymers

2.1.1 Polymerization of Acrylic Acid (AA), Methacrylic Acid (MAA) and Their Salts

AA and MAA are both potentially hazardous in the pure, anhydrous state, since they are liable to polymerize spontaneously and explosively. Thus, they should be kept in that state as briefly as possible, and certainly stored above their melting points ($\sim$13°C). Use of a copper-gauze filled fractionation column is recommended in the reduced-pressure distillation of the inhibited monomers, together with other safety procedures.[17] Both polymers are insoluble in their monomers and in a wide range of organic solvents, from which they normally precipitate during preparation as fine powders. Solubility in water increases with temperature in the case of poly(acrylic acid), PAA, but decreases in the case of poly(methacrylic acid), PMAA. The monomers are soluble in a wide range of solvents.

Several reviews of the polymerization and copolymerization of AA and MAA have appeared in recent years.[18-24] Polymerization may be carried out under homogeneous or heterogeneous conditions in either aqueous or non-aqueous media. Free-radical initiation is the usual route employed. Conventional initiators, such as organic azo compounds[25] (often water-soluble),[26] inorganic and organic peroxides and related compounds, alone[27] or coupled with suitable reducing agents or catalytic ions, are usually effective.[28,29]

2.1.2 Solution Polymerization

Initiator systems have been reviewed recently by several authors[24,30,31] and although these are not specific to the polymerization of AA and MAA, there is little doubt that they will usually be effective.

The following are among systems recently reported for the initiation of polymerization of AA and MAA.

The use of potassium persulphate as initiator[32] for the polymerization of concentrated sodium methacrylate ($3 \cdot 15$–$4 \cdot 67 \, \text{mol dm}^{-3}$) solutions has indicated second-order termination with an initiator exponent of $0 \cdot 5$ but the rate of polymerization (R_p) was zero order in monomer. The overall activation energy appeared rather high ($91 \cdot 5 \, \text{kJ mol}^{-1}$).

A study of AA and MAA polymerization using the same initiator was reported by Venkatarao et al.[33] These workers deduced that initiation involves interaction between persulphate ions and monomer, which could be in either the ionized or un-ionized form, depending upon the solution pH:

$$\left. \begin{array}{l} S_2O_8^{2-} + M_1^- \rightarrow M_1^{\cdot r} + SO_4^{\cdot -} \\ S_2O_8^{2-} + M_1H \rightarrow M_1^{\cdot} + SO_4^{\cdot -} \end{array} \right\} \quad \textit{Initiation}$$

$$\left. \begin{array}{l} M_1^{\cdot} + M_1H \rightarrow M_2H^{\cdot} \\[6pt] M_1^{\cdot r} + M_1H \rightarrow M_2H^{\cdot r} \\[6pt] M_1^{\cdot} + M_1^- \rightarrow M_2^{\cdot r} \\[6pt] M_1^{\cdot r} + M_1^- \rightarrow M_2^{\cdot r} \end{array} \right\} \quad \textit{Propagation}$$

or

or

The values of $[M_1^-]$ and $[M_1H]$ thus control the overall rate of initiation and therefore R_p would be expected to display a dependence upon the pH of the system.

Ag^+ ions were found to catalyze the persulphate initiation, an effect ascribed to reactions of the type:

$$Ag^+ + S_2O_8^{2-} \rightarrow AgS_2O_8^-$$
$$Ag^+ + MH \rightarrow AgM + H^+$$
$$S_2O_8^{2-} + MH \rightarrow R^{\cdot} + SO_4^{\cdot -}$$
$$AgS_2O_8^- + MH \rightarrow R^{\cdot} + Ag^+ + SO_4^{\cdot -}$$
$$S_2O_8^{2-} + AgM \rightarrow R^{\cdot} + Ag^{\cdot} + SO_4^{\cdot -}$$
$$AgS_2O_8^- + AgM \rightarrow R^{\cdot} + 2Ag^+ + SO_4^{\cdot -}$$

This mechanism explains the relatively minor salt effects observed. For example, Cu^{2+} ions were found to retard the rate of polymerization. This effect was ascribed to oxidative transfer with the growing radicals, leading to reduced rate and degree of polymerization:

$$\text{\textasciitilde\textasciitilde CH}_2 - \dot{\text{C}}\text{HR} + M_{(\text{aq})}^{n+} \rightarrow \text{\textasciitilde\textasciitilde CH}{=}\text{CHR} + M^{(n-1)+} + H^+$$

The catalytic effect of Ag^+ ions was also observed[34] in the polymerization of MAA by potassium peroxodiphosphate in aqueous H_2SO_4, reducing the overall energy of activation from the already low value of $34 \cdot 5 \, kJ \, mol^{-1}$ to $19 \cdot 3 \, kJ \, mol^{-1}$.

An interesting procedure was reported[35] for controlling product degree of polymerization in systems using redox initiators in the presence of multivalent metal ions. Addition of a powerful ion-chelating agent such as EDTA was found to cause the cessation of polymerization, whereupon the monomer concentration could be increased before addition of further metal ions to re-start the initiation process.

Aldehydes such as butanal, pentanal and hexanal in dioxane[36] and benzil dimethyl ketal (**I**) in aqueous solution at high pH^{37} were

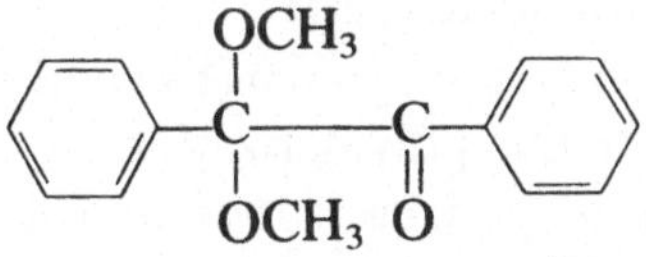

I

reported to be effective initiators. Photochemical decomposition of the latter in a mixture of acrylamide (ACM) and AA was claimed to give a more rapidly dissolving copolymer, which was considerably richer in AA units, than did **II**.

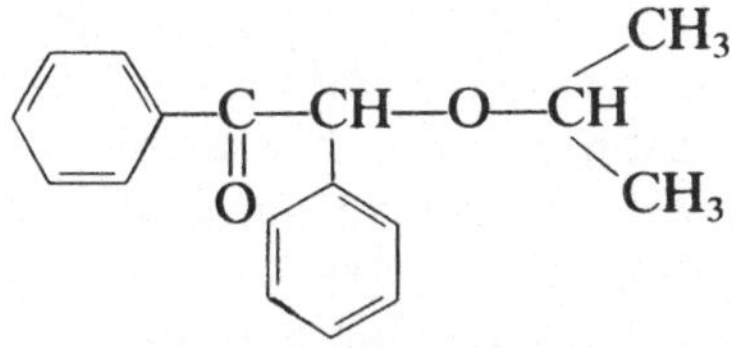

II

Nitrogen dioxide in tetrahydrofuran (THF) or 1,4-dioxane has been shown[38] to be an effective free-radical initiator (overall energy of activation $68 \, kJ \, mol^{-1}$). Peroxysilanes of the general structure **III**,

$$(CH_3)_{(4-n)}Si((CH_3)_3COO)_n$$

III

where $n = 1 - 3$, were used as initiators.[39] For $n = 3$, initiation with primary radicals formed by induced decomposition of the initiator yielded polymers containing peroxy groups. Poly(γ-

mercaptopropylsiloxane–lanthanide) complexes were also shown to initiate polymerization of AA by a radical mechanism.[40]

Manganese(III) has proved to be an efficient constituent of polymerization initiators. Manganese(III) benzenesulphonatobis(acetylacetonate), **IV**, where $C_5H_7O_2$ = ligand, R = H, Me, CH=CH$_2$, OH,

$$(C_5H_7O_2)_2MnSO_3-\!\!\left\langle\!\!\bigcirc\!\!\right\rangle\!\!-R$$

IV

COOH or $C_{10-14}H_{21-29}$, was found[41] to be an efficient initiator at 10–20°C in the presence of atmospheric oxygen. Initiation with manganese(III) acetate in aqueous H_2SO_4 displayed[42] a decrease in R_p with increase in ionic strength and [H$^+$]. $Mn(OH)^{2+}$ was postulated as the initiating species and manganese(III) as the terminating species.

Comparison[43] of the manganese system (**IV**) and persulphate/ascorbic acid initiating systems in the polymerization of AA showed that in the former case the rate of polymerization and degree of polymerization were greater by factors of about 80 and 2 respectively. It was also shown that a small quantity of ethanol acted as a chain transfer agent and controlled the rate of chelate decomposition and therefore polymerization.

Several redox systems involving manganese(III) have been used for the polymerization of AA. $Mn(CH_3CO_2)_3$/diglycolic acid[44] and $Mn(CH_3CO_2)_3$/isobutyric acid[45] in aqueous media both function by oxidation of the organic acid via complex mechanisms to yield reactive free radicals. Termination was ascribed to the interaction of growing chains with manganese(III) and manganese(III)–acid complexes, leading to regeneration of manganese(II).

Other systems have also been studied. These include the hydrogen peroxide/ascorbic acid system,[46] which gave overall energies of activation of 75·7 and 62·6 kJ mol^{-1} for AA polymerization in the pH range 7–9·4; these energies correspond to ionized monomer reacting with virtually un-ionized and completely ionized chains respectively. In the absence of monomer, the main product is oxalic acid. Sodium hypophosphite/2,2′-azobis(2-amidinopropane)/HCl is claimed to yield high molecular weight products.[47] Presumably, the initiator is redox in the early stages, after which thermal decomposition of the azo compound becomes the main initiation process.

A systematic study[48] of MAA polymerization initiated by perpropionic acid/metal acetate in benzene showed that the reaction proceeds in the presence of oxygen and that the activity of the catalyst increases in the order:

$$Ce < Fe < Cu < Co < Mn$$

Finally, the hydrogen peroxide/3,4,5,6-tetrahydroxy-2-oxohexanoic acid lactone(**V**) was shown[49] to be very effective in the polymerization of AA in alkaline solution, a temperature rise of 2°C being observed over the first 30 s of reaction.

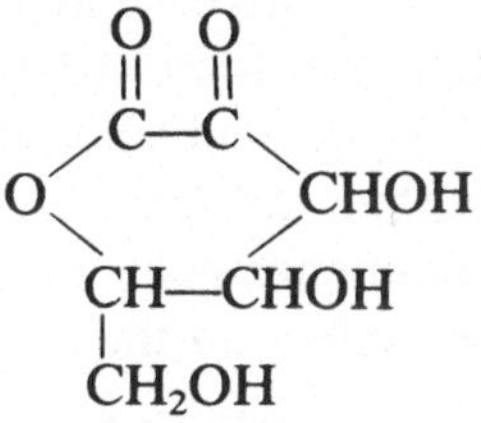

V

2.1.3 Emulsion Polymerization[50,51]

Homogeneous polymerization of AA and MAA in aqueous solution will normally yield a viscous solution or gel which may cause problems in handling. Use of water-miscible organic solvents often leads to precipitation of the polymer which, under suitable conditions, may yield an easily handled product. An alternative and widely practised technique is inverse emulsion or suspension (water-in-oil) polymerization.

This involves the production of submicroscopic water-swollen hydrophilic spheres by suitable mixing of aqueous monomer solution, hydrophobic phase, emulsifier, water-soluble or oil-soluble initiator and a wide range of other additives. These form the principal polymerization loci, but, in contrast to conventional oil-in-water emulsions, they are much more comparable in size with the aqueous monomer droplets which constitute the monomer reservoir; hence the situation is more complex. Inverse emulsions are also less stable and may tend to settle out over a period of days.

Dimonie et al.[52] compared inverse suspension with solution polymerization of ACM, and it is probable that their general conclusions can be extended to most water-soluble polymers. They examined the effect of method of aqueous phase addition, type and concentration of emulsifier, salt concentration, time intervals between mixing, and other variables, on the rate of polymerization and polymer molecular weight and nature.

They found that polyacrylamides prepared in inverse suspension are of considerably lower molecular weight than those prepared using the same initiator concentration in solution. These workers also found that in the former case, product molecular weight is independent of initiator concentration whereas in the latter, the normal rate-dependence was observed.

It was also found that in suspension polymerization, product degree of polymerization decreases with increase in surfactant concentration.

Addition of inorganic salts was found to have little effect whereas even small amounts of carboxylic acid salts caused a marked increase in product molar mass. Neither the length of time elapsing between mixing the phases and commencement of polymerization nor the intensity of stirring has much effect on the product molar mass.

On commencement of polymerization, phase-inversion was found to occur, yielding a continuous aqueous phase of increased viscosity.

Table 1 gives brief details of a few of the many recipes that have been reported in the literature for the polymerization or copolymerization of AA and/or MAA. A vast patent literature exists for the emulsion polymerization of other water-soluble monomers and such recipes will probably be effective in the present case. Detailed discussion is outside the scope of this chapter.

2.1.4 Factors Affecting the Polymerization Process

(a) Aqueous solution. The nature of the solvent medium[64] has been shown to exert a considerable effect on the polymerization of AA, MAA and other inorganic monomers. This is most clearly understood in studies of homogeneous polymerization under a range of conditions. In particular, variation of pH, ionic strength, nature of counterion and choice of different solvent compositions has provided a valuable insight into polymerization mechanisms.

Kabanov et al.[65] studied the effect of pH on the aqueous-phase polymerization of AA and MAA and their results are shown in Fig. 1.

AA monomer is a typical weak acid which undergoes ionization with a pK_a of 4·2. Ionization of PAA is a more complicated process, since partial ionization imparts a negative charge to the polyion and increases the free energy of ionization, thereby hindering further release of hydrogen ions. This results in a pK_a value which is usually considerably higher than the single pH-independent value for the

TABLE 1

SOME INVERSE EMULSION POLYMERIZATION SYSTEMS

Initiator	Monomer	Non-aqueous phase	Emulsifiers	Other components	Reference
$S_2O_8^{2-}$	Na acrylate	Water/Me_2CO MeOH, EtOH	Ethoxylated nonylphenol	Solid CO_2, triethanolamine	53
$S_2O_8^{2-}$	Na acrylate	Cyclohexane, methylcyclohexane	Sorbitol monostearate		54
$S_2O_8^{2-}$	Na acrylate	Hexane	Alkenyl (C_{16-18}) succinic acids, anhydrides, hydrogenation products		55
4,4'-Azobis-4-cyano-valeric acid	Na acrylate/ACM	Isoparaffins (Isopar)	Mono- and tri-glycerides poly-ethoxylated sorbitol hexaoleate	NaCl, NaBr, LiCl, LiBr Versenex (heavy-metal chelating agent)	56, 57
$S_2O_8^{2-}$/ $NaHSO_3$,	Na acrylate/ACM	2-Ethylhexyl phthalate	Esters of oleic or stearic acids with glycols or polyols, nonylphenol or polyethoxylated stearin		58
Azo compounds peroxides	AA	$CHCl_3$/ligroin (1:4)	Permethylated β-cyclodextrins (initiator-carrier)		59, 60
$S_2O_8^{2-}$	AA Na acrylate	Aliphatic ketones	Poly(ethylene glycol)	NaOH	61
Various soluble azo initiators	ACM etc.	Cyclohexane/isodecanol	Copolymer of dicyclopentadiene, vinyl acetate, maleic anhydride, etc.		62
Bis-(2-ethylhexyl)peroxydicarbonate	AA	CH_2Cl_2	Polyoxyethylene alkyl ether and/or sorbitan poly(oxyethylene monoester)		63

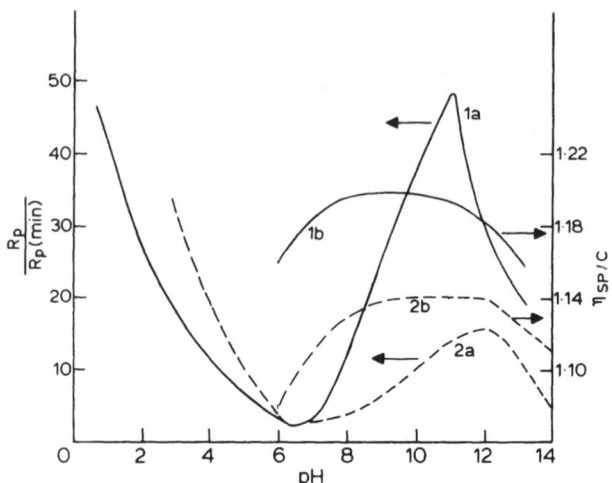

FIG. 1. Effect of pH on R_p and η_{sp}/C of model mixtures: 1a, $R_p/R_{p(min)}$ for AA; 1b, η_{sp}/C for PAA/AA/NaOH model mixture; 2a, $R_p/R_{p(min)}$, for MAA; 2b, η_{sp}/C for PMAA/MAA/NaOH model mixture. (Redrawn from ref. 65.)

ionogenic monomer and which increases with increasing extent of neutralization and therefore pH. The pK value is ca 5·8–6·4 in the case of 0·01 M PAA, and increases with decreasing ionic strength, increasing radius of the counterion and decreasing dielectric constant.[66]

Titration behaviour of polyacids obeys the modified Henderson–Hasselbach Equation:

$$pH = pK'_a + n \log \frac{\alpha}{1-\alpha}$$

where α is the degree of neutralization, pK'_a is the average value of pK_a and n is a measure of the configurational entropy (related in turn to coil dimension).

The time required for an ionization step ($\sim 10^{-9}$ s) is several orders of magnitude less than that for an addition step in free-radical vinyl polymerization ($\sim 10^{-4}$ s). It is likely therefore that the behaviour of a polyacid radical will be similar to that of a 'dead' polymer molecule. These ionization phenomena lead to progressive changes in charge and therefore reactivity of the reacting species.[67] As the pH increases from 1 to 7, the degree of ionization of AA monomer increases from 0% to almost 100% whereas the polyradicals are much less dissociated. The rate reduction is attributed to the lower reactivity of the monomer ions, in which form an increasing proportion of the monomer exists. The lower reactivity of AA ions compared with undissociated AA has been explained in terms of the enhancement of electron-density on the C=C double bond by the electron-donating carboxylate group.[68] Molecular orbital calculations[69] of the possible reactions between dissociated or undissociated monomer molecules and radicals showed that the greatest gain in resonance stabilization occurred on the addition of undissociated monomer. As the pH increases from 1 to 7, the proportion of monomer in this form decreases, as explained above; hence the overall rate would be expected to display a significant decrease.

Following a somewhat flat minimum around pH 6–7, the rate of polymerization (Fig. 1) increases sharply as the pH increases from 8 to 11, a trend which is a consequence of parallel variation in the propagation rate constant, k_p, the termination rate constant k_t being virtually independent of pH over the range ~3–13·6.[65] This increase occurs over the pH range in which PAA chains are displaying a marked increase in the degree of ionization. The predominant addition step will involve interaction of a polyion and a monomeric ion of the same charge with consequent repulsion effects. These are thought to be overcome by cation binding (VI)[65] which is a consequence of the

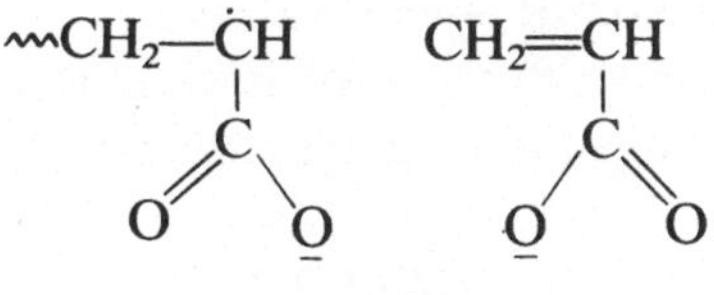

$$\text{wCH}_2\text{—CH} \qquad \text{CH}_2\text{=CH}$$

Cation

VI

addition of the base to raise the pH to the appropriate level. This reduces the charge repulsion between the reacting species by a shielding effect and also assists the reaction processes by affecting steric placement, thereby enhancing reaction rate.

Above pH 11, R_p decreases sharply in parallel with a corresponding decrease in k_p. Kabanov[65] ascribed this effect to 'chain collapse' under the influence of the high ionic strength of the system as a consequence of the amount of alkali required to give the high pH attained. Over this region, ionization is complete and substantial shielding of adjacent carboxylate groups will permit radical coiling. Access of monomer will be hindered, leading to reduction of k_p and R_p (see Table 2).

Kabonov et al.[65] reported convincing evidence for their conclusions. Addition of high concentrations of sodium salts increased the value of k_p by a factor of 5 at pH 7·9 but reduced R_p considerably at pH 11. Sodium ion binding is presumably effective in favouring chain growth in the former case; chain collapse is enhanced in the latter. Also included in Fig. 1 are viscosity data which are relevant. With an increase in pH, chain extension would seem to accompany chain ionization. The observed decrease is attributed to chain collapse. Addition of neutral electrolyte at pH 11 causes a sharp reduction in intrinsic viscosity and presumably the size of the polymer chain. However, Mandel[70] has criticized the chain-collapse explanation on the grounds that the ionic strengths are not sufficiently high to cause such effects. He has speculated on the possibility of high pH affecting the quality of the solvent by bringing it closer to theta conditions.

The results obtained with AA have, in general, been parallel with those for MAA. The main differences are due to the lower reactivity of the PMAA radical and to the normally extended conformation of the chain, a consequence of the presence of α-methyl groups.

Several workers have investigated pH effects by copolymerization studies on systems such as AA/ACM,[69,71] AA/N-vinylpyrrolidone[72] and MAA/N-vinylpyrrolidone,[73] and have reached similar conclusions to those made for the homopolymerizations.

Cation binding and its effects have been studied in some detail by several workers.[74-76] Titration of poly(acrylic acid) with LiOH, NaOH

TABLE 2

KINETIC DATA FOR EFFECT OF pH ON THE POLYMERIZATION OF ACRYLIC AND METHACRYLIC ACIDS[65] (23°C)

Monomer	pH	$k_p/10^4$	$k_t/10^8$
Acrylic acid	Acidic[64]	~3·0	1·8
	7·9	0·065	2·6
	7·9 (1·5 M NaCl)	0·315	2·6
	11·0	0·66	2·1
	13·6	0·25	2·25
Methacrylic acid	Acidic[64]	0·500	0·1
	8·8	0·067	2·1
	13·6	1·95	2·25

and KOH respectively satisfies the modified Henderson–Hasselbach Equation. The apparent dissociation constant, pK_a', increases and the chain-coil dimension decreases in that sequence.

This trend has been attributed to the decreasing extent of cation binding (i.e. $Li^+ > Na^+ > K^+$) as the cation size increases, ($Li^+ < Na^+ < K^+$). This results in the same sequence of decreasingly effective shielding of the negatively-charged —COO^- groups. Because of the lower charge concentration, ionization of the remaining —COOH groups is increasingly hindered in the same sequence as is chain expansion. An alternative explanation is that the larger cations are less able than the small ions to approach close to the polyion chains, so chain extension is less likely.

In the copolymerization of ACM (M_1) with Li, Na and K acrylates (M_2), monomer reactivity ratios (r_2) of 0·37, 0·30 and 0·24 respectively were obtained.[74a] This was explained by the reduction in the extent of cation binding in the sequence $Li > Na > K$ and changes in radical conformation, leading to reduction in values of k_{p22} (the rate constant of the reaction between monomer M_2 and radical $M_2^.$) by virtue of increased monomer–radical repulsion. On the other hand, studies[75] of the copolymerization of methyl methacrylate (M_1) with Li, Na and K methacrylates (M_2) gave 'best' values for r_2 of 0·073, 0·126 and 0·143 respectively. The same workers also found that the rate of homopolymerization of the methacrylate salts increases in the same sequence. Both observations are said to be consistent with an increase in k_{p22} with increased size of counterion.

In another study,[74b] copolymerization of ACM (M_1) with divalent metal salts of AA (M_2) gave r_2 values of 0·09, 0·05 and 0·33 for Ca^{2+}, Sr^{2+} and Ba^{2+} respectively. Clearly, the correlation with size observed with monovalent ions is not obeyed in this case. The trend ($Ba^{2+} > Ca^{2+} \sim Sr^{2+}$) reflects the variation in charge fraction (number of charges carried by the polymeric ion/degree of polymerization) of polyacrylate ions neutralized to 80% with $Ca(OH)_2$, $Sr(OH)_2$ and $Ba(OH)_2$ respectively.

There is no doubt that in polymerizations and copolymerizations involving different metal salts of acrylic and methacrylic acids, variation of polymerization rates and monomer reactivity ratios is caused by alteration in chain conformation and in forces of repulsion between the monomer ion and the polyion as a whole and not merely the growing radical end. Unfortunately, studies to date have concentrated on measurement of R_p, and monomer reactivity ratios r_1 and r_2.

As a result, it is difficult to identify, unequivocally, the effects of propagation, cross-propagation, termination and cross-termination. Thus, apparently contradictory results have been obtained. Only when values of the individual velocity coefficients for these processes are available will it be possible to obtain reliable comparisons.

A further consequence of the nature of the solvent medium is its effect on copolymer composition, which becomes a function of extent of reaction. If during the course of the reaction, the pH and ionic strength change with the ionogenic monomer concentration, then the values of the monomer reactivity ratios will also change, thereby contributing significantly to a drift in the copolymer composition, with probable effects on the properties of the products of the reaction. This effect has been amply demonstrated by Myagchenkov et al.[77,78] for the systems ACM/sodium maleate, maleic acid and citraconic acid. The drift in copolymer composition is most easily reduced, or even eliminated, if sufficient electrolyte is added to ensure a constant ionic strength and, if possible, pH.

(b) Non-aqueous solution. Polymerization and copolymerization of AA and MAA are highly sensitive to the composition of the solvent medium. This is apparent from a number of copolymerization studies in which, for example, r_1 values of 1·3, 0·48 and 0·67 have been observed[79] for AA $(M_1)/N$-vinylpyrrolidone (M_2) in bulk, toluene and dimethylformamide (DMF). Abkin et al.[64,79,80] (Table 3) demonstrated very significant solvent effects by measuring k_p and k_t in aqueous solution and in dimethylsulphoxide (DMSO).

TABLE 3

EFFECT OF DMSO ON ACRYLIC ACID AND FLUOROACRYLIC ACID POLYMERIZATION[64]

	Acrylic acid		Fluoroacrylic acid	
	Water	DMSO	Water	DMSO
$k_p/10^4$ (30°C)	3·19	0·076	0·36	0·11
$k_t/10^8$ (30°C)	1·8	0·2	0·9	0·5
$A_p/10^8$	0·6	47	0·6	0·66
$E_p/$(kJ)	13·0	33·4	18·8	21·7
$A_t/10^8$	1·8	0·2	2·0	1·0
$E_t/$(kJ)	0	0	2·5	1·7

Symbols: k = rate constant; A = Arrhenius factor; E = activation energy; suffix p = propagation step; suffix t = termination step.

The differences are thought to be due to monomer dimerization and radical–solvent complex formation with DMSO, and also to chain conformation effects. The presence of the highly electronegative fluorine atoms in fluoroacrylic acid (FAA) is thought to reduce radical reactivity by delocalization of the unpaired electron, to hinder formation of complexes with DMSO and to eliminate dimer formation. In addition, the fluorine atoms give rise to very stiff radical chains because of their mutual repulsion; consequently, the kinetic constants are much less sensitive to the nature of the solvent. Activation energies and frequency factors support these conclusions.

Chapiro[81-83] has provided considerable evidence that hydrogen bonding may markedly influence solvent effects in the polymerization of AA. He observed that R_p, auto-acceleration and stereoregularity of the polymer produced in bulk polymerization are virtually unaffected by dilution with up to 20% by volume of hydrogen bonding solvents (e.g. water, methanol and dioxane), whilst non-hydrogen bonding solvents (e.g. hexane, toluene) cause these quantities to decrease abruptly. He ascribed this effect to the influence of the solvent on a dimer–oligomer equilibrium of AA molecules in solution.

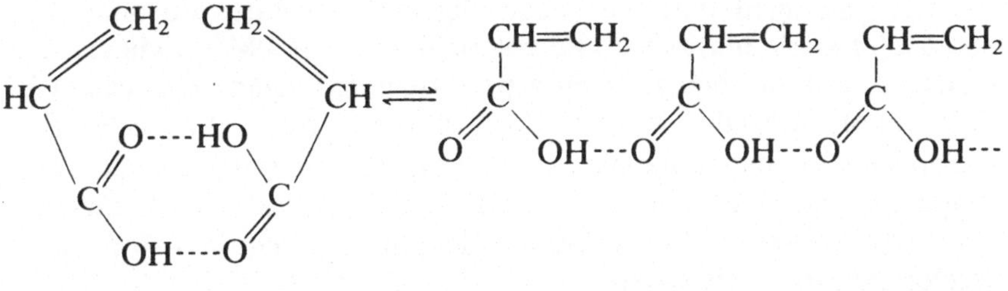

Dilution in non-polar solvents[83] and the decrease in temperature[84] both shift the equilibrium in favour of the dimer. Polar solvents, on the other hand, favour the formation of the 'pluri-molecular aggregates'. In the latter, the double bonds are in an organized narrow zone, so that when initiation occurs, a rapid polymerization and high degree of stereoregularity ensue. Chapiro[83] also proposed a matrix effect. He suggested that as PAA chains are formed, these act as 'templates' against which the monomer structures align themselves (VII), thereby enhancing the 'zip' character of the polymerization, causing auto-acceleration.

This mechanism is consistent with the observation that the resulting polymer probably consists of syndiotactic blocks interspersed with

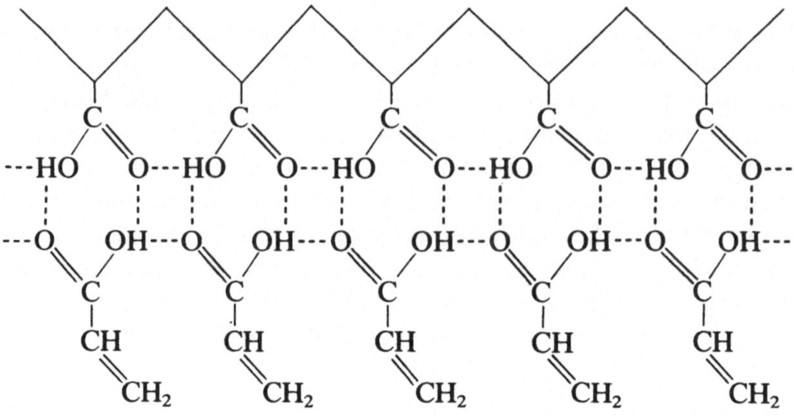

$$
\begin{array}{ccccc}
\text{C} & \text{C} & \text{C} & \text{C} & \text{C} \\
\text{---HO \quad O---HO} & \text{O---HO} & \text{O---HO} & \text{O---HO} & \text{O---} \\
\text{---O \quad OH---O} & \text{OH---O} & \text{OH---O} & \text{OH---O} & \text{OH---} \\
\text{C} & \text{C} & \text{C} & \text{C} & \text{C} \\
\text{CH} & \text{CH} & \text{CH} & \text{CH} & \text{CH} \\
\text{CH}_2 & \text{CH}_2 & \text{CH}_2 & \text{CH}_2 & \text{CH}_2
\end{array}
$$

VII

atactic segments. It is also significant that at very low concentrations of AA in non-polar solvents, auto-acceleration is in fact observed.[83] It is attributed to linear oligomers formed by association of 'free' monomer with polymer chains already formed.

By contrast, although MAA forms similar molecular associations, no auto-acceleration is observed[85] in any of a wide range of solvents. It has been assumed that the linear oligomer structures do arise, but because of steric hindrance and the rigidity of the PMAA chains, the monomer cannot align to form a pre-orientated complex as can AA. Bulk copolymerization of AA/MAA showed[86] that auto-acceleration disappeared in the monomer mixture which generated a copolymer containing one MAA unit to each AA unit. This was taken to indicate that AA sequences of a certain length are required for auto-acceleration to be displayed.

Several workers have studied the effect of preformed polymer chains on the polymerization of AA and MAA ('template polymerization'). Ferguson *et al.*[87a] showed that a significant increase in inhibition period, R_p and degree of polymerization occurred in the presence of poly(vinylpyrrolidone). In poly(4-vinylpyridine) (P4VP), the magnitude of the effect was shown to increase with the degree of polymerization of the added polymer[87b] and the degree of polymerization of the PAA produced was similar to that of the P4VP present. This was ascribed to the ability of longer P4VP chains to provide longer 'template' molecules which could bind longer chains of PAA and thereby retain them in solution for a longer period of their growth. This in turn was expected to result in the observed effects.

Random copolymers of P4VP with styrene and with ACM were shown to enhance the overall rate of polymerization of MAA, the rate maximum depending on the copolymer composition.[88] The results obtained in these studies were in parallel with the findings of Challa *et al.*[89] on the MAA/P2VP system. These workers suggested that two possible template mechanisms were operative:

Type I: Strong monomer–template interaction in which monomer is pre-adsorbed and propagation takes place exclusively along the template ('zip' mechanism), or

Type II: Weak interaction in which propagation commences in the bulk of the solution until a critical chain length is attained after which polymer complexation is favoured and propagation takes place along the template chain by adding monomer from the bulk. The magnitude of the various effects is therefore a function of the nature of the adsorption.

Other polymers have been used as templates in AA and MAA polymerization and interesting results have been obtained. The effectiveness of isotactic PMMA as a template in the polymerization of MAA was found[90] to be a function of the solvent and temperature used. In DMF between 0 and 40°C, R_p and the kinetic chain lifetimes were unaffected by the presence of PMMA. This was ascribed to weak interaction between the growing chains and the template which enabled termination to take place in bulk solution before association had occurred. At $-10°C$, the rate in the presence of the template was increased by a factor of 1·5. This was explained by proposing that the growing chains associated with the template displayed hindered termination by virtue of a reduced entropy of activation caused by the conformational limitations imposed by the association.

The absence of a template effect in 2-methoxyethanol and in $C_2H_5OH:H_2O$ (83:17), in which complexation is moderately strong, is explained[90] by the poor solubility of the polymer. It is noteworthy that with a more associating solvent in which the polymer is more soluble (30% dioxane in 2-methoxyethanol at $-10°C$), R_p was enhanced by a factor of 2·2.

In the copolymerization of MAA (M_1) with MMA (M_2), the monomer reactivity ratio r_1 increased from 1·5 to 2·1 and r_2 decreased from 0·55 to 0·1 on the addition of poly(ethylene glycol) (PEG).[91] The implied formation of longer AA sequences was supported by NMR studies, thus suggesting that PEG acts as a template for the polymerization. It was also observed that R_p for homopolymerization of AA is greater in the presence of PEG, and the greater the molecular weight of the PEG, the greater is the enhancement of R_p.

A study of the polymerization of MAA in the presence of persulphate ions and chitosan acetate, a β-D-$(1\rightarrow 4)$-linked polymer of **VIII** yielded interesting conclusions.[92] Initiation appears to occur at

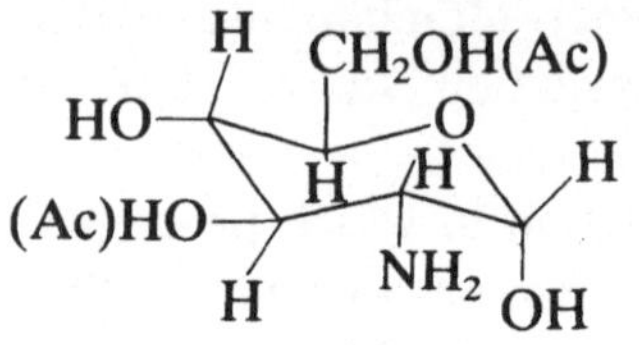

VIII

the reducing end groups of the chitosan acetate and propagation takes place via methacrylate monomer molecules bound to the chitosan molecules by ionic and hydrophobic bonding to yield living polymer radicals. It was also observed that the degree of polymerization and polydispersity of the polymer were almost identical to that of the chitosan—evidence, again, in favour of a template effect.

(*c*) *Molar mass control.* The initiator systems described earlier normally lead to high molar mass products, especially in polymerization of AA, for which $k_p/k_t^{0.5}$ has unexpectedly high values of 2·2 or 0·4, the value depending upon whether the reagents are in the un-ionized or ionized form respectively.[64,65]

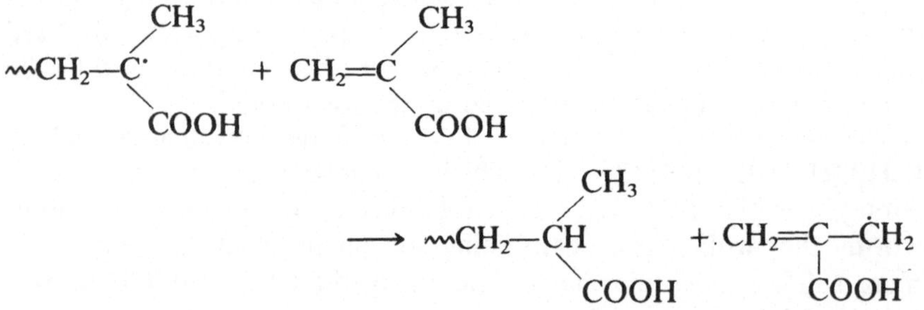

The relevant values[64,65] in the case of MAA over a comparable pH range are 1·6 and 0·05. This, together with the greater likelihood of monomer transfer because of resonance stabilization in the product radical will lead to considerably lower polymer molecular weight.

For effective flocculation behaviour, high molar masses are required. Effective dispersant behaviour is favoured, however, by low molecular weight. Chain transfer is the principal route to such materials. Ethanoic acid,[93] isopropanol, isobutanol, isopentanol, isohexanol/Cu, Zn or Mn acetates,[94] tertiary amines such as $Me_2NCH_2CH_2OH$[95] and $Me_2NCH_2CH_2NMe_2$[96,97] and thiols such as *tert*-hexadecyl mercaptan[98] and octyl mercaptan[99] have recently been cited as chain transfer agents. Another approach has been the design of continuous processes, which allow better control of reactant concentration and thereby polymer molar mass.[100,101]

2.1.5 Other Polyacids

Maleic anhydride polymers, usually as copolymers with various acrylic, allyl, alkylvinyl and vinyl esters and itaconic acid (**IX**), are probably

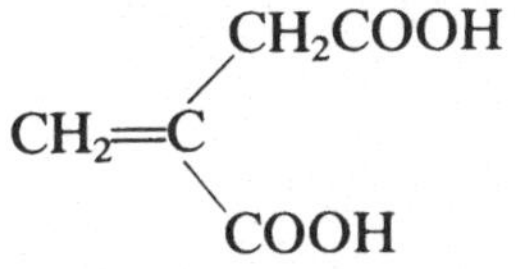

$$CH_2 = C \overset{\displaystyle CH_2COOH}{\underset{\displaystyle COOH}{}}$$

IX

the most important carboxylic polymers after PAA and PMAA; on treatment with alkali, these yield a range of polyelectrolytes. To the author's knowledge, few reviews which are relevant to this chapter have appeared in recent years.[14]

2.1.6 Formation of Polycarboxylic Acids by Polymer Modification (Hydrolysis)

Hydrolysis of suitable polymers represents an alternative route to polyacids which eliminates the hazards of handling the monomeric acids and which may, in fact, yield a more desirable product. Polyacrylates, polymethacrylates and especially polyacrylamides are the polymers which are usually subjected to such modification.

Kinetic studies of such reactions are difficult because of the wide range of possible microenvironments of a given unit in a chain. These are a consequence of the various possible tacticities of the relevant triad and the possible previous reaction of one or both neighbouring groups.[102] The effect of these variables is further complicated by their response to attack by different reactants.

Klesper *et al.*[103] demonstrated that in the hydrolysis of syndiotactic PMMA with relatively concentrated KOH, the product tended to contain alternating acid and ester units, whereas with a weak base such as pyridine, blocks of ester or acid units resulted. These findings were ascribed to the effect of charge repulsion on ester hydrolysis by OH^- ions in the first case, but to the predominance of a neighbouring-group effect in the latter, whereby an already hydrolyzed and ionized group assists the hydrolysis of its neighbour.

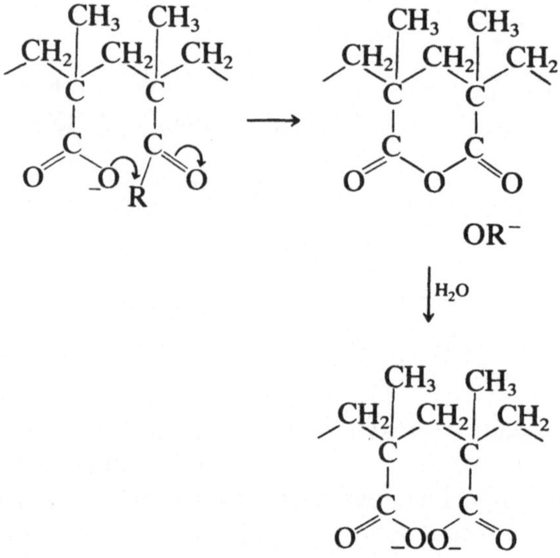

Kinetic approaches, often based on the analysis of McQuarrie,[104] have been used to evaluate the rate constants k_0, k_1 and k_2, where the subscripts represent the number of adjacent groups already reacted. Platé[105] obtained absolute values for these rate constants for a number of polyacrylates and polymethacrylates. His calculations gave values for k_0, k_1 and k_2 in the ratio $1:0.4:0.4$ for hydrolysis in $0.2M$ KOH but $1:8:100$ in pyridine/water (95:5), thereby supporting Klesper's findings.

More recently, it has been shown that neighbouring-group effects are displayed in both acid- and base-catalyzed hydrolysis of poly([N,N-dimethylaminoethyl] methacrylate) **X**.[106]

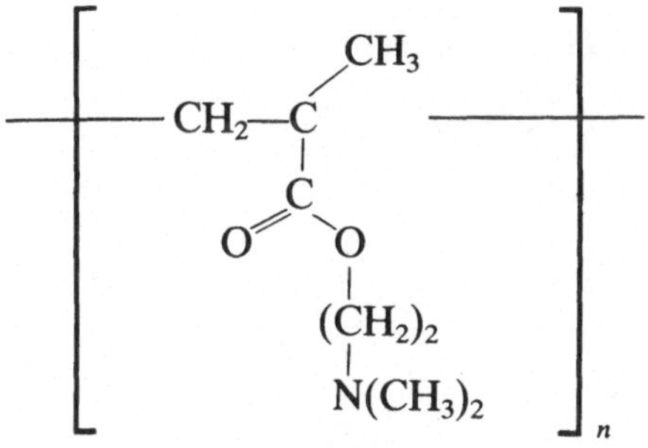

$$
\left[
\begin{array}{c}
\text{CH}_2\text{—C} \overset{\displaystyle \text{CH}_3}{\underset{\displaystyle \underset{\displaystyle \underset{\displaystyle (\text{CH}_2)_2}{|}}{\overset{\displaystyle \text{C}}{O \diagup\diagdown O}}}{|}}
\end{array}
\right]_n
$$

X

A much more important route to polyacids is the hydrolysis of polyACM and polymethacrylamide (polyMACM).

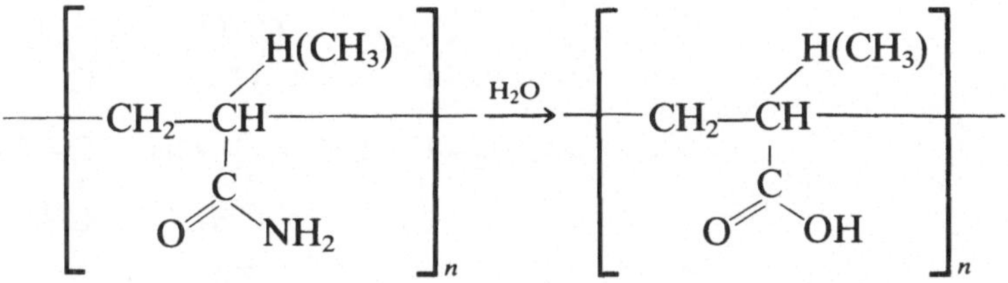

$$
\left[
\begin{array}{c}
\text{—CH}_2\text{—CH—} \\
\overset{\displaystyle \text{H(CH}_3)}{} \\
\underset{\displaystyle O \diagup\diagdown \text{NH}_2}{\overset{\displaystyle \text{C}}{}}
\end{array}
\right]_n
\xrightarrow{\text{H}_2\text{O}}
\left[
\begin{array}{c}
\text{—CH}_2\text{—CH—} \\
\overset{\displaystyle \text{H(CH}_3)}{} \\
\underset{\displaystyle O \diagup\diagdown \text{OH}}{\overset{\displaystyle \text{C}}{}}
\end{array}
\right]_n
$$

This may be carried out during or after polymerization, although it is, in fact, difficult to avoid 1–2% hydrolysis occurring during polymerization under normal conditions.

It is thought that polyMACM resists hydrolysis beyond 72% under alkaline conditions.[107] This conversion limit is a consequence of the kinetic relationship:

$$ k_0 \sim k_1 \gg k_2 $$

which in this case is probably the result of OH⁻ ion repulsion by the two charged neighbouring groups which have already undergone hydrolysis, reinforced by hydrogen bonding which causes the polymer chains **XI** to adopt a more compact and therefore less accessible conformation, **XII**.

The effect is thought to be increased by an unfavourable neighbouring group effect as a result of hydrogen bonding, in which the N–H electron-pair is partly released towards the N-atom and a

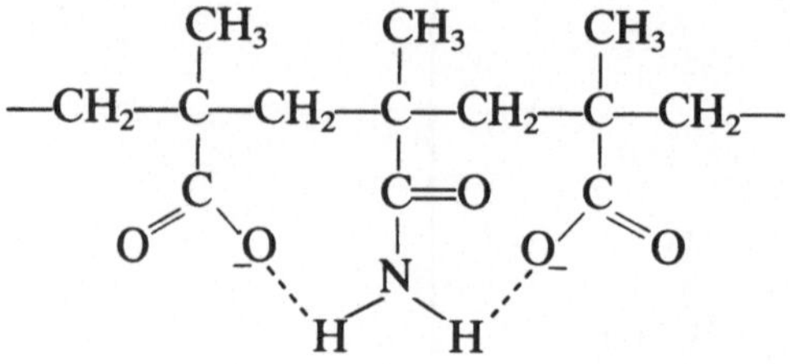

XI

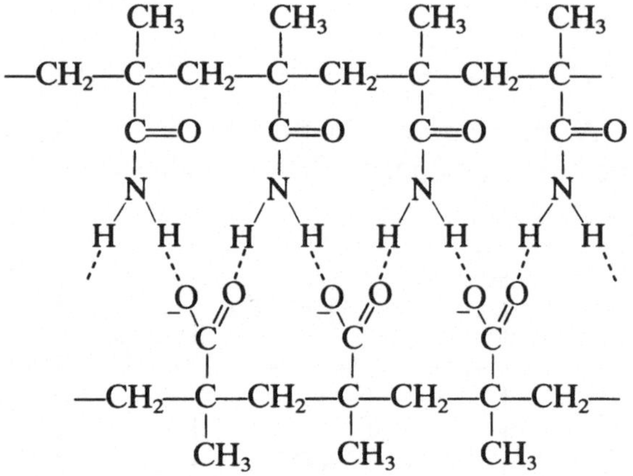

XII

fractional negative charge is transmitted to the amide C-atom, thereby hindering the approach of the OH⁻ ion.

With poly(*N*-methylacrylamide) (**XIII**) and poly(*N,N*-dimethylacrylamide) (**XIV**), limiting conversions of 55% and 35% respectively are attributed to shielding.[108] In line with this, it is found that poly[*N,N*-dimethylmethacrylamide] **XV**, is unattacked by dilute alkali. In the presence of an *N*-carboxy group, on the other hand, auto-acceleration of hydrolysis occurs.[109]

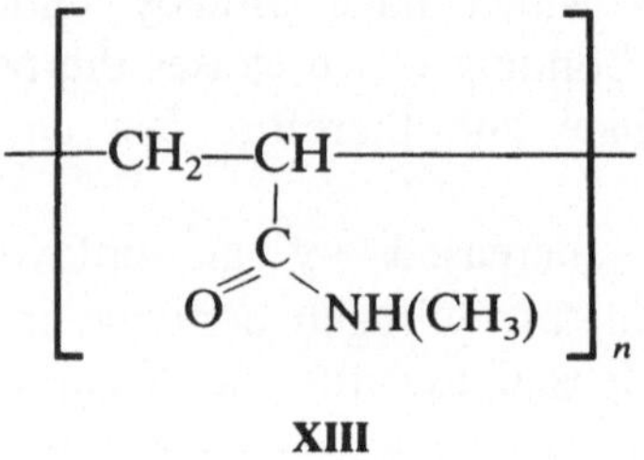

XIII 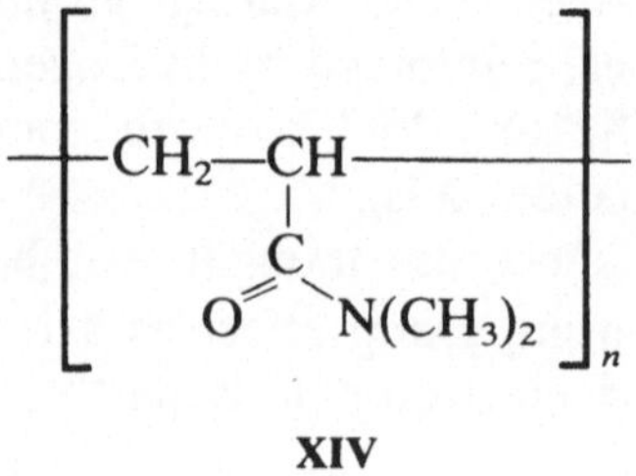   **XIV**

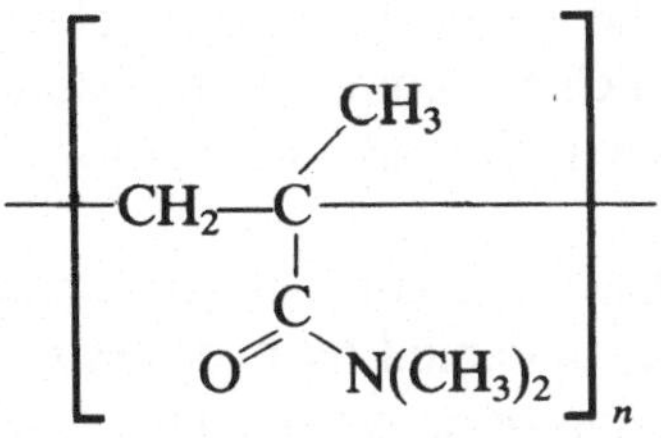

$$\left[-CH_2-\underset{\underset{\underset{O}{\diagup}\underset{N(CH_3)_2}{\diagdown}}{\overset{|}{\underset{C}{|}}}{\overset{CH_3}{\overset{|}{C}}}- \right]_n$$

XV

In the case of polyACM, hydrolysis takes place in two distinct stages, but complete conversion may be attained.[110]

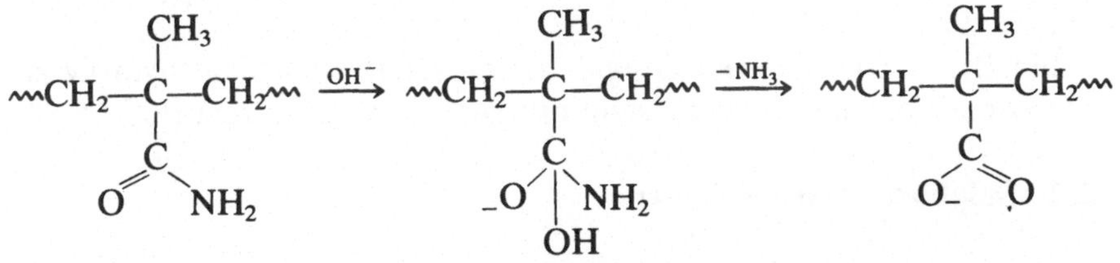

$$\text{wCH}_2-\underset{\underset{O\diagdown NH_2}{\underset{C}{|}}}{\overset{CH_3}{\overset{|}{C}}}-CH_2\text{w} \xrightarrow{OH^-} \text{wCH}_2-\underset{\underset{\underset{OH}{{}_-O\diagdown|NH_2}}{\overset{C}{|}}}{\overset{CH_3}{\overset{|}{C}}}-CH_2\text{w} \xrightarrow{-NH_3} \text{wCH}_2-\underset{\underset{O_-\diagdown \diagup O}{\overset{C}{|}}}{\overset{CH_3}{\overset{|}{C}}}-CH_2\text{w}$$

The first ~40% conversion takes place considerably more rapidly than the second stage, and beyond 70% the rate is very slow indeed. The transition is thought to correspond to chain extension which occurs when a critical polymer charge concentration has been reached and the polyACM chains suddenly uncoil. The absence of shielding by methyl groups is presumably responsible for the possibility of complete hydrolysis.

In more recent studies,[111] it has been claimed that the rate of alkaline hydrolysis decreases sharply after 10–16% conversion, the decrease being ascribed to similar causes and to aggregate formation of the polyACM.

The patent literature contains many references to alkaline hydrolysis of polyACM and polyMACM, particularly in emulsion. The reaction of a water-in-oil emulsion of the hydrolyzing agent (e.g. NaOH) with a similar emulsion of the polymer is claimed to give a stable emulsion of the required hydrolyzed polymer.[112] Treatment of an emulsion copolymer of MAA/N-alkyl acrylamide with NaOH gives a polymer suitable for a wide range of uses.[113] Conditions for achieving 20–30% hydrolysis of polyACM with NaOH have been described;[114,115] the product is claimed to be highly effective for enhanced oil recovery. Rapid hydrolysis with less than the stoichiometric amount of NaOH or other hydrolyzing agent occurs at 125°C under pressure,[116] and at 150–220°C rapid hydrolysis occurs in the absence of acids or bases.[117–118]

Hydrolysis with NaOH in aqueous acetone solutions gives powdered hydrolyzed polymers which are conveniently handled.[119] Evidence has been presented,[120] however, that alkaline hydrolysis also leads to some degree of chain rupture.

Polymerization of ACM[121] in concentrated solution (>30%) in the presence of NaOH is accompanied by hydrolysis of the monomer and polymer, and the formation of ammonia. This reacts with unconverted ACM to form $N(CH_2CH_2CONH_2)_3$ which induces decomposition of the persulphate initiator leading to an increase in R_p. If [ACM] ~ 4–10%, formation of the tertiary amine is negligible; consequently R_p and product molar mass are controlled by normal polymerization parameters.

Finally, various process designs for polymerization, and hydrolysis of various polymers to give useful products, have been reported.[122–126]

2.2 Sulphonic-based Polymers

2.2.1 Polymerization of Styrene Sulphonic Acid and its Salts (XVI)
Polymerization of styrene sulphonic acid (SSA), sodium styrene sulphonate (NaSS) and potassium styrene sulphonate (KSS) alone or

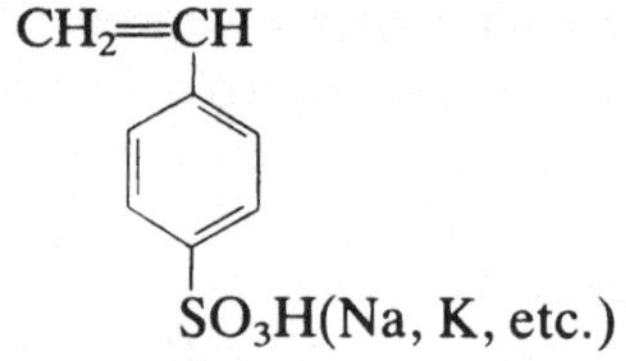

$$CH_2{=}CH$$

$$SO_3H(Na, K, etc.)$$

XVI

with comonomers yields polyelectrolytes which are considerably stronger acids (pK_a ~ 1) and whose degree of ionization is therefore less pH-dependent than that of polycarboxylic acids (pK_a ~ 6).

Kangas[127] has reviewed the polymerization reaction of a range of sulphonated monomers and has discussed solvent and related effects in some detail. Polymerization occurs via free-radical, anionic and co-ordination routes. Free-radical initiation is the most common method, the processes being similar to those for the polymerization of carboxylic monomers, using thermally sensitive or redox initiators in solution[128–130] and in emulsion.

Because of their permanent charge, styrene sulphonate units in a copolymer chain stabilize the particles formed on precipitation from a non-solvent or from an inverse emulsion. The amount of surfactant required is therefore reduced or its use may be eliminated altogether.[131–135] The addition of even small amounts of styrene sulphonate salts to ACM in a water–toluene system, for instance, was found to stabilize the surfactant-free emulsion,[136] and lower the overall energy of activation from 47·8 to 40·2 kJ mol⁻¹; this is in agreement, incidentally, with the earlier studies of van der Hoff.[137] Various processes have been described for emulsion polymerization of these monomers involving techniques such as evaporative cooling[138] and recycling.[139,140]

2.2.2 Factors Affecting the Polymerization Process

Kinetic studies of the polymerization of these monomers have been reviewed by Kangas.[127] The high value of $k_p/k_t^{0.5}$ (0·4 at 45°C) for NaSS polymerization in DMSO,[141] compared with 0·03 for styrene[142] at 50°C, was attributed to the inductive effect of the sulphonate group on the reactivity of the vinyl double bond. Estimated values[127,143,144] of 4·5 (at 45°C) and 2 (at 50°C) were obtained for the aqueous-phase polymerization of NaSS. The considerably larger value compared with that in DMSO was ascribed to the different ionic repulsion effects on propagation and termination in the two solvents.

More recently, Kurenkov and Myagchenkov[145] reported a detailed study of the polymerization of NaSS and KSS in a series of solvents. In

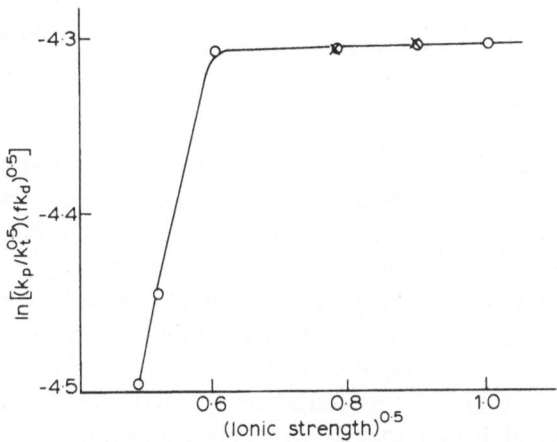

FIG. 2. Effect of ionic strength on $(k_p/k_t^{0.5})(fk_d)^{0.5}$ for polymerization of KSS in water at 70°C: ×, without addition of KCl; ○, with addition of KCl.[145]

aqueous solution, the value of $(k_p/k_t^{0.5})(fk_d)^{0.5}$ for KSS polymerization was found to vary with ionic strength as shown in Fig. 2 (k_d is the dissociation rate constant for the initiator and f is the fraction of radicals formed by dissociation, which initiate propagation). An increase in ionic strength would be expected to reduce the extent of ionic repulsion between charged reactants, thereby increasing both k_p and k_t. These workers claimed that the increase in k_p was the dominant factor and that this was the result of variation of electrostatic repulsion and chain conformational effects.

The effect of variation of dielectric constant was studied in dioxane/water mixtures. Results were explained on the basis of the simplified ionic equilibria:

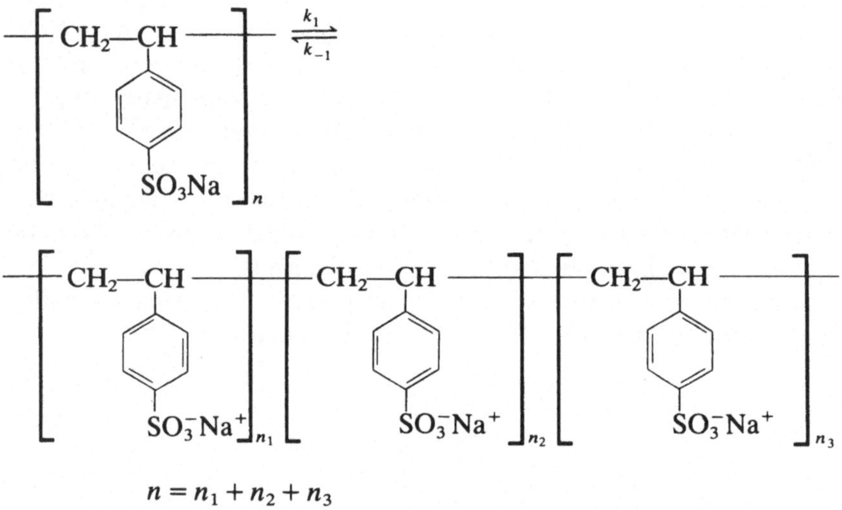

$$\left[\begin{array}{c} CH_2-CH \\ | \\ \bigcirc \\ | \\ SO_3Na \end{array}\right]_n \quad \underset{k_{-1}}{\overset{k_1}{\rightleftharpoons}}$$

$$\left[\begin{array}{c} CH_2-CH \\ | \\ \bigcirc \\ | \\ SO_3^-Na^+ \end{array}\right]_{n_1} \left[\begin{array}{c} CH_2-CH \\ | \\ \bigcirc \\ | \\ SO_3^-Na^+ \end{array}\right]_{n_2} \left[\begin{array}{c} CH_2-CH \\ | \\ \bigcirc \\ | \\ SO_3^-Na^+ \end{array}\right]_{n_3}$$

$$n = n_1 + n_2 + n_3$$

A decrease in dielectric constant would be expected to enhance electrostatic interactions between radicals and ions and, by the same token, to lower the degree of ionization of the chains. These conclusions were borne out by measurements of R_p, viscosity and electrical conductivity of the polymer solution. Unfortunately, the rate of initiation is a function of the composition of the water/dioxane mixture, so $k_p/k_t^{0.5}$ was not obtainable.

The rate of initiation was found to be unaffected by DMSO but measurements indicated that $k_p/k_t^{0.5}$ decreases from 0·62 to 0·37 on changing the solvent from water:DMSO, 3:1, to pure DMSO, with its lower dielectric constant. This was ascribed largely to an increase in k_t, and evidence from conductivity and viscosity measurements was presented which indicates that the effective degree of ionization of the polyradical and monomer decreases with the decrease in dielectric constant. It was also deduced that the polyradical is less ionized than the monomer by virtue of sodium-ion binding. Other conformational and reactivity factors may well be involved, however. These workers also found that for equal reactant concentrations, R_p for KSS is greater than that for NaSS. Surprisingly, this was ascribed to more significant binding to the polyions of K^+ ions compared with Na^+ ions.

Kurenkov et al.[146] also studied the effect of divalent cations (Ca^{2+}, Sr^{2+} and Ba^{2+}) on the aqueous-phase polymerization of SSA. They found that the viscosities of solutions of fixed concentrations of these polymers decreased in the order Ca–KSS > Sr–KSS > Ba–KSS and that the effect was more pronounced at lower ionic strengths.

The effect was attributed to greater coiling of the polymer chains, as a result of increased ion-binding with the increase in cation radius ($Ca^{2+} < Sr^{2+} < Ba^{2+}$). They obtained values of 0·43, 0·87 and 1·18 for $k_p/k_t^{0.5}$ (70°C) in the case of Ca–SS, Sr–SS and Ba–SS respectively. It was proposed that these values were a consequence of the decreased electrostatic repulsion between monomer and growing polymer radical by virtue of ion binding. This in turn would lead to an increase in k_p and k_t. The observed increase in viscosity-average molar mass was taken to indicate that k_p increases by a greater factor than does k_t.

Matrix polymerization of NaSS, usually on polymeric cationic backbones, has been reported by various workers. Examples include the polymerization on a matrix consisting of the ionene formed from 1,4-diazabicyclo[2,2,2]octane and 1,4-dibromobutane in the presence of NaX (X = F, Cl, Br, ClO_4). The alteration of the ionic-selectivity sequence on the addition of isopropanol was explained by hydration of the counterions.[147]

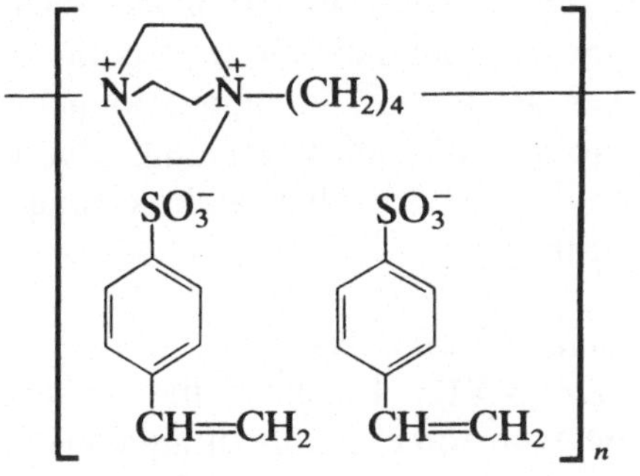

The influence of the matrix on R_p was measured[148] for the separate polymerization of NaSS and bicyclo[2,2,2]octane dimethylammonium styrene sulphonate:

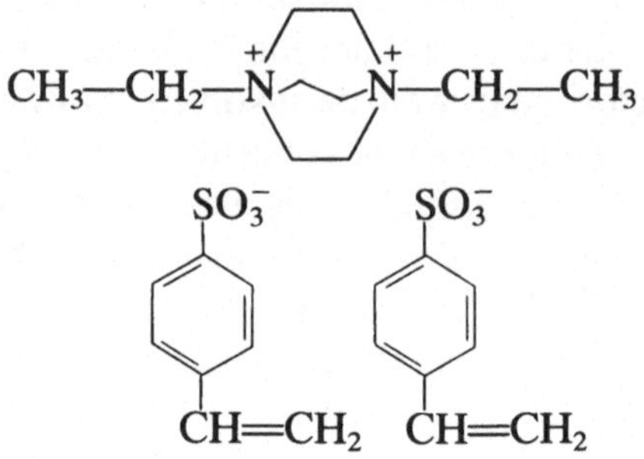

$$CH_3—CH_2—\overset{+}{N}\diagdown\diagup\overset{+}{N}—CH_2—CH_3$$

Each rate was found to be much lower than that for the polymerization of a salt of SSA in the presence of the above polymeric ionene. This indicated that the rate enhancement is very strong when passing from monomeric or dimeric ions to polymeric ions (DP ~ 12). In the latter case, the rate was found to increase with the decrease in dielectric constant of the solvent.

However, when the solvent composition approached that of pure water, the dependence was found to be reversed, and it was concluded that there was a weakening of the electrostatic interaction between matrix and monomer, and possible enhancement of hydrophobic forces. Measurement[149] of the variation of conductance of the above system was used to explain the effect of ionized polymeric counterions on the matrix polymerization of polyelectrolytes, using counterion condensation theory.

Copolymerization studies have been reported by many workers and the most interesting point to emerge is the complicated effect of the non-isoionic character observed in many,[145,150–152] but not all,[153] such systems. This is a result of the change in factors such as ionic strength and pH as the ionogenic monomers are incorporated in polymers and copolymers. Another group has attempted to gain a better understanding of the 'blockiness' and alternating character of different copolymers involving ACM together with a range of sulphonated monomers and dextran.[154]

2.2.3 Other Sulphonated Monomers

Ethyl sulphonic acid (**XVII**). Kangas[127] has reviewed the polymerization of this monomer in some detail. In aqueous solution, rate and maximum extent of conversion vary considerably according to the

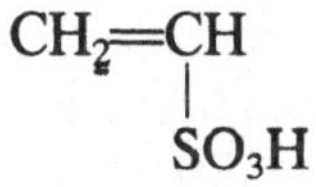

$$CH_2\!\!=\!\!CH$$
$$|$$
$$SO_3H$$

XVII

monomer concentration range, ionic strength and, surprisingly for a strong acid, pH.[155] Electrostatic repulsion, immobility of growing chains at high monomer concentration and heterophase polymerization have been suggested to explain the various phenomena.

Sulphoethyl methacrylate[127] (**XVIII**). In contrast with SSA, the value of $k_p/k_t^{0.5}$ for this monomer was found[156] to increase with

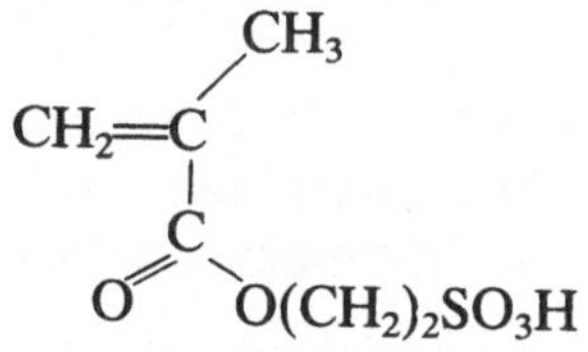

$$CH_2\!\!=\!\!C\overset{\displaystyle CH_3}{\underset{\displaystyle \underset{O}{\overset{\displaystyle \|}{C}}\,O(CH_2)_2SO_3H}{}}$$

XVIII

decreasing ionic strength. This was ascribed to a greater increase in repulsion between the radicals than between radical and monomer. The value of $k_p/k_t^{0.5}$ at 60°C was estimated[127] to be 3·9 compared with 0·14 for methyl methacrylate in organic media,[157] a similar relationship to that observed in the case of SSA and styrene.

Various other sulphonated polymers including sulphonated polyacrylamides, polyacrylates and poly(α-methacrylates) are described in the literature and are included in Kangas's review.[127] Routes to rather more complex sulphonated polymers involving polycondensation methods have also been described, but these are outside the scope of this chapter.

2.2.4 Formation of Polysulphonates by Polymer Modification
The most important commercial route to poly(styrene sulphonate) and its salts is probably the sulphonation of polystyrene.

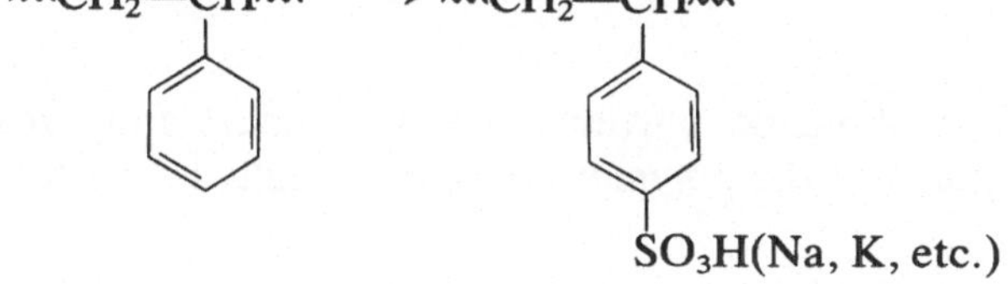

Unless special precautions are taken, concentrated sulphuric acid and oleum are not good sulphonating agents since they cause crosslinking and thereby insolubility of the product. Sulphur trioxide in halogenated solvents is much more satisfactory, provided certain conditions are satisfied. These include low reactant concentration with concurrent supply, efficient agitation, low temperature and pure solvents.[158] Various processes which avoid crosslinking are outlined by Goethals[159] and by Vorchheimer.[10] More recent patents refer to the use of carefully controlled reactant ratios for sulphonation of poly-styrene and subsequent neutralization,[160] the sulphonation[161] of poly-styrene surfaces by anhydrous H_2SO_4 and the reaction of a hydrocarbon-soluble acyl sulphate sulphonating agent with a polystyrene/hydrocarbon cement.[162,163]

Other polysulphonates are obtained by various polymer modifications. Sulphonation of low molar mass styrene/maleic anhydride copolymers[164,165] yields an essentially alternating copolymer of the type **XIX**.

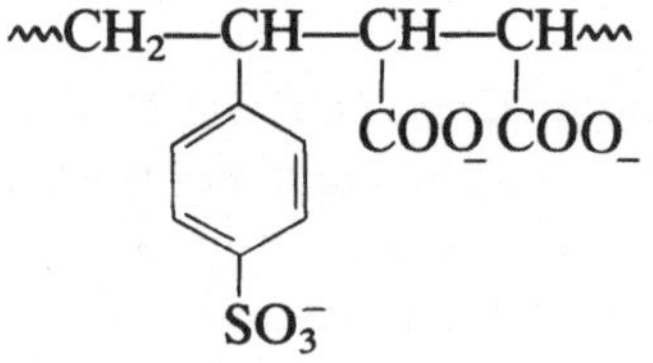

XIX

Poly(vinyl-*p*-toluene sulphonate) is obtained by reaction of poly(vinyl alcohol) with *p*-toluene thionyl chloride in pyridine.[166] The

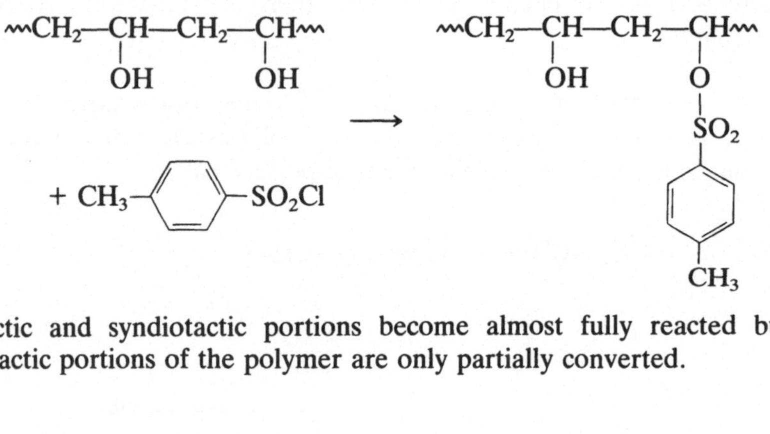

atactic and syndiotactic portions become almost fully reacted but isotactic portions of the polymer are only partially converted.

NaHSO$_3$ has been used in the sulphonation of polymers and copolymers of (di)cyclopentadiene containing a residual double

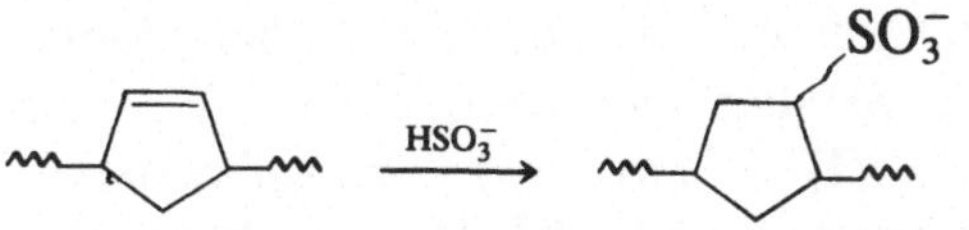

bond[167] and in the formation of adducts of acrolein–acrylic acid polymers:[168]

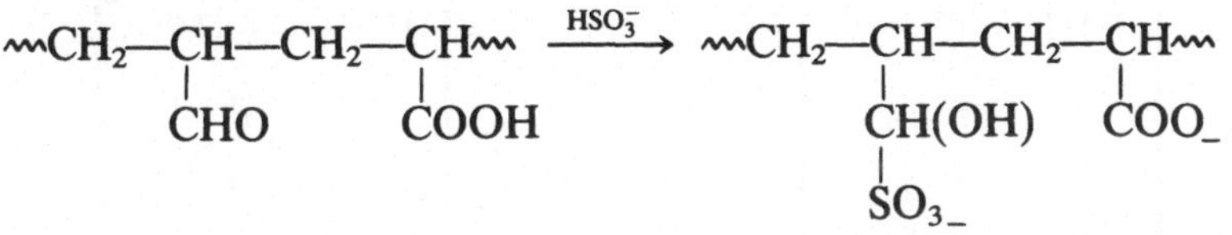

It has also been reported[169] that selective sulphonation occurs when H$_2$SO$_4$ is used under suitable conditions to treat copolymers of the type **XX**.

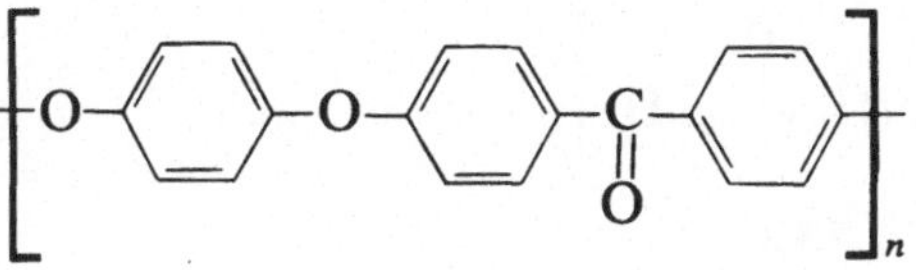

A

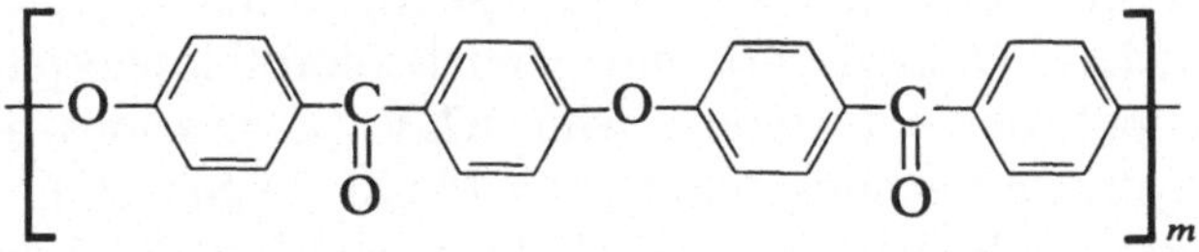

B

XX

It is claimed that at least 70% of A units are sulphonated and that B units are unaffected.

Finally, corn starch has been grafted[170] with a copolymer of CH$_2$=CH·CO·NH·CMe$_2$·CH$_2$·SO$_3$H and acrylonitrile, and is then hydrolyzed to form a material capable of absorbing large quantities of water from emulsions, suspensions and dispersions; one application is for solidifying sewage.

2.3 Phosphorus-containing Polymers

These polyelectrolytes are comparatively recent developments in water treatment chemicals and have therefore received less coverage in the literature. Vorchheimer[10] has described their formation by various routes, as have Sandler and Karo.[171] A wide range of poly(vinyl phosphates) (**XXI**), poly(vinyl phosphonates) (**XXII**) and poly(vinylphosphonic acids) (**XXIII**) has been described.[171] Free-radical polymerization, initiated by conventional azo and peroxide

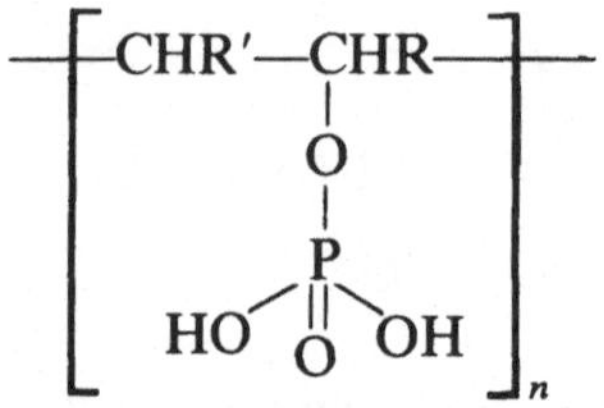

XXI

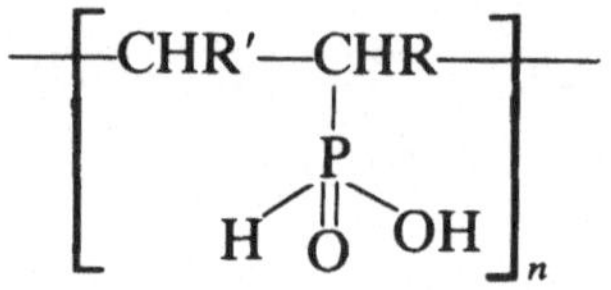

XXII

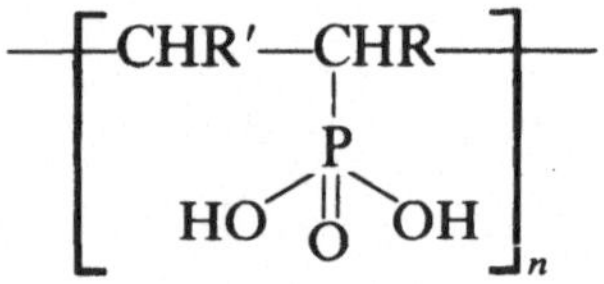

XXIII

initiators, is the usual mechanism. In recent years, attention has been concentrated on complex phosphorus-containing monomers. Examples include O-ethyl-O-(p-vinylphenyl)chlorothiophosphate (**XXIV**)[172] and copolymers of styrene with **XXV** (Z = various aromatic and alkyl hydrocarbon substituents; R = Me_3Si, Et_3Si, etc.);[173] the trimethylsilyl ester hydrolyzes more easily than the triethylsilyl derivative.

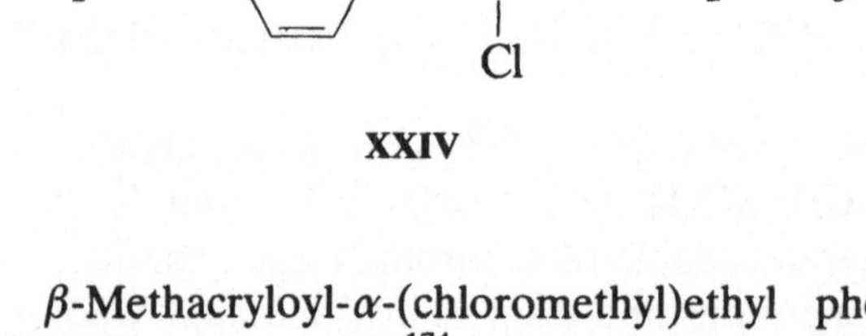

XXIV

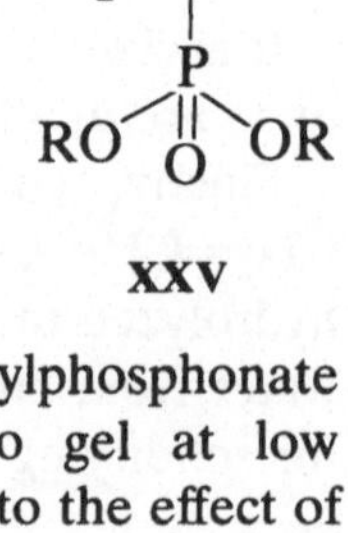

XXV

β-Methacryloyl-α-(chloromethyl)ethyl phenyl methylphosphonate (**XXVI**) was shown[174] to polymerize rapidly and to gel at low conversions. The high polymerization rate was ascribed to the effect of

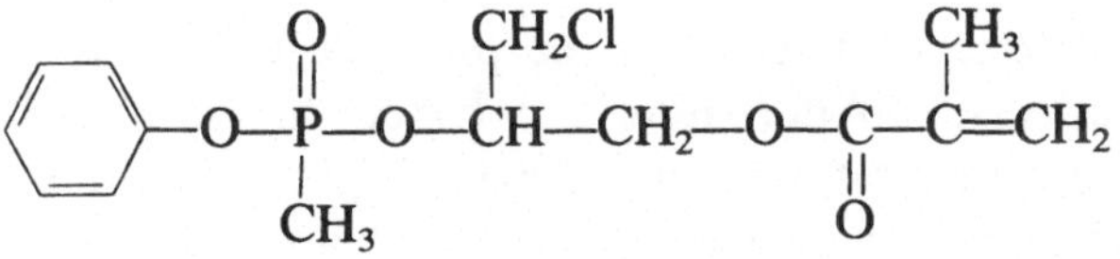

$$\text{C}_6\text{H}_5\text{—O—P—O—CH—CH}_2\text{—O—C—C=CH}_2$$

XXVI

steric hindrance of the bulky phosphorus-containing group on the termination reaction.

Hexachlorocyclotriphosphazene (**XXVII**) was polymerized[175] using boron halides or their triaryl phosphate complexes. The reaction is thought to involve the formation of mixed chlorophosphates and phenyl borates.

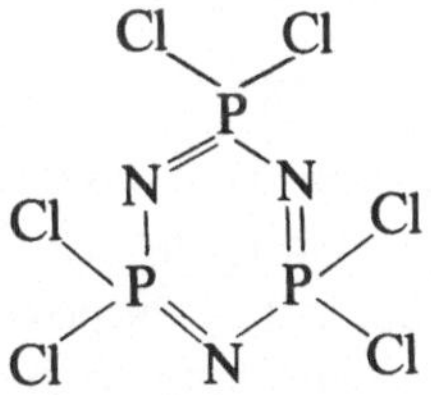

XXVII

Copolymerization provides a popular route to phosphorus-containing polymers. These often involve alkyl phosphonates, but clearly, hydrolysis to the acid is usually feasible. Examples include the copolymerization of dibutyl or di-isobutyl vinylphosphonate with acrylic acid,[176] in which the sequence distribution of the chains depends on the length and branching of the alkyl group in the vinylphosphonate. In the copolymerization of $CH_2\text{=}C(R)\cdot PO(OR')_2$ with styrene, methyl methacrylate, vinyl acetate and $CH_2\text{=}CH\cdot PO\cdot(OEt)_2$, it was found that the comonomers displaying the greatest tendency to become incorporated in the copolymer are those with the weakest conjugation.[177]

2-Acrylamido-2-methyl-1-propanephosphonic acid (**XXVIII**) and its

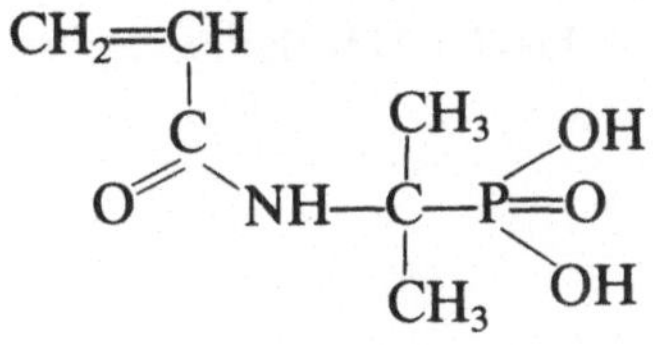

XXVIII

salts copolymerize with ACM in the presence of strong acid to give versatile products.[178] Alternating copolymers are formed by the un-catalyzed polymerization of 2-phenyl-1,3,6,2-trioxophosphocane (**XXIX**) with acrylic, methacrylic acid or α-keto-acids.[179]

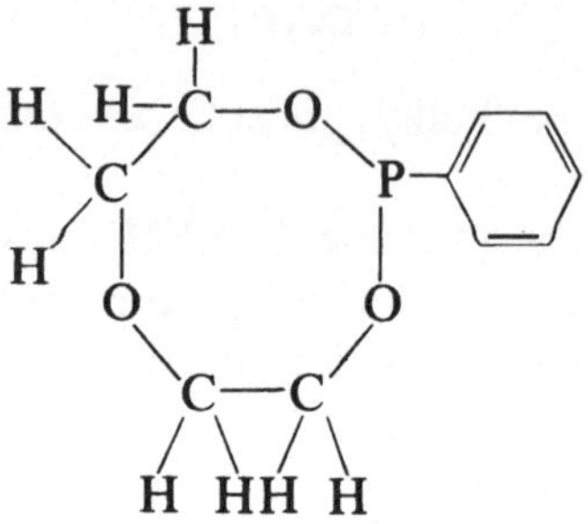

XXIX

Condensation polymerization is a common route to phosphorus-containing polyelectrolytes.[10] An approach that is frequently used is to react a polyfunctional amine with a diepoxide, dihalide or epihalohydrin to give linear nitrogen-containing chains:

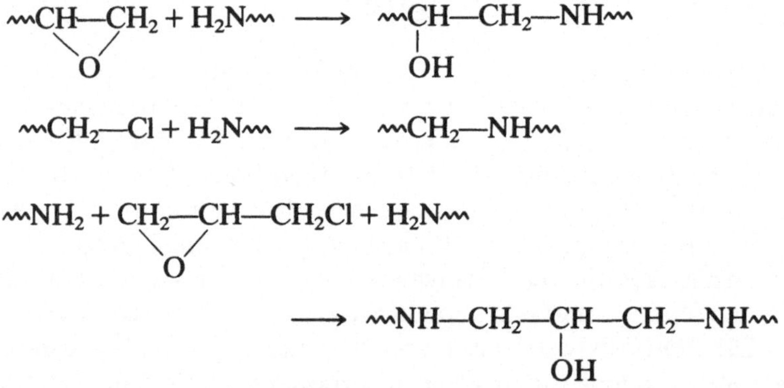

Phosphonic or other phosphorus-containing groups are then introduced by the Mannich Reaction, e.g.

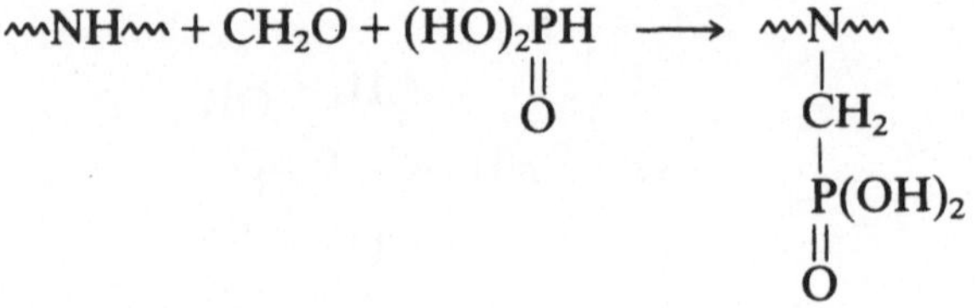

Various condensation reactions involving reactants such as $C_6H_5PO(Cl)_2$ and $C_6H_5PO(OH)_2$,[180] CH$_2$—CH—CH$_2$Cl, POCl$_3$ and

CH$_2$—CH—CH$_3$,[181] [RSi(OH)$_2$O]$_3$M with R$_2'$PO·OH,[182] (EtO)$_2$PCl

with benzoquinone[183] and 2-phenyl-4H-1,3,2-benzodioxaphosphorin-4-one (**XXX**) with 4,4'-diphenoquinones[184] yield phosphorus-containing chain backbones.

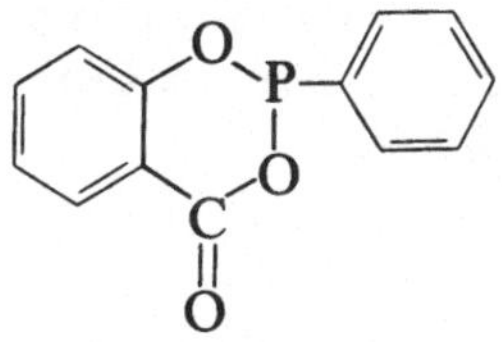

XXX

Modification of existing polymers using phosphonating agents is a useful route to phosphonic polymers. Poly(vinyl alcohol) is a widely used polymeric substrate.[171] Poly(ethylene glycol) is also widely used,[185] for example in:

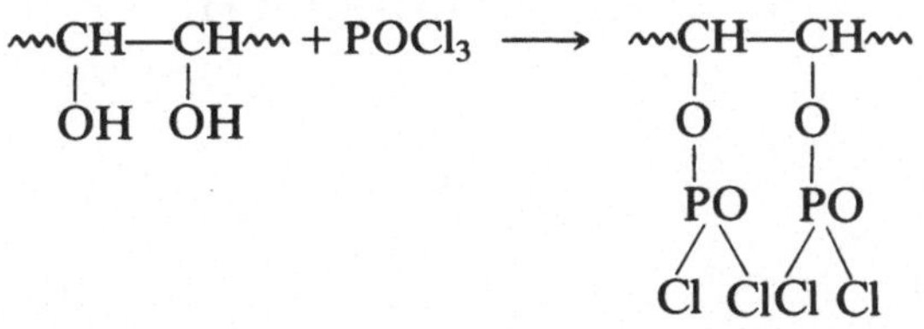

$$\text{wCH—CHw} + POCl_3 \longrightarrow \text{wCH—CHw}$$

3 CATIONIC POLYELECTROLYTES

Anionic polymers are normally prepared by free-radical polymerization of suitable monomers or by relatively simple modification of conventional polymers. Cationic polymers normally involve addition or condensation polymerization of much more complex monomers, often already converted to the cationic form. As before, an alternative approach is to modify existing polymers, but usually by more complex routes. Reference has already been made to the distinction between 'weak' and 'strong' polyelectrolytes. In the former case the degree of ionization is a function of pH (increasing due to protonation at low pH), and in the latter charge is independent of pH by virtue of the formation of quaternary ammonium salts.

3.1 Nitrogen-containing Polyelectrolytes

A number of reviews of nitrogen-containing polyelectrolytes have appeared. Tomalia[186] and Luskin[187] have described the preparation, polymerization and copolymerization of monomers such as alkenyl-2-imidazolines (**XXXI**), alkyl-vinylpyridines (**XXXII**), 2-aziridinyl ethyl methacrylate (**XXXIII**), aminoalkylacrylates and methacrylates such as diethylaminoethyl methacrylate (**XXXIV**), aminoalkylacrylamides and methacrylamides (**XXXV**).

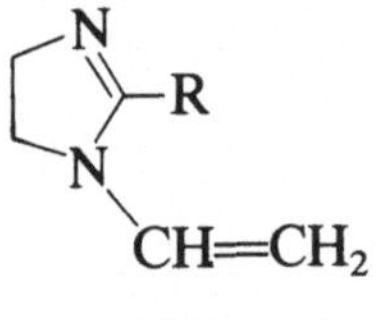

XXXI

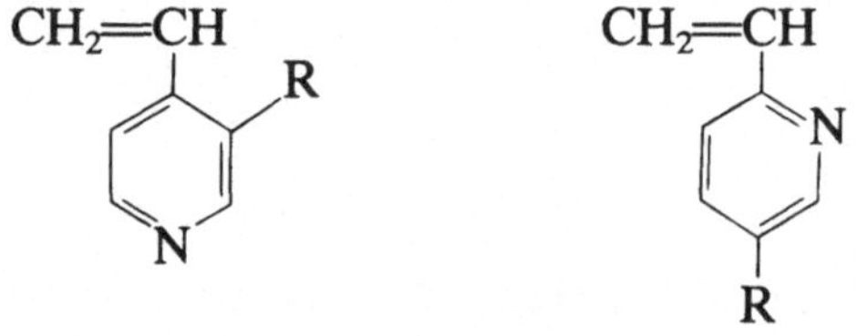

XXXII

etc.

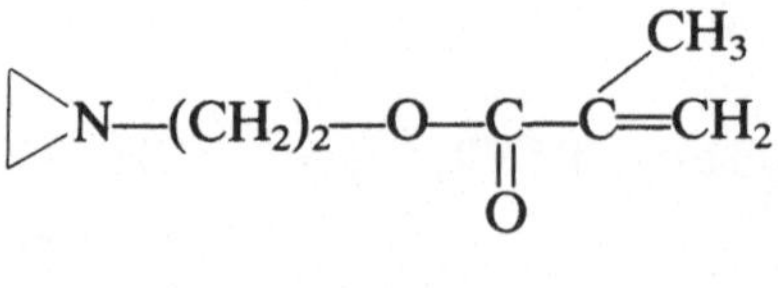

XXXIII

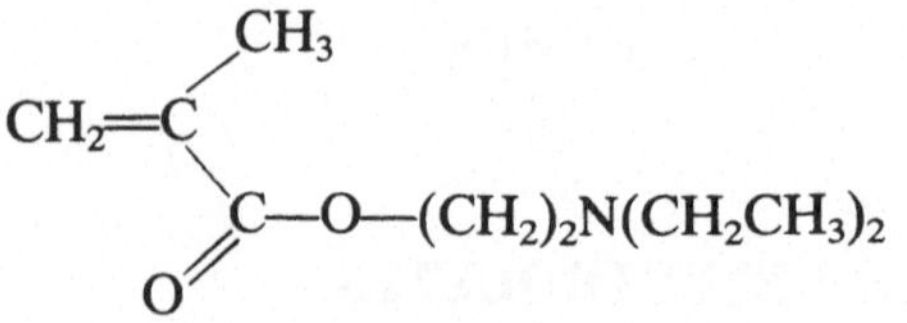

XXXIV

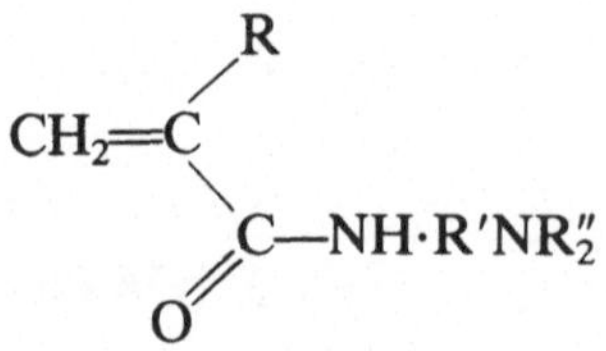

XXXV

Vorchheimer[10] has reviewed a range of more complex nitrogen-containing polyelectrolytes in some detail, as have Hoover et al.[4-6]

Because of the high cost of nitrogen-containing monomers suitable for polyelectrolyte formation, these are usually incorporated in relatively low amounts (typically ~10 mol %) as comonomers with cheaper monomers such as ACM. Addition polymerization and copolymerization usually take place by free-radical or ionic routes, the former being the more important. Initiation is usually by conventional systems, although peroxides may react with the amino group of the monomer via a redox reaction. This is usually to be avoided since it normally leads to very rapid initiator loss and low extents of conversion. Under some circumstances, however, it may serve as the initiator system. In the polymerization[188] of N,N-dimethylaminoethyl methacrylate (DMAEMA; **XXXVI**) in the presence of lauroyl perox-

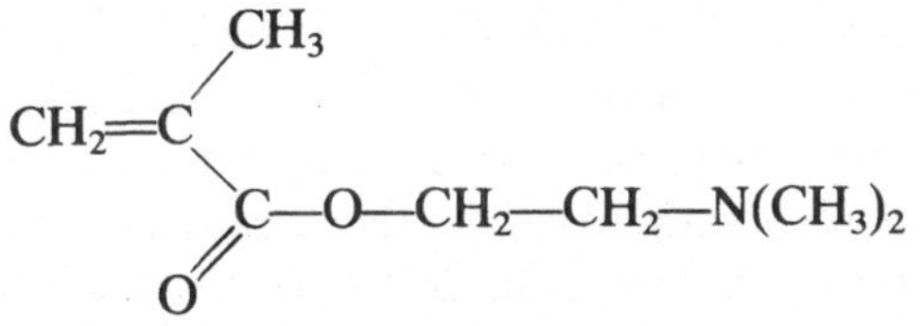

$$CH_2=C \overset{CH_3}{\underset{\underset{O}{\overset{\|}{C}}-O-CH_2-CH_2-N(CH_3)_2}{}}$$

XXXVI

ide, the rate equation was found to be:

$$R_p = [DMAEMA]^{1 \cdot 5}[\text{peroxide}]^{0 \cdot 5}$$

and the overall energy of activation was $40 \cdot 6 \, \text{kJ mol}^{-1}$. It is likely therefore that initiation takes place via a redox reaction between the peroxide and the amino group of the monomer.[186]

3.1.1 Polymerization of Aminoalkyl Acrylates and Aminoalkyl Methacrylates

The most commercially important members of this class are *tert*-butylaminoethyl and dimethylaminoethyl methacrylates and amino-ethyl methacrylate hydrochloride. Luskin's review[187] deals in considerable detail with the preparation and properties of the monomers and the resulting polymers. These monomers may be polymerized in an organic or aqueous phase. In the latter case, low pH is usually maintained, to ensure solubility by means of the formation of the protonated form of the amine.

Bulk polymerization[189] yielded values of $0 \cdot 260$ and $0 \cdot 195$ for $k_p/k_t^{0 \cdot 5}$ at $44 \cdot 1^\circ C$, indicating the likelihood of reasonably high values of R_p and product molar mass. The gel effect is also apparent,[190] thereby leading to higher values of these quantities. Copolymerization characteristics are strongly influenced by the nature of the solvent. For example, with styrene as the comonomer, polar solvents such as DMF give monomer reactivity ratios which are close in value to those obtained in bulk, whereas in non-polar solvents such as toluene and tetrachloromethane, the reactivity of the aminoester is greatly enhanced. This was ascribed to the effect of chain transfer.[191]

In a more recent study,[192] the copolymerization of (phenylamino)ethyl methacrylate with AA and MAA was studied in acetic acid and dimethylsulphoxide (DMSO). In both cases, R_p was found to increase with the increase of AA and MAA concentration in the feed. In acetic acid, this was ascribed to the higher reactivity of AA and MAA compared with DMAEMA. In DMSO, the reactivity of DMAEMA is greater. The increased R_p is ascribed to intramolecular hydrogen bonding in the acids, which leads to changes in k_p and k_t.

The decrease[193] in R_p with the increase in pH for the copolymerization of the aminoesters with MAA under acidic conditions displays similar characteristics to those of the homopolymerization of MAA.[65] Hydrolysis under alkaline conditions prevented extension of the studies to cover the same pH range as was covered with AA and MAA. Various patents for polymerization of aminoacrylate esters in acid solution, or of the corresponding hydrochlorides, have been taken out. It is claimed,[194] for instance, that addition of NaCl to the $H_2O_2/HO·CH_2SO_2Na$-initiated copolymerization yields a polyelectrolyte whose water solubility is enhanced. It is also claimed[195] that in the adiabatic polymerization of DMAEMA in strongly acid solution using a water-soluble azo initiator, rapid cooling following attainment of the maximum temperature considerably lessens thermal degradation of the polymer, thereby increasing product molar mass and cationic content.

Other workers[196] have shown that lowering of the pH by addition of butyric acid to the system leads to increased R_p and product molar mass in the polymerization of DMAEMA. This was attributed to a reduction in the rate of termination. The same workers also observed a matrix effect giving enhanced R_p and molar mass on the addition of poly(methacrylic acid).

A further effect of variation of pH and ionic strength is undoubtedly the resultant change in polymer chain conformation. It has been reported[197] that copolymers of DMAEMA and MAA undergo chain uncoiling when the pH of the solvent medium decreases from 8·65 (the iso-electric point) to 3. This is a predictable consequence of the intrachain electrostatic repulsion arising from increased protonation of the amino groups with lowering of pH. It is significant that the uncoiling over this pH range is greater in water than in $0.1M$ KCl. The increased ionic strength would be expected to reduce electrostatic repulsion, thereby causing less chain expansion. In view of the results obtained for AA and MAA, such conformation effects would be expected to exert a significant effect on polymerization behaviour of these monomers.

Many papers describing suitable conditions for copolymerization have appeared in recent years. Heavy metal salts and inorganic phosphorus compounds are claimed[198] to increase R_p in the azo-initiated aqueous-phase copolymerization of methacrylamide, MAA, aminoalkyl acrylates and methacrylates and their water-soluble compounds. A two-step aqueous phase process involving UV irradiation to give 95% conversion before concentration and finishing off with ^{60}Co γ-rays has been reported.[199] The concentration of the additives was found to be critical in ensuring very high conversion.

Self-heating during initial photosensitized polymerization is used in a process for the production of ACM/DMAEMA copolymers.[200] After an irradiation period of 10 min, thermal decomposition of the azo initiator in the unstirred system took place. Under conditions of low heat loss, a peak temperature of 90°C was attained and high molar mass polymer was obtained.

In the dispersion polymerization[201] of an aqueous solution of DMAEMA/ACM in $C_{12}-C_{18}$ alkanes, poly(oxyethylene)oleate is claimed to be a necessary ingredient for the production of a copolymer which gives a smooth paste when dissolved in water. In its absence, flocculation results.

A large number of workers have measured monomer reactivity ratios for copolymerizations involving DMAEMA.[187] More recently, values for diethylaminoethyl methacrylate with ACM and MACM[202,203] and DMAEMA with N,N-diethylacrylamide[204] have been reported. It is clear from the results that solvent effects and, especially, pH effects exert a profound influence. Conversion of DMAEMA to the protonated form clearly increases monomer reactivity very considerably.[203]

3.1.2 Polymerization of Aminoalkyl Acrylamides, Aminoalkyl Methacrylamides and Other Amines

Several authors[4-6,10,14] have reviewed the formation and polymerization of basic monomers.

Major routes for the synthesis of such monomers involve the Mannich Reaction, for example:

$$RCONH_2 + CH_2O + R'NH_2 \longrightarrow RCONHCH_2NHR'$$

$$RCOCH_3 + CH_2O + R'_2NH \longrightarrow RCO(CH_2)_2NR'_2$$

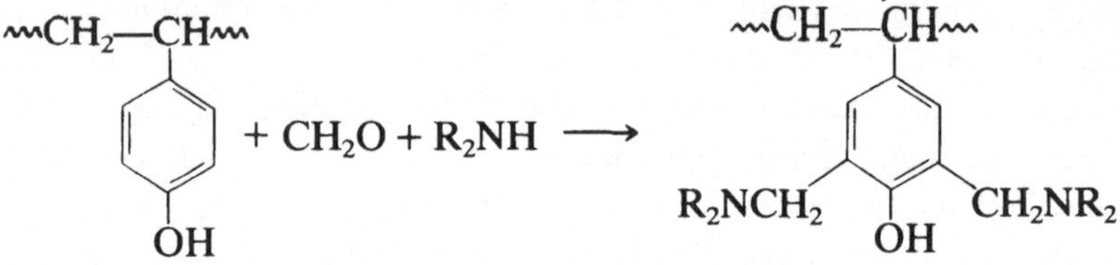

The Hoffmann Degradation is also widely used in the conversion of various acrylamides to the corresponding amines:

$$RCONR'_2 \xrightarrow{\text{NaOBr/NaOH}} RNR'_2$$

Many examples of the application of such reactions are quoted in the literature.[10] Studies of the polymerization of the uncharged monomers are uncommon, most work being done with protonated amines or quaternary ammonium salts.

3.1.3 Polymerization of Aminostyrene and its Derivatives
The trichloroacetic acid initiated polymerization of 4-(N,N-dimethylamino)styrene (4-DMAS; **XXXVII**) was studied[205] in solvents

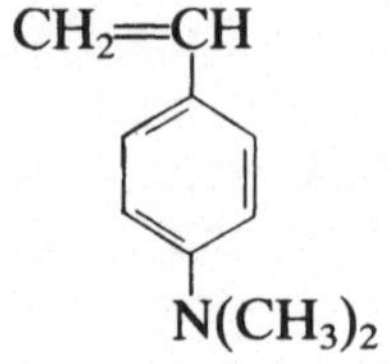

XXXVII

of different electron donating and accepting properties and of different relative permittivities. The relative permittivity was shown to correlate well with R_p. Results in benzene were very different from those in nitrobenzene. It was also found[206] that although the rate of the benzoyl peroxide or lauroyl peroxide initiated polymerization is very slow, low concentrations of the monomer form a redox couple with benzoyl peroxide capable of initiating the polymerization of MMA. The rate equation is:

$$R_p = K_p[\text{MMA}][4\text{-DMAS}]^{0.5}[\text{Bz}_2\text{O}_2]^{0.5}$$

The overall energy of activation was found to be fairly low (30·9 kJ mol^{-1}) and the 4-DMAS became incorporated in the polymer chain.

Copolymerization[207] of 4-DMAS with styrene initiated by AIBN or trichloroacetic acid yielded a homopolymer of pure 4-DMAS. In benzene, the trichloroacetic acid initiated polymerization obeyed the rate equation:

$$R_p = K_p[4\text{-DMAS}]^{0.99}[\text{CCl}_3\text{COOH}]^{1.10}$$

and the reaction was virtually unaffected by inhibitors. The radical nature of the process is, thus, not certain.

The position of the ring-substituent was found to be important in determining reactivity. Copolymerization parameters[208] for **XXXIX** are similar to those for styrene and α-methylstyrene, whereas the amino group has a significant effect on **XXXVIII**, whose parameters are similar to those of **XXXVII**.

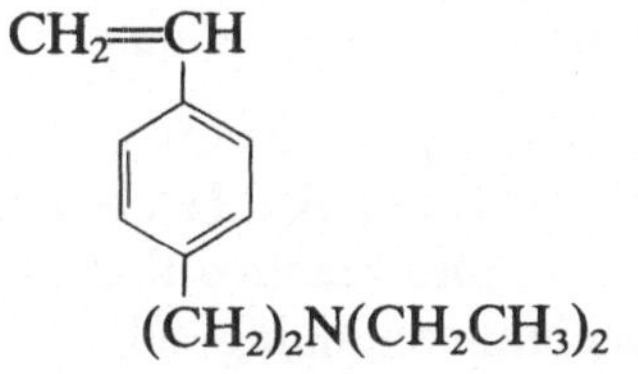

CH$_2$=CH

(CH$_2$)$_2$N(CH$_2$CH$_3$)$_2$

XXXVIII

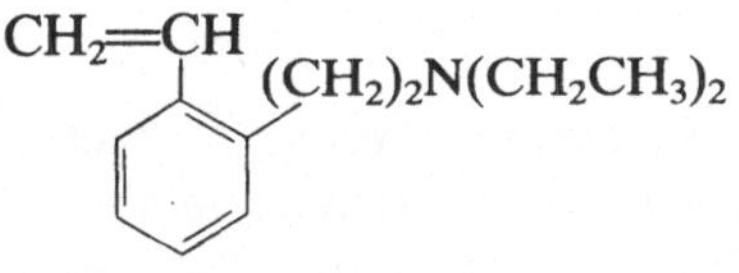

CH$_2$=CH
(CH$_2$)$_2$N(CH$_2$CH$_3$)$_2$

XXXIX

3.1.4 Polymerization of Vinylpyridine and its Derivatives

Luskin[187] has reviewed the synthesis of these monomers in considerable detail. They undergo polymerization by free-radical, ionic and co-ordination mechanisms.

Vinylpyridines and styrene are of comparable reactivity to attack by free radicals and the 3-isomer (**XL**) is normally somewhat more reactive than the 2- (**XLI**) or 4-isomer (**XLII**).

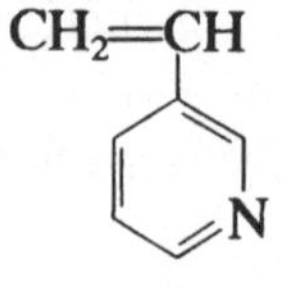

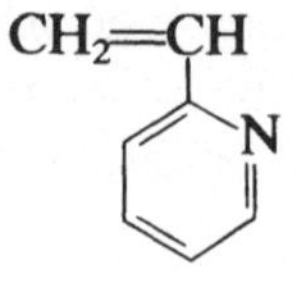

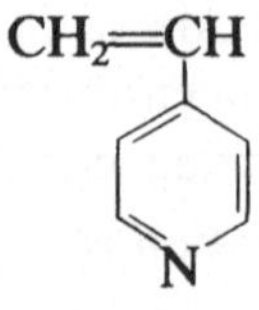

The relative reactivity of the two types of monomer to attack by ions is very different, however, and this is attributed to interaction between the weakly nucleophilic N-atom and the charged initiating species.[209]

In the free-radical polymerization of 4-vinylpyridine, conventional initiators are effective. Values of k_p and k_t for bulk polymerization (12×10^6 and 3×10^6 $dm^3\,mol^{-1}\,s^{-1}$ respectively at 25°C[210]) are both rather low compared with most polymerizations. However, the ratio of $k_p/k_t^{0.5}$ has a value of 4, signifying that rapid polymerization occurs. The surprisingly high value of the monomer transfer constant ($\sim 7 \times 10^{-4}$ at 25°C) will of course lower the product molar mass. Chain transfer in the polymerization of both 2- and 4-vinyl-pyridine with a number of transfer agents has been studied.[211] Both radicals display similar reactivities and the substrate reactivities are in the order:

ethanol < octanol < toluene < benzyl alcohol

< chloroform < nitroethane[211]

In another study,[212] transfer constants for polymerization of the same two monomers in the presence of R_3GeH and R_3SiH showed that, towards both substrates, the 2-vinyl- was more reactive than the 4-vinyl-pyridine radical and that the germanium-containing substrate was more reactive than the silicon-containing substrate towards both radicals. This was taken to indicate that donor–acceptor interactions were influential in the reactions.

NMR studies have indicated[213] that linear relationships between chemical shift for 2-methyl-5-vinylpyridine and the reactivity parameters displayed by it in vinyl polymerization were in agreement with those for other monomers.

Polymerization is also carried out in solution, suspension and emulsion,[187] but in the aqueous solutions low pH must be used to ensure monomer and polymer solubility, in which case the nature of the species changes considerably by virtue of partial ionization. As well as reactant reactivity being altered, chain conformation may also be influenced, thereby affecting monomer–radical and radical–radical interaction. In the aqueous-phase copolymerization[214] of 2-methyl-5-vinylpyridine with MAA, for instance, product viscosity was found to display a minimum at pH 4·5. The viscosity maxima observed at pH ~2 and ~10 presumably reflect the chain expansion occurring due to increased ionization at the two extreme pH values.

Among studies reported recently are the copolymerization[215] of 2-vinylpyridine with methyl acrylate or methacrylate in benzene solution. NMR measurements indicated an alternating co-isotactic microstructure for the methyl acrylate/2-vinylpyridine copolymer but a random structure for the methyl methacrylate/2-vinylpyridine copolymer.

In the persulphate-initiated polymerization of 2-methyl-5-vinylpyridine,[216] formation of monomer radicals is said to occur by oxidative coupling of $S_2O_8^{2-}$ with the monomer. This involves formation of an intermediate between the monomer and cetylpyridinium chloride, present as an emulsifier, and results in a comparatively low overall energy of activation (39·5 kJ mol^{-1}).

Charge-transfer polymerization of 2-vinylpyridine has been reported[217] in the presence of n-butylamine/tetrachloromethane; mechanisms considered are:

$$RNH_2 + M \underset{k_{-1}}{\overset{k_1}{\rightleftharpoons}} A$$

$$A + CCl_4 \overset{k_2}{\longrightarrow} M\dot{C}Cl_3 + R\overset{+}{N}H_3\bar{C}l$$

(RNH_2 = donor; M = acceptor), and

$$RNH_2 + CCl_4 \underset{k_{-3}}{\overset{k_3}{\rightleftharpoons}} B$$

$$B + M \overset{k_4}{\longrightarrow} M\dot{C}Cl_3 + R\overset{+}{N}H_3\bar{C}l$$

(M = 2-vinylpyridine; R = C_4H_9; A, B = charge-transfer complexes).

Various groups have studied anionic polymerization of these monomers. The effect of alkaline earth counterions has been investigated,[218] and also of their triphenylmethyl derivatives,[219] $Ph_3CMX(THF)_n$ (M = Ca, Ba, Sr; X = Br, Cl; n = 2, 4, 5) in tetrahydrofuran (THF),

1,2-dimethoxyethane or in bulk. THF gives greater tacticity in poly-(2-vinylpyridine), and this decreases with change in initiator in the order $Ca > Sr > Ba$ and is also greater when $X = Br$.

Complex bases of the type $NaNH_2$–$RONa$ ($R = MeO(CH_2)_2$, $Et(OCH_2CH_2)_2$ or $CH_2 = CHCH_2$) are reported[220] to be good initiators of 2-vinylpyridine polymerization in THF and toluene. Variation of the alkoxide and the solvent gave polymers of different molar mass.

Initiation of living anionic polymerization of 2-vinylpyridine by benzylpicolylmagnesium (**XLIII**) in hydrocarbon solvents to give

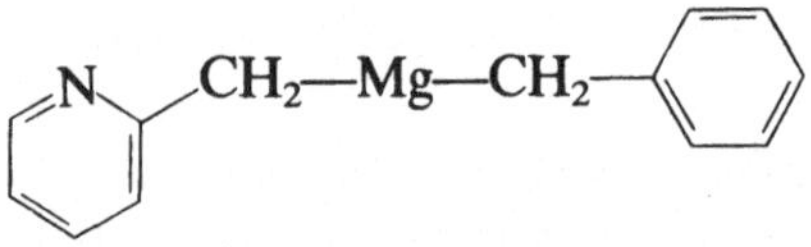

XLIII

isotactic polymer has been reported.[221] It is suggested that the stereo-regulating mechanism involves displacement by the monomer molecule of the last pyridine ring co-ordinated with Mg, and complexation at vacant sites on the Mg^{2+} cation. The monomer is subsequently incorporated in the growing chain following a planar four-centre transition state, with simultaneous electron transfer between the C—Mg bond and the vinyl bond, and retention of configuration.

Finally, studies[222] of the polymerization of 4-vinylpyridine in the presence of Ziegler–Natta catalysts ($TiCl_3$, Al alkyls or Zn alkyls) indicate a mechanism which is different from a free-radical route.

3.1.5 Preparation of Quaternized Nitrogen-containing Polymers

It has already been pointed out that the degree of ionization of basic nitrogen-containing polymers is a function of pH. Conversion to quaternary ammonium salts gives products whose degree of ionization is pH-independent. Such polymers are prepared by polymerization of the previously quaternized monomer or by quaternization of the uncharged polymer.

3.1.6 Quaternization of Nitrogen-containing Monomers

Quaternization usually involves the reaction of a tertiary amine-containing monomer with alkyl halides or sulphates to produce ionic

salt-like products:

$$RNR_2' + R''Cl \rightarrow R\overset{+}{N}R_2'R''Cl^-$$

$$RNR_2' + R''SO_3H \rightarrow R\overset{+}{N}R_2'R''\overset{-}{S}O_3H$$

Commonly used monomers are vinyl pyridine and its derivatives, (*tert*-aminoalkyl)acrylates, (*tert*-aminoalkyl)methacrylates, (*tert*-aminoalkyl)acrylamides and (*tert*-aminoalkyl)methacrylamides. In general, the latter are easier to use. They are usually dry, crystalline solids which are more reactive in polymerization and more hydrolytically stable than the usually liquid aminoesters.

The quaternization process occurs readily. For example, the addition of methyl chloride to DMAEMA over a period of 4 h followed by stirring at 35°C and then at 70°C gives a large yield[223] of quaternary methacryloyloxyethylammonium chloride **XLIV**. The addition of hydroquinone is necessary to avoid spontaneous polymerization.

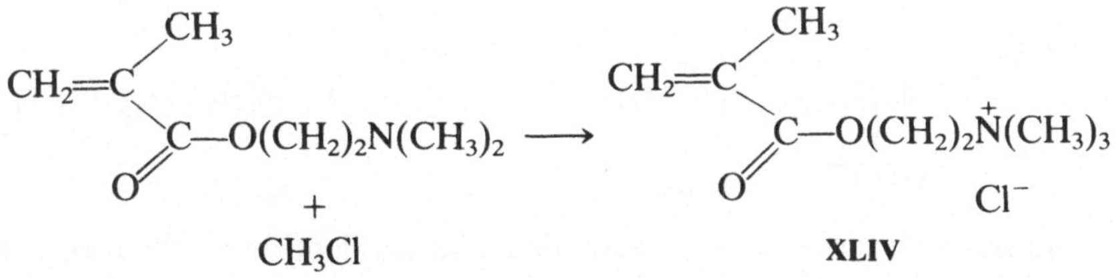

Kabanov *et al.*[224] reported that in the quaternization of DMAEMA and 2-diethylaminoethyl methacrylate, spontaneous polymerization followed an initial period of quaternization. This did not occur in the absence of DMAEMA and ethyl bromide.

Complex quaternary compounds are also readily formed. A typical example[225] is the formation of salts with the structure **XLV** ($R^3 = H$,

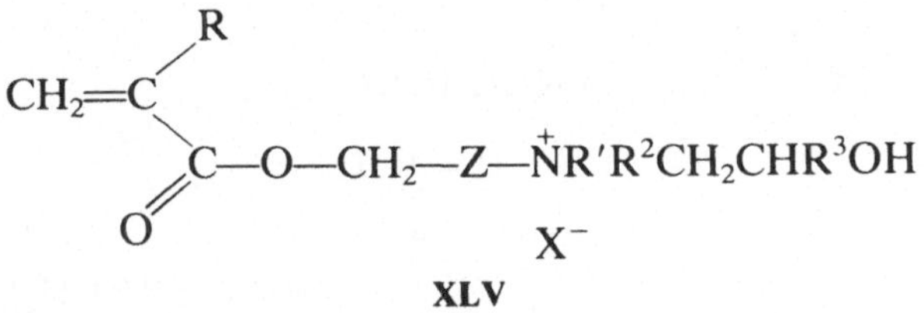

Me; R^1, $R^2 = C_{1-3}$ alkyl; $Z = CH_2$, hydroxyethylene, hydroxypropylene; $X = Cl$, Br, ClO_4, NO_3, p-Me·C_6H_4·SO_3 or AcO). Typically these salts are formed by reaction with reactants such as

N,N-diethylethanolamine, neutralized with gaseous HCl in propane-2-ol/toluene, and treated for 5 h at 60°C with glycidyl methacrylate (**XLVI**).

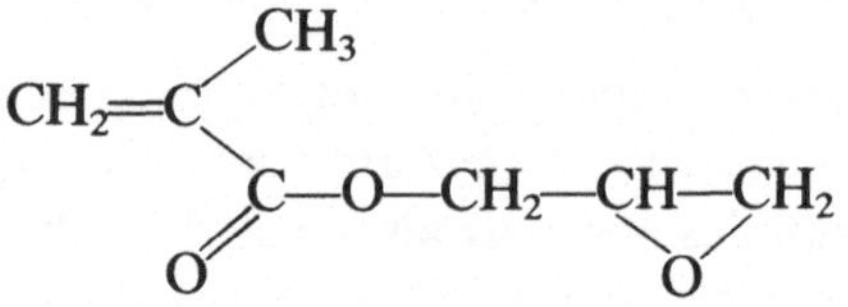

$$CH_2{=}C{\overset{CH_3}{\underset{\underset{O}{\parallel}}{\;}}}C{-}O{-}CH_2{-}CH{-}CH_2$$

XLVI

Acrylamide derivatives are also easily converted to the quaternary ammonium form. N-(3-Dimethylaminopropyl) methacrylamide (**XLVII**) in methanol solution at 40°C is quantitatively converted[226] in 3–6 h by reaction with methyl chloride to (3-methacrylamidopropyl)-trimethylammonium chloride (**XLVIII**).

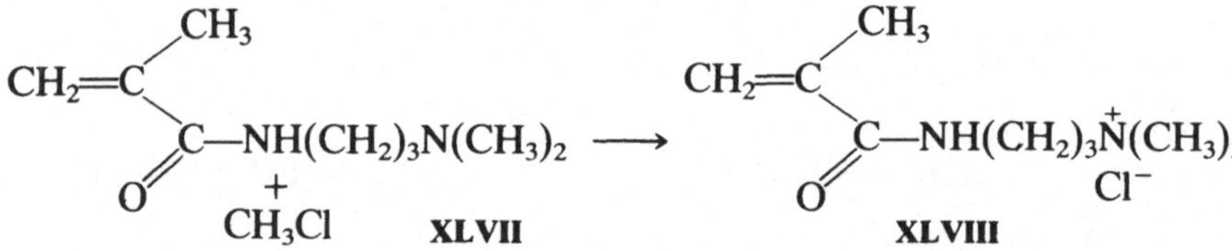

$$CH_2{=}C{\overset{CH_3}{\underset{\underset{O}{\parallel}}{\;}}}C{-}NH(CH_2)_3N(CH_3)_2 \longrightarrow CH_2{=}C{\overset{CH_3}{\underset{\underset{O}{\parallel}}{\;}}}C{-}NH(CH_2)_3\overset{+}{N}(CH_3)_3$$

CH$_3$Cl **XLVII**   Cl$^-$ **XLVIII**

Similarly, treatment of acrylamide, methacrylamide or their derivatives with C_{1-4} alkyl chlorides or dialkyl sulphate at 40–45°C and a pressure of 3–3·5 bar gives 99% conversion after 50 min.[227]

An alternative approach to the reaction of an alkyl halide with an unsaturated tertiary amino compound is to react a tertiary amine with a halogenated monomer. A good example[228] of this is the reaction:

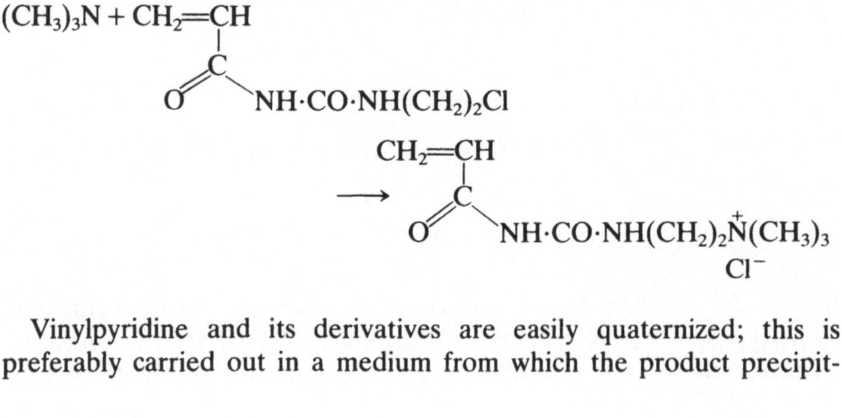

$$(CH_3)_3N + CH_2{=}CH{\underset{C}{\overset{\vert}{\;}}}{\overset{O}{\;}}{}NH{\cdot}CO{\cdot}NH(CH_2)_2Cl$$

$$\longrightarrow CH_2{=}CH{\underset{C}{\overset{\vert}{\;}}}{\overset{O}{\;}}{}NH{\cdot}CO{\cdot}NH(CH_2)_2\overset{+}{N}(CH_3)_3$$

Cl$^-$

Vinylpyridine and its derivatives are easily quaternized; this is preferably carried out in a medium from which the product precipitates on formation. Spontaneous polymerization of the charged polymer occurs at high concentrations in polar solvents (see later).

Further complex routes, in addition to those mentioned in other reviews,[4–6,10] include the reaction between an epoxide, N-propyl methacrylamide and propane-2-ol in the presence of acetic acid at 50°C for 1 h to give **XLIX**.[229] $CH_2 = C(Me) \cdot CO \cdot NH \cdot (CH_2)_3 \cdot NMe_2$ reacted[230]

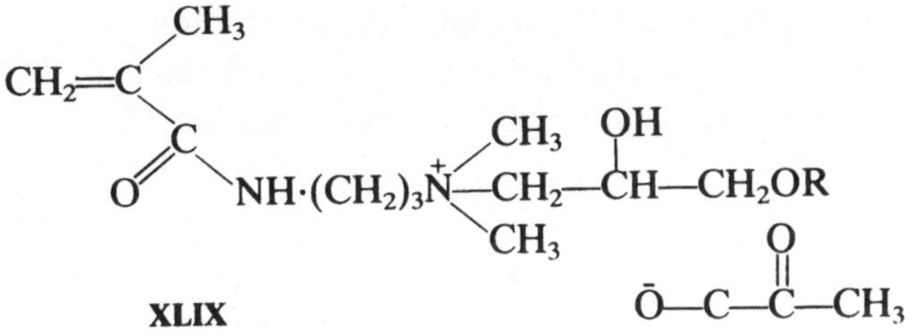

$$CH_2{=}C \overset{\displaystyle CH_3}{\underset{\displaystyle C}{\Big|}}$$

XLIX structure:
$CH_2{=}C(CH_3){-}C({=}O){-}NH{\cdot}(CH_2)_3{-}\overset{+}{N}(CH_3)(CH_3){-}CH_2{-}\underset{OH}{CH}{-}CH_2OR$

$\bar{O}{-}C{-}\overset{O}{\overset{\|}{C}}{-}CH_3$

XLIX

with ethylene oxide, water and acetic acid for 1 h at 50°C gives a 95% yield of **L**.

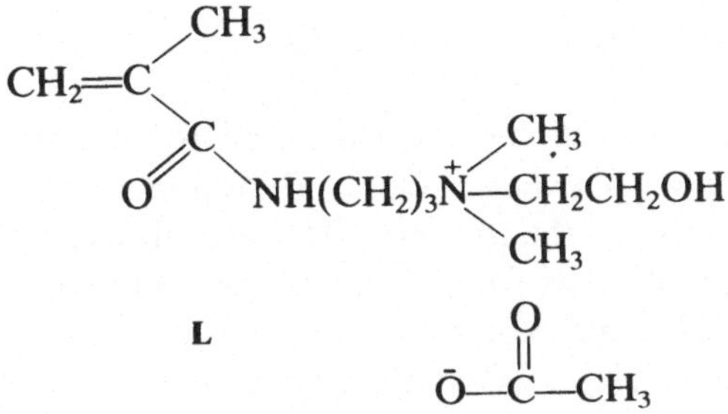

L structure:
$CH_2{=}C(CH_3){-}C({=}O){-}NH(CH_2)_3{-}\overset{+}{N}(CH_3)(CH_3){-}CH_2CH_2OH$

$\bar{O}{-}\overset{O}{\overset{\|}{C}}{-}CH_3$

L

3.1.7 Polymerization of Quaternized Aminoalkyl Acrylates and Aminoalkyl Methacrylates

Several general reviews of the formation and quaternization of these monomers are available.[6,10,187] Polymerization occurs readily by free-radical mechanisms using a varied selection of initiators.

Sulphite and bisulphite are claimed[231] to initiate polymerization rapidly in the presence of salts of Fe, Cu, Sn, Mn, and Ce. Other additives which increase the rate of polymerization include sodium benzene sulphinate (**LI**)[232] which, on addition to a solution of

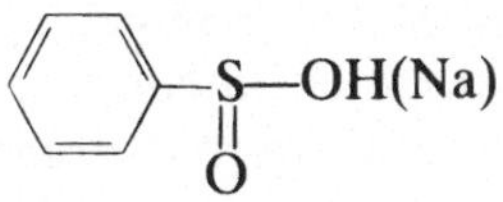

$C_6H_5{-}\overset{O}{\underset{O}{\overset{\|}{\underset{\|}{S}}}}{-}OH(Na)$

LI

water-soluble initiator and monomer, considerably reduces the induction period, increases R_p and the solubility of the resulting polymer. Ascorbic acid and sodium formaldehyde sulphoxylate are said[233] to exert a similar effect.

The addition of a small quantity of a peroxide (which alone is insufficient to cause discernible polymerization) has been shown[234] to increase considerably the rate of initiation by water-soluble azo compounds, presumably by induced decomposition.

Polymerization of methacryloyloxy (dimethylaminoethyl) bromide (**LII**) in aqueous solution obeys the rate equation:[235]

$$R_p = K_p[M]^{1.83}[S_2O_8^{2-}]^{0.5}$$

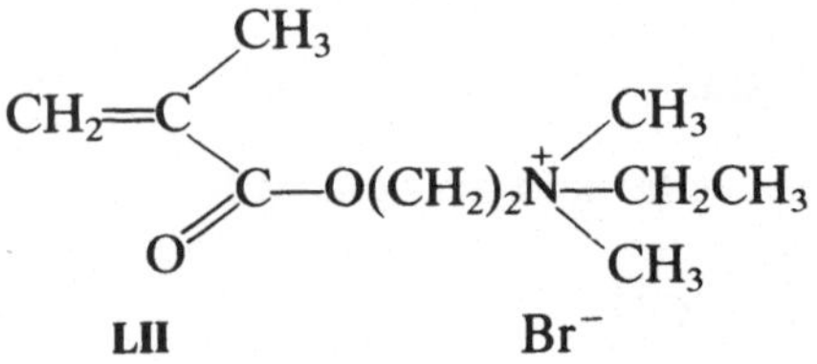

LII Br^-

The high order in monomer is attributed to the formation of an intermediate complex between $S_2O_8^{2-}$ and monomer, so increasing the rate of decomposition of the initiator. The overall energy of activation for the polymerization was found[236] to be $58.2\,kJ\,mol^{-1}$. Polymerization of the chloride salt was found[237] to increase in rate with the increase in pressure. This was ascribed to an increase in k_p.

Combinations of initiators and initiation techniques are claimed to give improved yields of more soluble polymers. Examples include the use of photosensitive initiators such as benzoin together with redox systems such as $S_2O_8^{2-}$/ethylenediamine.[238] For the first three hours, redox initiation enables 90% conversion to be attained. Thereafter, benzoin decomposition by UV irradiation completes the conversion in a 5 min period.

Another example[239,240] is the combination of similar redox systems with an azo initiator. Under the conditions used, 80% conversion was attained in 4 h at 5–40°C. Thereafter, heating for 1 h at 70–90°C caused expansion of the polymer molecules and enabled almost complete conversion to a soluble, non-gelling polymer.

The polymerization process is, in common with the polymerization of other ionic monomers, sensitive to the reaction medium. Radiation-induced copolymerization of the methyl chloride quaternized salt of DMAEMA with ACM in the presence of only 5–20% water gave a copolymer which could be ground to a powder.[241] The addition[242] of a small quantity of the prepared polymer markedly reduces the induction period and probably increases R_p.

The presence of reactive transfer agents (e.g. propane-2-ol) was claimed to eliminate crosslinking, thereby giving a water-soluble polymer.[241] Other chain-transfer agents employed include 2-imidazolidinone (**LIII**),[242] tetrahydro-2-pyrimidinone (**LIV**),[243] sodium bisulphite[244] and formic acid.[245]

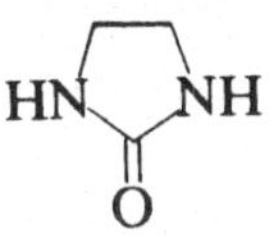

LIII 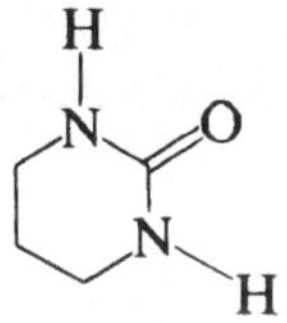　　　　　　　　**LIV**

Suspension and emulsion copolymerizations are also popular. Typical inverse phase systems for the polymerization and copolymerization of quaternized alkylaminoesters and the relevant stabilizers which have been reported include xylene/sorbitan mono-oleate,[246] toluene/(copolymer of AA and 2-ethylhexyl acrylate)[247] and cyclohexane/(sorbitan ethers or ethylcellulose);[248] the last gives larger beads and slightly higher solution viscosity. An alternative approach is the evaporation of the cyclohexane until the residual water content is 15%, when the product is obtained as a dry powder.[249,250]

Finally, aqueous solutions of quaternary aminoesters have been extracted with tetrachloromethane.[251] The resulting extract was freed of solvent by evaporation and the residual monomer polymerized using H_2O_2 and $S_2O_8^{2-}$ to give water-soluble polyelectrolytes.

3.1.8 Polymerization of Quaternized Aminoalkyl Acrylamides and Aminoalkyl Methacrylamides

Procedures are similar to those for polymerization of quaternized DMAEMA monomers. Recent examples include[228] copolymerization of **LV** (R = H, Me; X = Cl, I) with ACM at 15°C and pH 3. The

$$CH_2{=}C \qquad R$$
$$NH{\cdot}CO{\cdot}NH(CH_2)_n\overset{+}{N}(CH_3)_3$$

LV　　　X^-

polymerization was initiated by a complex system consisting of AIBN, $K_2S_2O_8$, $Na_2S_2O_4$ and $FeSO_4$. Thorough purification of the monomers was shown to be of great importance in the production of a polymer free of insoluble residue[252] and the use of ethyl cellulose as a stabilizer for emulsion polymerization is recommended.[253]

Studies of the conformational transitions occurring on counterion exchange at the quaternary ammonium sites and protonation of the residual tertiary amino groups, and their effect on chiro-optical properties, have been reported[254] for partially quaternized, optically active poly[thio-1-(N,N-diethylaminomethyl)ethylene], LVI,

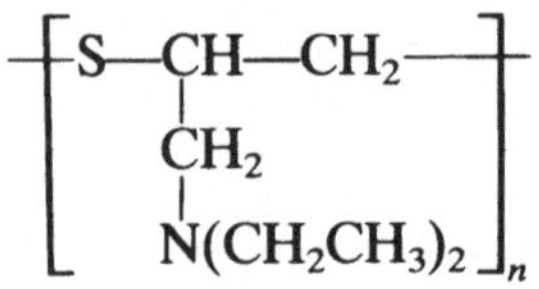

$$\left[\begin{array}{c} S-CH-CH_2 \\ | \\ CH_2 \\ | \\ N(CH_2CH_3)_2 \end{array} \right]_n$$

LVI

In water, the partially protonated chains are said to adopt a globular conformation and protonation proceeds through a globular → extended coil co-operative transition. The quaternized sample, on the other hand, behaved conventionally.

3.1.9 Polymerization of Quaternized Diallylamines (Diallyldialkyl Ammonium Halides)

Diallyldialkyl ammonium halides[255] of the general structure LVII

$$(CH_2{=}CH-CH_2)_2 \overset{+}{N}(CH_3)_2$$

LVII X^-

are of increasing importance in the production of ion-containing polymers and copolymers.[10,256]

Difunctional vinyl monomers would normally be expected to yield branched and subsequently crosslinked polymers. Furthermore, the presence of the allyl group would be expected to favour frequent degradative chain transfer with monomer and therefore low rates of polymerization, because of the probable stability of the allyl radical produced. In this case, however, rapid polymerization produces a water-soluble product with little evidence of exhibiting any of these characteristics. This has been attributed to the prevalence of a cyclization mechanism[257] although some debate has arisen as to the size of the rings formed.[258]

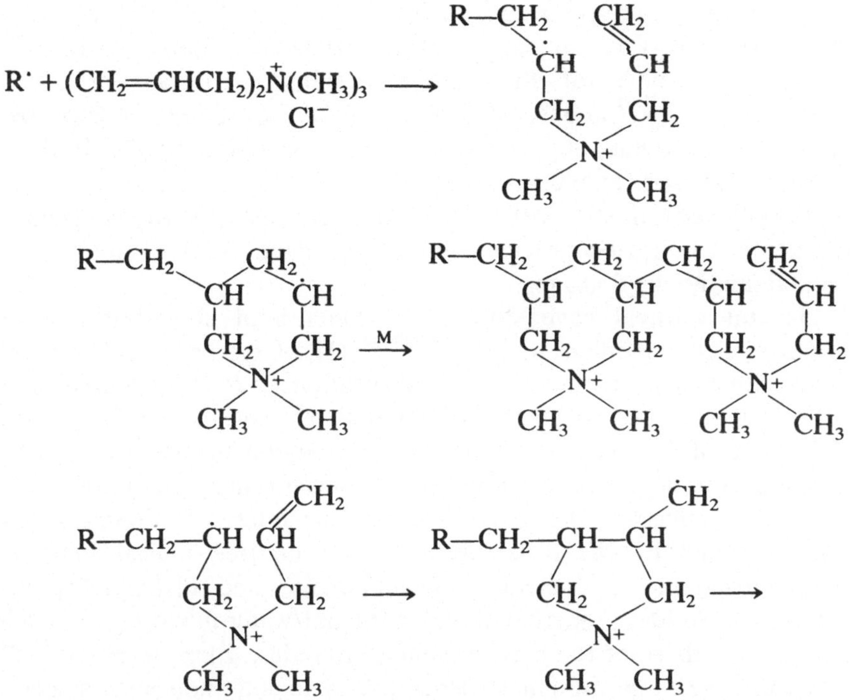

$$R^{\cdot} + (CH_2{=}CHCH_2)_2\overset{+}{N}(CH_3)_3 \quad\longrightarrow$$
$$Cl^-$$

Recent work supports the cyclization mechanism. In the $S_2O_8^{2-}$-initiated polymerization of diallyldimethylammonium chloride, comparison of the experimentally observed rate equation:[259]

$$R_p = K_p[M]^{2\cdot3}[S_2O_8^{2-}]^{0\cdot47}$$

with the theoretical equation for the polymerization of unconjugated dienes indicated partial cyclization during propagation and termination by the combination of cyclized radicals.[259] Further evidence has been provided by Kabanov et al.,[260] who suggested that rearrangement of the allyl radical (**LVIII**) formed by monomer transfer, gave a more reactive radical (**LIX**) which reduced the extent of degradative transfer and chain branching.

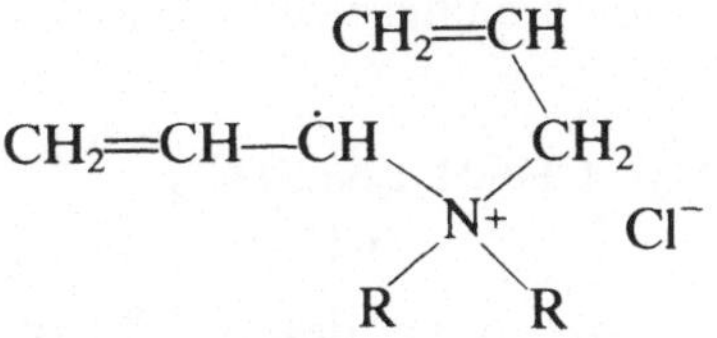

LVIII

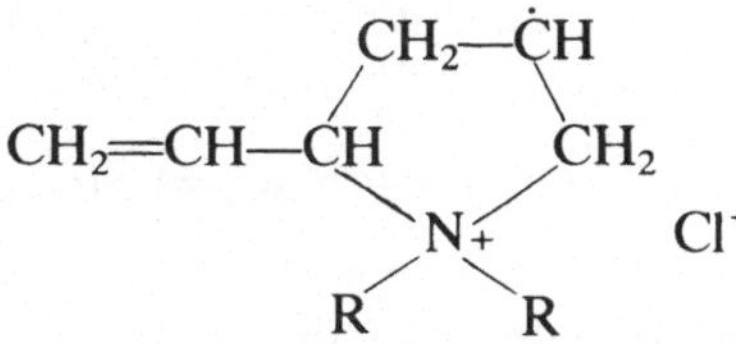

LIX

The presence of one, rather than two, terminal double bonds and a terminal methyl group support this mechanism. Other workers[261] obtained a value for the monomer transfer constant of $2 \cdot 5 \times 10^{-3} \, dm^3 \, mol^{-1} \, s^{-1}$ at 35°C. This is typical of the value to be expected for a conventional allyl monomer which would normally lead to degradative transfer and rate reduction.

Copolymerization[262] with DMAEMA satisfied the Kelen–Tuedos equation for cyclocopolymerization and gave soluble copolymers containing no vinyl groups.

Several workers have studied the kinetics of polymerization of diallyldialkylammonium halides.[261,263–266] The large order of the reaction with respect to monomer concentration (>2) is attributed as a rule to its involvement in the initiation step[263] and to its influence on the value of k_p. Addition[261] of NaCl was found to increase R_p and decrease the dependence on monomer concentration. The implication of charge effects is therefore evident. The effect of counterions is demonstrated by the observation[267] that the free radical initiated polymerizations of diallyldimethylammonium and diallyldiethylammonium chlorides in methanol and methanol/water mixtures are much faster than those of the corresponding bromides. These workers[267–269] attribute the large monomer dependences to intermolecular association and to the decrease in values of k_t with increase in monomer concentration. The latter phenomenon was ascribed[269] to the high viscosities of the methanolic solutions of monomer and the consequent hindrance of radical diffusion.

Various workers have reported suitable conditions for the satisfactory polymerization of diallyldialkylammonium halides. Control of pH appears to be important although acid and alkaline conditions are recommended in different procedures.

Polymerization using $S_2O_8^{2-}$ initiation in $0 \cdot 07M$ NaOH and $0 \cdot 02M$ $(C_2H_5)_3N$ produces good yields of polymer in the presence of air.[270] A continuous process involving $S_2O_8^{2-}$ initiation in an aqueous medium buffered at pH $6 \cdot 7$–$10 \cdot 3$ is claimed to have low sensitivity to air.[271,272]

Reagent purity is once more essential for the production of good yields of gel-free polymer. For instance, when a quaternized monomer is synthesized by the reaction:[273]

$$(CH_3)_2NH + 2CH_2{=}CHCH_2Cl \rightarrow (CH_2{=}CHCH_2)_2\overset{+}{N}(CH_3)_2$$
$$Cl^-$$

contamination of the secondary amine by small quantities ($<2\%$) of the primary amine $MeNH_2$ leads to quaternized allyl salts of functionality >2:

$$CH_3NH_2 + 3CH_2\!\!=\!\!CHCH_2Cl \rightarrow (CH_2\!\!=\!\!CHCH_2)_3\overset{+}{N}(CH_3)$$
$$Cl^-$$

On polymerization, this leads to crosslinking and gelling. Monoallyl ammonium salts have also been shown[274,275] to undergo free-radical polymerization readily, often in the presence of phosphates. It would appear[275] that rather low molar mass products are obtained, indicating the likelihood of monomer transfer.

N-Cetyl-N,N-diallyl(dodecyloxycarbonylmethyl)ammonium bromide (**LX**) has been polymerized using ultrasound and the products used for enzyme immobilization.[276]

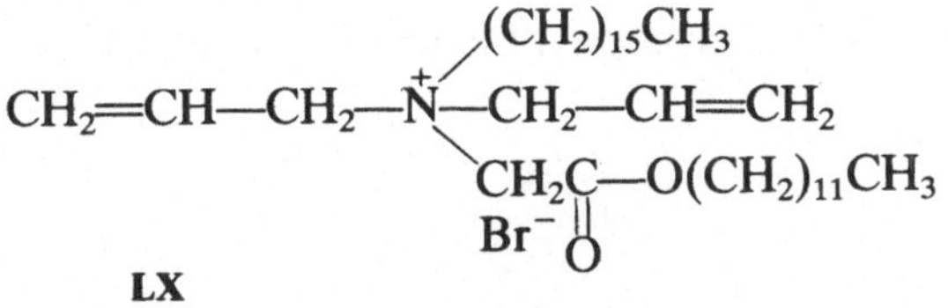

$$CH_2\!\!=\!\!CH\!-\!CH_2\!-\!\overset{+}{N}\overset{\displaystyle (CH_2)_{15}CH_3}{\underset{\displaystyle \underset{Br^-}{CH_2C}\overset{\displaystyle }{\underset{O}{\parallel}}-O(CH_2)_{11}CH_3}{-CH_2\!-\!CH\!=\!CH_2}}$$

LX

3.1.10 Polymerization of Quaternized Pyridine Salts (Vinylpyridinium Compounds)

In common with quaternized aminoalkylacrylates, aminoalkylmethacrylates, aminoalkylacrylamides and aminoalkylmethacrylamides, 2- and 4-vinylpyridine display a marked tendency to undergo spontaneous polymerization on quaternization with alkyl halides or on protonation in concentrated solution. Two general mechanisms have been suggested[277-283] and the preferred route depends upon the nucleophilicity of the counterion in the system. If this is strong, the probable mechanism of the attack is:

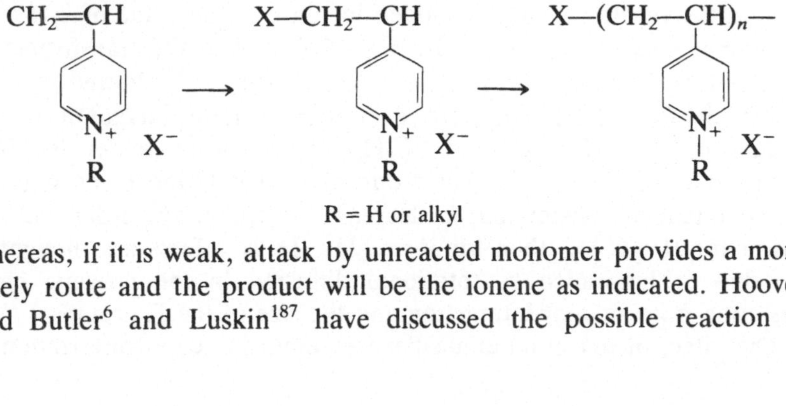

R = H or alkyl

whereas, if it is weak, attack by unreacted monomer provides a more likely route and the product will be the ionene as indicated. Hoover and Butler[6] and Luskin[187] have discussed the possible reaction in

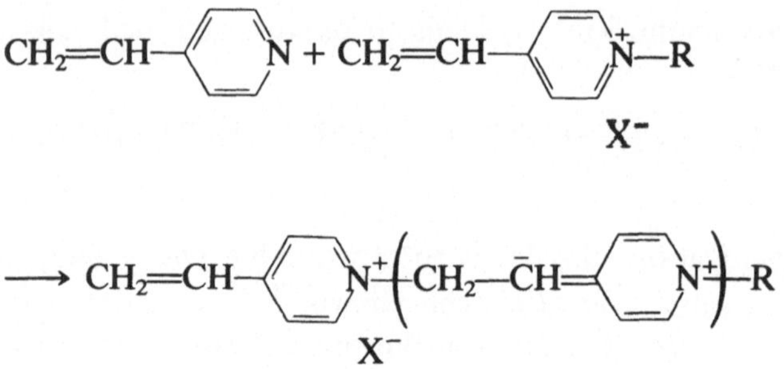

$$CH_2{=}CH{-}\langle\text{pyridine}\rangle N + CH_2{=}CH{-}\langle\text{pyridine}\rangle \overset{+}{N}{-}R$$

$$X^-$$

$$\longrightarrow CH_2{=}CH{-}\langle\text{pyridine}\rangle \overset{+}{N}{+}\left(CH_2{-}\overset{-}{CH}{=}\langle\text{pyridine}\rangle \overset{+}{N}{\pm}\right){-}R$$

$$X^-$$

R = H or alkyl

considerable detail in their reviews. Salamone,[284] in a more recent paper, has confirmed the existence of a free-radical process in the spontaneous polymerization.

Free-radical polymerization has been studied in various solvent systems. In the polymerization of **LXI** (R = C_4H_9, C_8H_{17}, $C_{10}H_{21}$)[285]

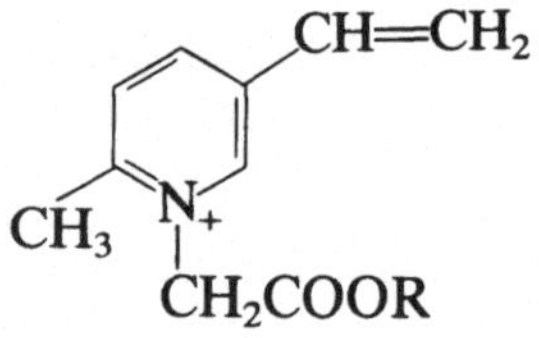

LXI

the rate of polymerization was shown to increase with the increase in the size of group R ($C_4H_9 < C_8H_{17} < C_{10}H_{21}$). This was ascribed to the increase in association of monomer molecules in solution because of the increasing surfactant properties of the monomer. The effect of micelle formation by the monomer was investigated.[286] Increasing the length of R (R = butyl, octyl, decyl) resulted in an increase in R_p. Monomer association will lead to an increase of monomer concentration in the reaction zone. A sudden increase in the value of $k_p/k_t^{0.5}$ at monomer concentrations $>0.3M$ was attributed to the transformation of spherical into lamellar micelles. Further studies[287] showed that the effect of solvent on the rate of polymerization increased in the sequence $CH_3OH < C_2H_5OH < C_4H_9OH$ and was greater for R = C_8H_{17} than for R = C_2H_5. The value of k_p was found to be virtually independent of solvent, but k_t decreased in the above order and was lower for R = C_8H_{17} than for R = C_2H_5. These effects are thought to be due to hindrance of rotational diffusion in the growing chain resulting from solvent interaction and the size of the R group.

The effect of pH in an emulsifier-free emulsion copolymerization of

4-vinylpyridine with styrene was studied.[288] At pH 11, the instantaneous copolymer composition was almost identical with that of the feed whereas, at pH 2, the 4-vinylpyridine entered the copolymer preferentially. The difference in behaviour was attributed to the different surface charges on the polymerization loci. Various recipes are available in the literature for copolymerization of these monomers.[289,290]

3.1.11 Formation of Nitrogen-containing Polyelectrolytes by Post-polymerization Modification

The principal methods employed involve amination accompanied by, or followed by, protonation or more often quaternization. The most common route involves modification of polyacrylamide or its C- and N-derivatives by the Mannich Reaction:

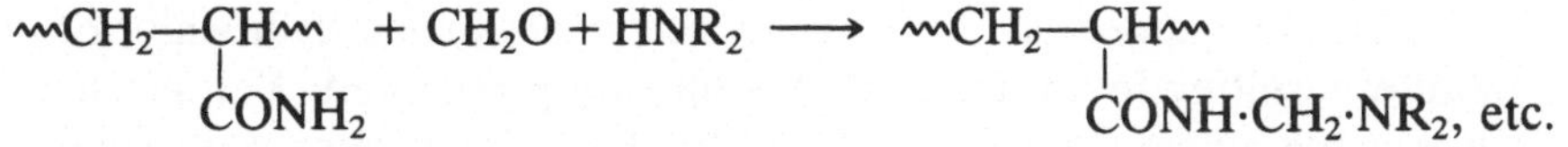

$$\text{wCH}_2\text{---CH}\text{w} + \text{CH}_2\text{O} + \text{HNR}_2 \longrightarrow \text{wCH}_2\text{---CH}\text{w}$$
$$\quad\quad\quad | \quad\quad\quad\quad\quad\quad\quad\quad\quad\quad\quad\quad\quad\quad | $$
$$\quad\quad\text{CONH}_2 \quad\quad\quad\quad\quad\quad\quad\quad\quad\quad\text{CONH·CH}_2\text{·NR}_2, \text{ etc.}$$

Recent examples of this include modification of polyACM with formaldehyde and Me_2NH in aqueous alkaline solution (pH 9–9·5),[291] and also of a copolymer of ACM/(methacrylamidopropyl)trimethylammonium chloride[292] at 40°C, for example:

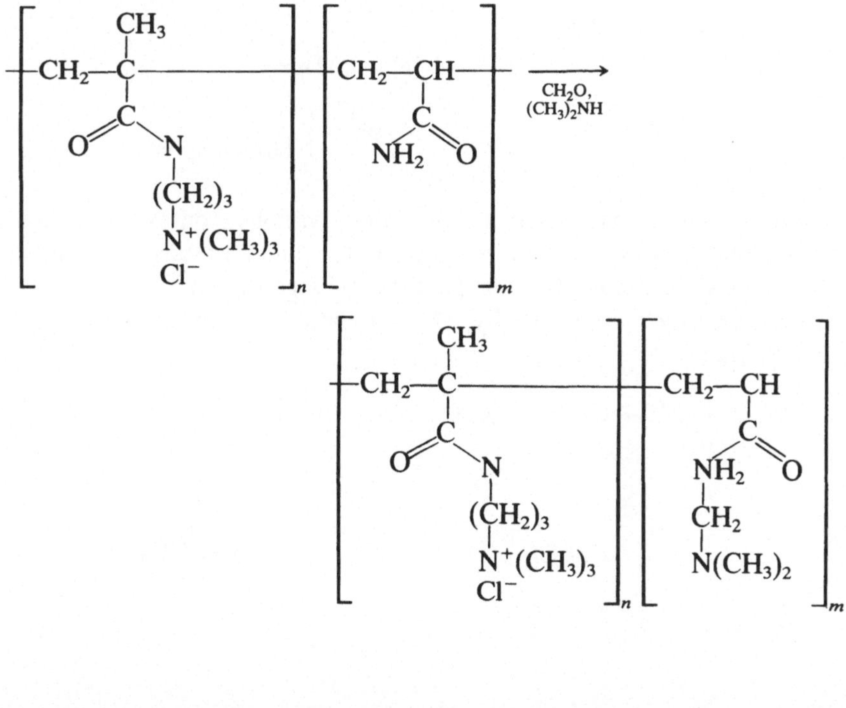

Copolymers of styrene/2-ethylhexyl acrylate/ACM or MACM[293] in dioxane have also been modified (**LXII**).

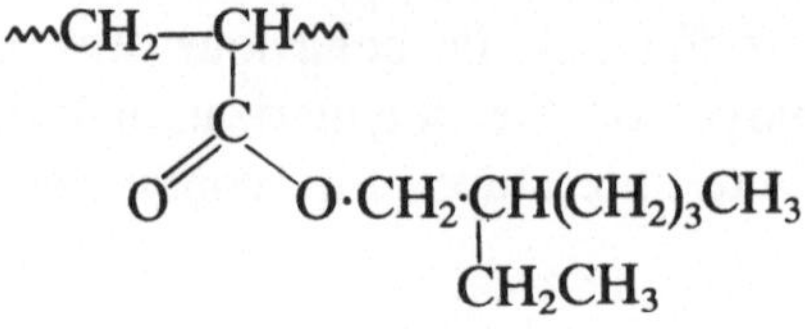

$$\text{\textasciitilde CH}_2\text{—CH\textasciitilde}$$

$$\underset{\underset{\displaystyle \text{CH}_2\text{CH}_3}{|}}{\overset{\displaystyle \text{C}}{\underset{}{}}}$$

O   O·CH$_2$·CH(CH$_2$)$_3$CH$_3$
                  CH$_2$CH$_3$

LXII

Product processability is claimed[294] to be improved by the addition of 1–30% poly(ethylene glycol) to the polyACM hydrogel before reaction with NaOH/Me$_2$NH/CH$_2$O.

Lowering of pH is essential to stabilize the polyamine product; otherwise crosslinking accompanied by gelation occurs rapidly.[295–298] An alternative approach[297] is to use a secondary amine hydrochloride, thereby resulting in an amine that is already protonated. Yet another route is the aqueous-phase reaction of polyacrylonitrile or copolymers of acrylonitrile and vinyl acetate with N,N-dialkylaminoalkylamines of the type H$_2$N(CH$_2$)$_m$NR$_2$ ($m = 2, 3$; R = Me, Bu, CH$_2$CH$_2$OH), to produce poly(dialkylaminoalkyl acrylamides):[299]

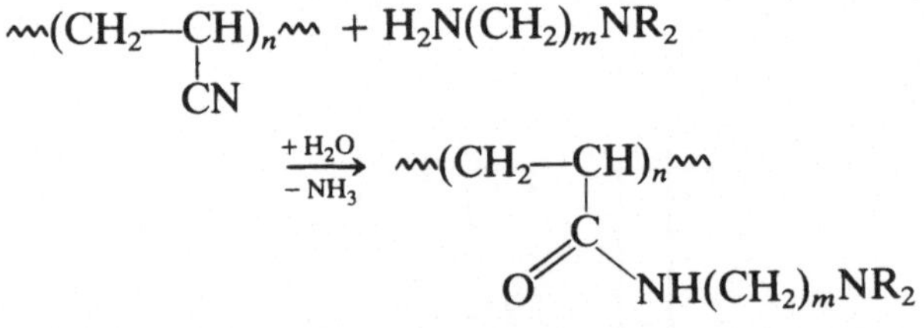

$$\text{\textasciitilde(CH}_2\text{—CH)}_n\text{\textasciitilde} + \text{H}_2\text{N(CH}_2\text{)}_m\text{NR}_2$$
$$\qquad\qquad \underset{\displaystyle |}{\text{CN}}$$

$$\xrightarrow[-\text{NH}_3]{+\text{H}_2\text{O}} \quad \text{\textasciitilde(CH}_2\text{—CH)}_n\text{\textasciitilde}$$

$$\underset{\displaystyle \text{O}}{\overset{\displaystyle \text{C}}{}}\quad \text{NH(CH}_2\text{)}_m\text{NR}_2$$

Vorchheimer[10] has discussed in detail various routes by which carboxamide groups may be converted to amino groups. The most important is probably via the Hoffmann Degradation[300] referred to earlier. Another route is by the amination of chloromethylated polystyrene:

$$\text{\textasciitilde CH}_2\text{—CH\textasciitilde} + \text{R}_2\text{NH} \longrightarrow \text{\textasciitilde CH}_2\text{—CH\textasciitilde}$$

CH$_2$Cl   CH$_2$NR$_2$

Rate constants were found to depend on the nature of the amine and the solvent. On amination with $BuNHCH_2CH_2OH$ or $MeNHCH_2CH_2OH$, rate acceleration occurred in dioxane,[301] but steric hindrance was observed after 75% conversion in DMF. Another group[302] showed that rate acceleration occurred in the reaction with $CH_3CH_2CH(NH_2)CH_2OH$ in dioxane but deceleration was observed with triethylamine in DMSO. These results were attributed to the formation of hydrogen bonds and to the electrostatic effect with neighbouring groups in the transition state. Interaction between positive charges on the chain and the nucleophilic amine is thought[303] to be responsible for the two-stage reaction with hydroxy-ethyldimethylamine, bis(hydroxyethyl)methylamine or triethylamine in DMF, in which case $k_1 < k_2$, where the subscript refers to the number of neighbouring groups already reacted.

The degree of conversion attained in the reaction with tri-alkylamines was found[304] to be much higher in DMF than in dioxane, THF or benzene, which have much lower dielectric constants. The reactivity of these amines in the quaternization reaction was found to decrease with increasing alkyl group size and to be lower than that of alkyldimethylamines with long-chain alkyl groups.

Amination of poly(vinyl chloroformate) has also been carried out with compounds containing labile hydrogen atoms (e.g. amines, alcohols, phenols):[305]

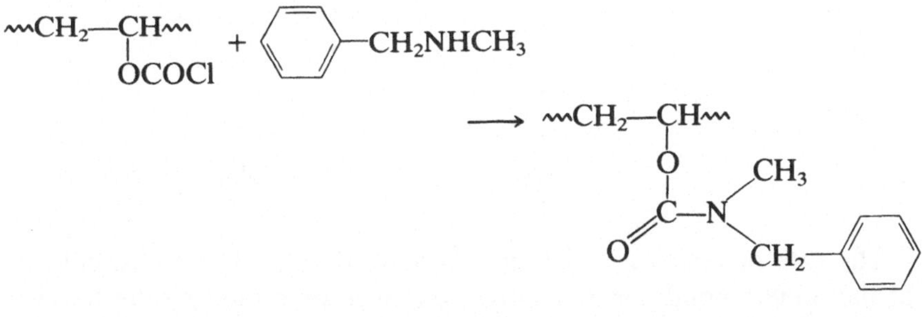

Results with ethylethanolamine demonstrate that the NH group is much more reactive than the hydroxyl group. Conditions are described which limit the amount of degradation and hydrolysis.[306] Amination reactions involving other halogenated compounds, e.g. copolymers of α-chloroacrylic acid with butyl acrylates, have also been described.[307]

Conditions for the quaternization of polyamino chains are similar to those for the quaternization of nitrogen-containing monomers.

Quaternization of poly(vinylpyridine) and its derivatives is probably the most intensively studied of the quaternization reactions.[187,308,309] The quaternization of poly(4-vinylpyridine) with various bromides[310] and iodides[311] has been discussed recently. A neighbouring-group effect controlled by steric, rather than charge, factors is thought to be responsible for the decrease in the quaternization rate constant with increase in conversion and increased size of alkyl group. Conversely, in the quaternization by methyl iodide in N-methyl-pyrrolidone of copolymers containing 2-methyl-5-vinylpyridine, the rate constant was found to be independent of conversion.[312] Other workers studied the quaternization of the same polymer with hexadecyl bromide in dioxane/water mixtures.[313] Kinetic behaviour was found to depend on the solvent composition, and participation of ionic species was suggested.

An alternative method[314] of quaternization of poly(vinylpyridine) and its derivatives, and of a wide range of other polyamines, is the reaction of the nucleophilic N-atom with an electrophilic double bond of the type $CH_2{=}CH(R)$, where R = COMe, COOH, COOMe or $CONH_2$.

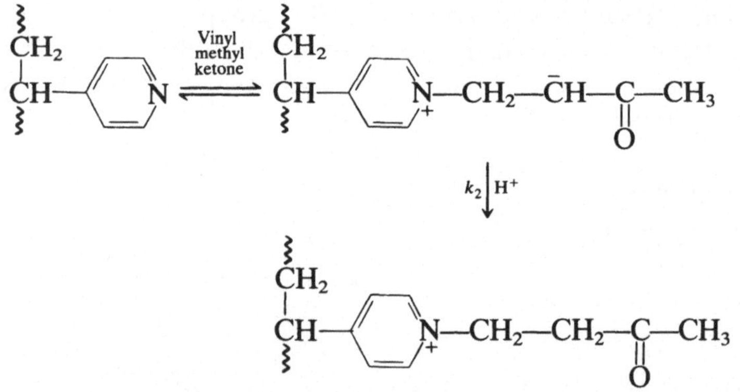

The rate of reaction was found to depend upon the electrophilicity of the double bond and the steric hindrance presented by the tertiary amine.[314]

Other workers who studied the poly(4-vinylpyridinium) ion reaction[315] showed that the reactivities of the vinyl monomers decrease in the sequence

$$-COOH > -CONH_2 > -COOCH_3 > -CN$$

This implies nucleophilic attack, since the double bond is increasingly electron-deficient in the same order.

Finally, the kaolin-flocculating ability of solutions of polyDMAEMA quaternized with C_{2-18} alkyl bromides has been measured.[316] The rate of sedimentation of the kaolin suspension was found to be independent of the degree of quaternization in the ethyl bromide/DMAEMA case. With bulkier alkyl bromides (hexyl, dodecyl, stearyl), the rate of sedimentation showed an increase with reduction in degree of quaternization. Reduction of pH was also found to favour flocculation.

3.2 Sulphonium Polyelectrolytes

Sulphonium monomers and their polymerization have been reviewed in detail by Hoover[5] and more recently, but in considerably less detail, by Vorchheimer.[10] These are the sulphur analogues of the quaternary ammonium compounds **LXIII**.

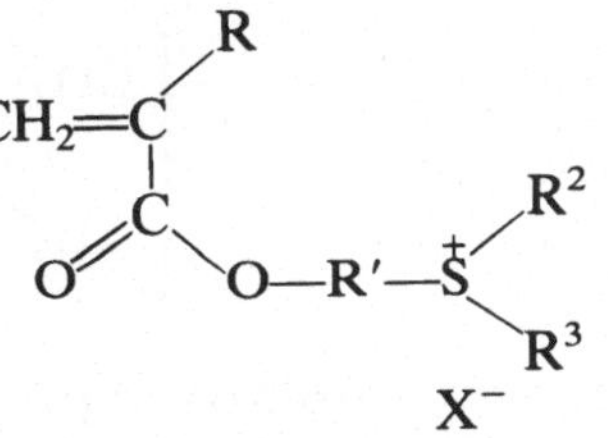

LXIII

In general, such polymers show a tendency to decompose fairly readily and form unpleasantly smelling sulphides. They have few, if any advantages over their nitrogen-containing analogues and are therefore of much less commercial importance. Typical examples include poly(2-acryloyloxyethyl dimethyl sulphonium methyl sulphate (**LXIV**), poly(vinylbenzylsulphonium) (**LXV**) and poly(diallylsulphonium) (**LXVI**).

3.3 Phosphonium Polyelectrolytes

Hoover[5] also reviewed polyphosphonium formation. As for sulphonium polyelectrolytes, comparatively little work appears in the

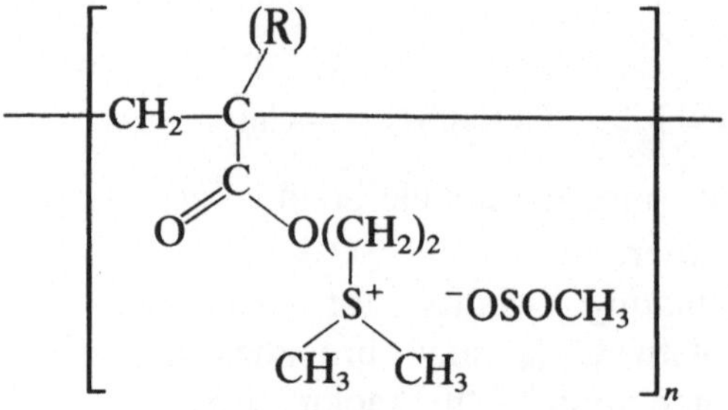

LXIV

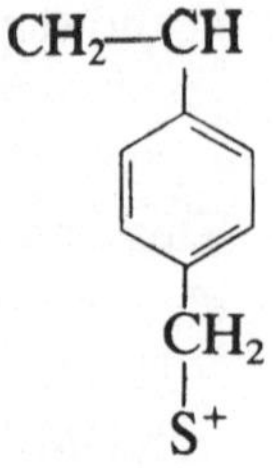

LXV

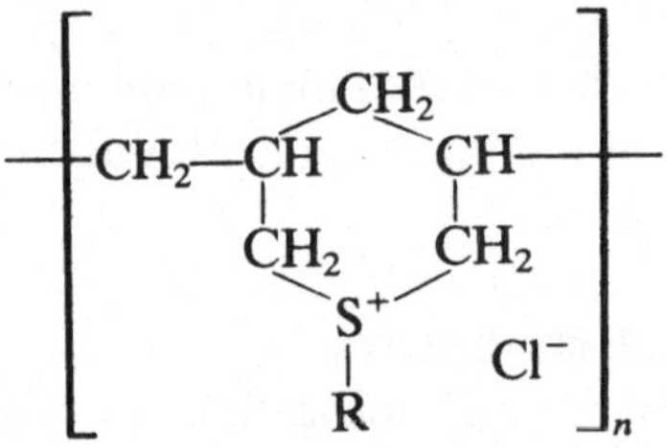

LXVI

literature. Unlike the N- and S-analogues, solutions of tributylvinyl-phosphonium bromide undergo polymerization on irradiation:

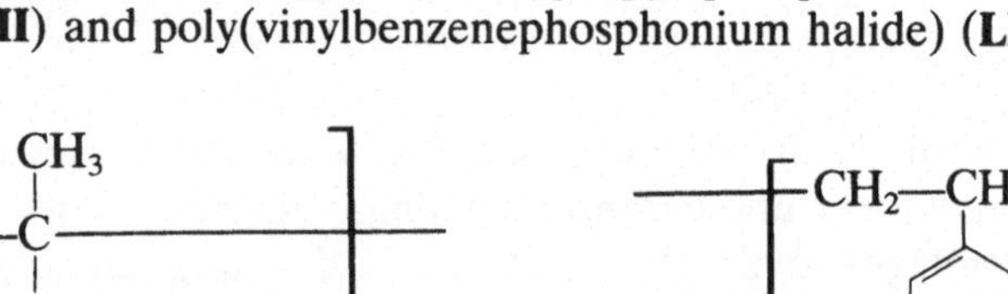

Other examples are poly(2-hydroxypropyl phosphonium methacrylate) (**LXVII**) and poly(vinylbenzenephosphonium halide) (**LXVIII**).

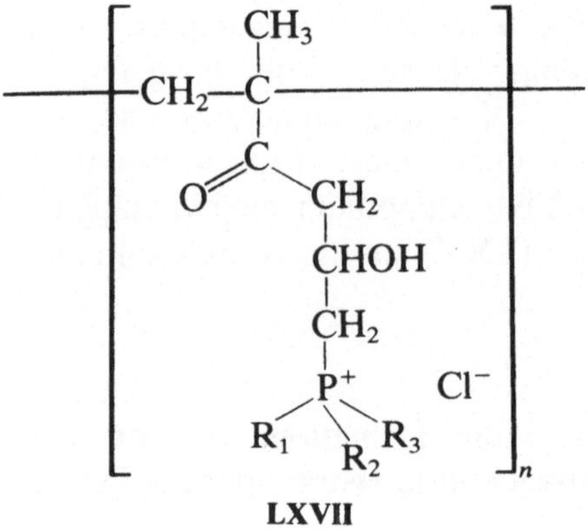

LXVII

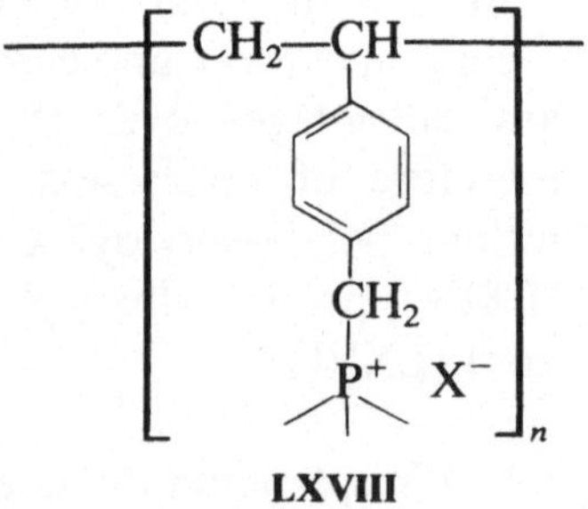

LXVIII

4 POLYAMPHOLYTES (AMPHOTERIC POLYELECTROLYTES)

These are polyelectrolytes containing charges of both signs. A major use appears to be as emulsifiers. They may be prepared by copolymerization of suitable anion-containing and cation-containing monomers or alternatively, by post-polymerization modification.[9,10]

4.1 Copolymerization

In principle, any pair of oppositely charged monomers capable of undergoing copolymerization will form a polyampholyte. Typical monomer pairs include unsaturated acids with substituted acrylamides or aminoalkylacrylamides or methacrylamides.

Recent examples include the copolymerization of AA/DMEAMA, initiated by benzoyl peroxide in the presence of lauroyl mercaptan as a chain-transfer agent and ethylene glycol monobutyl ether.[317] This yields a low molar mass polymer suitable for use as an emulsifier. Styrene, AA and N-(dimethylaminopropyl)methacrylamide were copolymerized[318] at 270°C and 28 bar pressure to produce a copolymer of molar mass ~3000 in the absence of moderator. This product was claimed to be a better polymerization emulsifier than the equivalent product prepared in an aqueous system.

A block copolymer[319] of MAA and 4-(N,N-dimethyl)aminostyrene was prepared by anionic polymerization of the latter with trimethylsilyl methacrylate using lithium naphthalene as the catalyst. Sodium was not a suitable counterion since it caused premature termination and incomplete conversion. Addition of the methacrylate followed by hydrolysis of the copolymer gave the required product which displayed micellar properties in solution.

Copolymers of quaternized alkylaminoacrylamides and alkylacrylamidosulphonates have been prepared and their properties in aqueous solution studied.[320] In contrast to normal polyelectrolytes, viscosities either increased or remained constant as the ionic strength of the solution was increased.

4.2 Polymer Modification Methods

Partial hydrolysis of polyDMEAMA under acid or alkaline conditions has been shown[321] to produce polyampholytes with considerable differences in solubility and titration characteristics from those formed by copolymerization of the appropriate monomers. These differences were attributed to the different distribution of the sequence of acid and basic units in the various cases.

Hydroxymethylation is an important polymer modification technique in these cases. Partial hydrolysis of polyACM followed by amination with $CH_2O/(CH_3)_2NH$ was claimed to give a stable product.[322] A dispersion of ACM/styrene copolymer in an emulsifier-free latex was modified by the Mannich Reaction and the Hoffmann Degradation to yield a copolymer containing primary and tertiary amino groups. Partial hydrolysis was unavoidable and gave amphoteric latices.[323,324]

ACKNOWLEDGEMENT

The author wishes to thank his daughter Kirsty for allowing him to monopolize her microcomputer, without which the writing of this chapter would not have been possible.

REFERENCES

1. HOLLIDAY, L., in *Ionic Polymers* (L. Holliday, Ed.), 1975, Applied Science Publishers, London, p. 1.
2. EISENBERG, A. and KING, M., *Ion-containing Polymers*, 1977, Academic Press, New York, Ch. 1.
3. WILSON, A. D. and PROSSER, H. J., *Developments in ionic polymers—1)* (A. D. Wilson and H. J. Prosser, Eds.), 1983, Applied Science Publishers, London, Ch. 1.
4. HOOVER, M. F., *ACS Polymer Prepr.*, 1969, **10**, 908.
5. HOOVER, M. F., *J Macromol. Sci., Chem.* 1970, **A4**, 1327.
6. HOOVER, M. F. and BUTLER, G. B., *J. Polym. Sci., Polymer Symp.* 1973, **45**, 1.
7. AKERS, R., *Rep. Prog. App. Chem.*, 1976, **60**, 605.
8. GUTCHO, S., *Waste Treatment with Poly-electrolytes and Other Flocculants*, 1977, Noyes Data Corpn., Park Ridge, N.J.
9. VOSTREIL, J. and JURACKA, F., *Commercial Organic Flocculants*, 1976, Noyes Data Corpn., Park Ridge, N.J.
10. VORCHHEIMER, N., in *Poly-electrolytes Waste-water Treatment* (W. L. C. Schwoyer, Ed.), 1981, C.R.C. Press, Boca Raton, Fla., p. 1.
11. ISE, N., *Macromol. Chem., Phys. Suppl.*, 1981, **5**, 1.
12. RICHARDS, D. H., *ACS Symposium Ser.*, 1981, **166** (Anionic Polym.), p. 343.
13. EISENBERG, A., *ACS Polymer Preprints* (Div. Polym. Chem.), 1979, **20**(1), 286.

14. Yocum, R. H. and Nyquist, E. B., (Eds.) *Functional monomers*, 1973 and 1974, Vols. 1 and 2, Dekker, New York.
15. Molyneux, P., *Water-soluble Synthetic Polymers, Properties and Behaviour*, 1983, Vols. 1 and 2, C.R.C. Press, Boca Raton, Fla.
16. Armstrong, R. W. and Strauss, U. P., *Encyc. of Polymer Science & Tech.*, 1964, **10**, 781.
17. *Storage and Handling of Acrylic and Methacrylic Esters and Acids*, Bull. CM-17, 1972, Rohm and Haas Inc., Philadelphia, Penn.
18. *Acrylic and Methacrylic Monomers. Typical Properties and Specifications*, Bull. CM-16, 1972, Rohm and Haas Inc., Philadelphia, Penn.
19. Davidson, R. L. (Ed.), *Handbook of Water-Soluble Gums and Resins*, 1980, McGraw-Hill, New York.
20. Sandler, S. R. and Karo, W., *Polymer Syntheses*, Vol. 2, 1977, Academic Press, New York, p. 264.
22. Nemec, J. W. and Bauer, W., 'Acrylic acid' in *Kirk–Othmer Encyc. of Chem. Tech.*, 3rd edn, (M. Grayson *et al.*, Eds.), Vol. 1, 1978, Wiley, New York, p. 330.
23. Nemec, J. W. and Kirch, L. S., 'Methacrylic acid and derivatives' in *Kirk–Othmer Encyc. of Chem. Tech.*, 3rd edn, (M. Grayson *et al.*, Eds.), Vol. 15, 1981, Wiley, New York, p. 346.
24. Kine, B. B. and Novak, R. W., 'Methacrylic polymers' in *Kirk–Othmer Encyc. of Chem. Tech.*, 3rd edn, (M. Grayson *et al.*, Eds.), Vol. 15, 1981, Wiley, New York, p. 377.
25. Zand, R., 'Azo catalysts' in *Encyc. of Polymer Sci. & Tech.*, (N. M. Bikales, Ed.), Vol. 2, 1965, Wiley–Interscience, New York, p. 278.
26. Nuyken, O. and Kerber, R., *Makromol. Chem.*, 1978, **179**, 2845.
27. (a) *Diacyl Peroxides*, (b) *Diallyl Peroxides*, (c) *Tertiary Alkyl Peroxides*, (d) *Peroxyesters*, Product Bulletins, 1977, Lucidol Division, Pennwalt Corpn., Buffalo, New York.
28. Nayak, P. L. and Lenka, S., *J. Macrol. Sci.—Rev. Macromol. Chem.*, 1980, **C19**(1), 83.
29. Misra, G. S. and Bajpai, U. D. N., *Prog. Polym. Sci.*, 1982, **8**(1–2), 61.
30. Sheppard, C. S. and Kamath, V., 'Initiators' in *Kirk–Othmer Encyc. of Chem. Tech.*, 3rd edn, (M. Grayson *et al.*, Eds.), Vol. 13, 1981, Wiley, New York, p. 355.
31. Mishra, M. K., *J. Macromol. Sci.—Rev. Macromol. Chem.*, 1981, **C20**(1), p. 149.
32. Loginova, T. F., Shubin, A. A., Vyalkov, V. V. and Kisel'nikov, V. N., *Izv. Vyssh. Uchebn. Zaved., Khim. Khim. Tekhnol.*, 1983, **26**(12), 1476.
33. (a) Manickam, S. P., Venkatarao, K. and Subbaratnam, N. R., *Eur. Polymer J.*, 1979, **15**(5), 483; (b) Manickam, S. P., Singh, U. C., Venkatarao, K. and Subbaratnam, N. R., *Polymer*, 1979, **20**(7), 917.
34. Sarasvathy, S. and Venkatarao, K., *Makromol. Chem., Rapid Commun.*, 1981, **2**(3), 219.

35. KNORR, R. S., CHANDLER, J. D. and TRAN, L. N. (Monsanto Co.), US Patent 4 433 122, 21 Feb. 1984.
36. OUCHI, T., MURAYAMA, N. and IMOTO, M., *Bull. Chem. Soc. (Jpn)*, 1980, **53**(3), 748.
37. BARTISSOL, A., BOUTINI, J. and WASCHOWSKI, F., (Rhone-Poulenc Ind., Ltd), Fr. Demande 2 489 336, 5 Mar. 1982.
38. MISHRA, M. K. and BHADANI, S. N., *Makromol Chem.*, 1983, **184**(5), 955.
39. NISTRATOVA, L. N., KOPYLOVA, N. A. and YABLOKOVA, N. V., Deposited Doc., 1982, SPSTL 993, Khp-D81 (Avail. SPSTL).
40. HUANG, M. and WU, R., *Ziran Zashi.*, 1982, **5**(12), 950; *Chem. Abstr.*, 996067.
41. NIKOLAEV, A. F., SHIBALOVICH, V. G., BONDARENKO, V. M. and KHOKHRIN, S. A., USSR 1 004 404, 15 Mar. 1983.
42. KALIYAMURTHY, K., ELAYAPERUMAL, P., BALAKRISHNAN, T. and SANTAPPA, M., (a) *Makromol. Chem.*, 1979, **180**(6), 1575; (b) *J. Macromol. Sci. Chem.*, 1982, **A18**(2), 219.
43. NIKOLAEV, A. F., BELOGORODSKAYA, K. V., BONDARENKO, S. G., BRATTER, M. A. and BANDYUK, O. V., *Zh. Prikl. Khim. (Leningrad)*, 1982, **55**(8), 1826.
44. ELAYAPERUMAL, P., BALAKRISHNAN, T., SANTAPPA, M. and LENZ, R. W., *J. Polym. Sci., Polymer Chem. Ed.*, 1982, **20**(12), 3325.
45. ELAYAPERUMAL, P., BALAKRISHNAN, T. and SANTAPPA, M., *J. Polym. Sci., Polym. Chem. Ed.*, 1980, **18**(8), 2471.
46. UHNIAT, M., SIKORSKII, R. T. and WOROSZYLO, L., *Vysokomol. Soedin., Ser. A*, 1981, **23**(11), 2420.
47. KOBAYASHI, G. and SEKIGUCHI, T. (Nippon Kayaku Co, Ltd), Jpn. Kokai, Tokkyo Koho 79 106 596, 21 Aug. 1979.
48. SUPRUN, V. YA., BORISLAYSKII, O. A. and MOKRVI, E. N., *Visn L'viv Politekh. Inst.*, 1983, **171**, 138; *Chem. Abstr.*, 99122981.
49. UHNIAT, M., RUBAI, M., WOROSZYLO, L., WASILEWSKI, J., BERES, J. and KOBYLARZ, S., Pol 101 050, 31 Dec. 1979; *Chem. Abstr.*, 938779.
50. MUNZER, M. and TROMMSDORFF, E., in *Polymer Processes, High Polymers*, Vol. 29, (C. E. Schildnecht/I. Skeist Eds.), 1977, Wiley, New York.
51. DUNN, A. S., 'Emulsion polymerization', in *Developments in Polymerization—2*, (R. N. Haward, Ed.), 1979, Applied Science Publishers, London, 45.
52. DIMONIE, M. V., BOGHINA, C. M., MARINESCU, N. N., MARINESCU, M. M., CINCU, C. I. and OPRESCU, C. G., *Eur. Polym. J.*, 1982, **18**(7), 639.
53. BERES, J. et al., Ciezkiej Syntezy Organicznej 'Blachownia', Pol. 105 454, 16 Dec. 1979; *Chem. Abstr.*, 92216036.
54. Seitetsu Kagaku Co. Ltd, Jpn. Kokai Tokkyo Koho 81 26 909, 16 Mar. 1981.
55. Kao Soap Co. Ltd, Jpn. Kokai Tokkyo Koho 82 49 602, 23 Mar. 1982.
56. ZECHER, D. C. (Hercules Inc.), US Patent 4 379 883, 12 Apr. 1983.
57. ZECHER, D. C. (Hercules Inc.), Brit. Pat. Appln. 2 093 464, 2 Sept. 1982.

58. DIMONIE, M., DINU, G. M., BOGHINA, C., OPRESCU, C., CINCU, C., POPESCU, G. and MORARU, I. (Intreprinderea Chimica Risnov), Rom 75 667, 28 Feb. 1981.
59. TAGUCHI, H., KUNEIDA, N. and KINOSHITA, M., *Makromol. Chem., Rapid Commun.*, 1982, **3**(7), 495.
60. TAGUCHI, H., KUNEIDA, N. and KINOSHITA, M., *Makromol. Chem.*, 1983, **184**(5), 925.
61. Seitetsu Kagaku Co. Ltd, Jpn. Kokai Tokkyo Koho 81 147 806, 17 Nov. 1981.
62. HEIDE, W., HARTMANN, H., BURKERT, H. and BUENSCH, H. (BASF A–G), Ger. Offen. 3 220 114, 1 Dec. 1983.
63. SEHM, E. J. (B. F. Goodrich Co.), US Patent 4 419 502, 6 Dec. 1983.
64. GROMOV, V. F., GAL'PERINA, N. I., OSMANOV, T. A., KHOMIKOVSKII, P. M. and ABKIN, A. D., *Eur. Polym. J.*, 1980, **16**, 529.
65. KABANOV, V. A., TOPCHIEV, D. A. and KARAPUTADZE, T. M., *J. Polym. Sci.*, 1973, **C42**, 173.
66. MILLER, M. L., *Encyc. of Polymer Sci. & Tech.* (N. M. Bekalis, Ed.), Vol. 1, 1964, Wiley–Interscience, New York, p. 197.
67. PLOCHOKA, K., *J. Macromol. Sci.—Rev. Macromol. Chem.* 1981, **C20**(1), 67.
68. TEDDER, J. M., 'The reactivity of free radicals' in *Reactivity, Mechanism and Structure* (A. D. Jenkins and A. Ledwith, Eds.), 1974, Wiley, Chichester.
69. CABANESS, W. R., YEN-CHIN LIN, T. and PARKANYI, C., *J. Polym. Sci., Part A*-1, 1971, **9**, 2155.
70. MANDEL, M., University of Leiden, private communication.
71. PONRATNAM, S. and KAPUR, S. L., *Makromolek. Chem.*, 1977, **178**, 1029.
72. PONRATNAM, S. and KAPUR, S. L., *J. Polym. Sci., Polymer Chem. Ed.*, 1976, **14**, 1987.
73. PONRATNAM, S., RAO, S. P., JOSHI, S. G. and KAPUR, S. L., *J. Makromol Sci.—Chem. Ed.*, 1976, **A-10**, 1055.
74. PLOCHOTKA, K. and WOJNAROWSKI, T. J., (a) *Eur. Polymer J.*, 1971, **7**, 797; (b) *Eur. Polym. J.*, 1972, **8**, 921.
75. HAMOUDI, A. and MCNEILL, I. C., *Eur. Polym. J.*, 1978, **14**, 177.
76. KUNIN, R. and FISHER, S., *J. Phys. Chem.*, 1962, **66**, 2275.
77. MYAGCHENKOV, V. A., KURENKOV, V. F. and FRENKEL, S. YA., *Vysokomol Soedin, Ser. A*, 1968, **10**, 1740; *Dokl. Akad. Nauk, SSSR*, 1969, **184**, 880; *Acta Polym.*, 1982, **33**(b), 388.
78. MYAGCHENKOV, V. A., KURENKOV, V. F., KUZNETZOV, E. N. and FRENKEL, S. YA., *Eur. Polym. J.*, 1970, **6**, 63.
79. GAL'PERINA, N. I., GUGANAVA, T. A., GROMOV, V. F., KHOMIKOVSKII, P. M. and ABKIN, A. D., *Vysokomol. Soedin., Ser. A*, 1975, **17**, 1670; *Chem. Abstr.*, 83164674.
80. GAL'PERINA, N. I., GUGANAVA, T. A., GROMOV, V. F., KHOMIKOVSKII, P. M. and ABKIN, A. D., *Vysokomol. Soedin., Ser. B*, 1976, **18**, 384; *Chem. Abstr.*, 8316467.
81. CHAPIRO, A., *Pure Appl. Chem.*, 1981, **53**, 643.

82. CHAPIRO, A. and SOMMERHATTE, T., *Eur. Polymer J.*, 1969, **5**, 707, 725.
83. CHAPIRO, A. and DULIEU, J., *Eur. Polymer J.*, 1977, **13**, 563.
84. CHAPIRO, A., GOLDFIELD-FREILISH, D. and PEMCHION, J., *Eur. Polymer J.*, 1975, **11**, 515.
85. CHAPIRO, A. and GOULONBANDI, R., *Eur. Polymer J.*, 1974, **10**, 1159.
86. CHAPIRO, A., MANKOWSKI, Z. and RENAULD, N., *Eur. Polymer J.*, 1977, **13**, 401.
87. (a) FERGUSON, J., GRANMAYEH, R. and AL-ALAWI, S., *Proc. IUPAC, Macromol. Symp. 28th.*, 1982, 227 (Eng.); (b) FERGUSON, J., AL-ALAWI, S. and GRANMAYEH, R., *Eur. Polymer J.*, 1983, **19**(6), 475.
88. FUJIMORI, K., *Polymer Bull. (Berlin)*, 1982, **8**(5–6), 207.
89. SMID, J., TAN, Y. Y. and CHALLA, G., *Eur. Polymer J.*, 1983, **19**(10/11), 853.
90. LOHMEYER, J. H. G. M., TAN, Y. Y. and CHALLA, G., *J. Macromol. Sci. Chem.*, 1980, **A14**(6), 945.
91. POLOWINSKI, S., *Eur. Polymer J.*, 1983, **19**(8), 679.
92. KATAOKA, S. and ANDO, T., *Kobunshi Ronbunshu,* (a) 1981, **38**(12), 821; (b) 1981, **38**(11), 797.
93. GAGNE, P. (Rhone-Poulenc Industries SA), Fr. Demande 2 463 780, 27 Feb. 1981.
94. SZALKAI, A. M., HENNING, S., LIANE, G. and BERGHECIAN, M. G. O. (Intreprinderea Chimica Risnov), Rom. RO 78 280, 30 Jan. 1982.
95. HUGHES, K. A., SWIFT, G. and KINE, B. B. (Rohm and Haas Co.), Braz. Pedido 8000 357, 7 Oct. 1980.
96. Sumitomo Chemical Co. Japan, Jpn. Kokai Tokkyo 57 31 210, 14 Aug. 1982.
97. Sumitomo Chemical Co. Japan, Jpn. Kokai Tokkyo 57 131 209 (82 131 209), 14 Aug. 1982.
98. SCHULTZ, G. O. and WILSON, D. M. (Johnson, S. C. & Son Inc.). Eur. Pat. Appn. 71 116, 9 Feb. 1983.
99. PEASCOE, W. J., WHITE, W. W. and SOMMA, L. E. (Uniroyal Inc.), Eur. Pat. Appln. 29 970, 10 Jun. 1981.
100. BERES, J. and BOGDALSKA, H., *Zesz. Nauk.—Wyzsza Szk. Pedagog. im. Powstancow Slask, Opolu, (Ser): Chem.* 1981, **5**, 89; *Chem. Abstr.*, 96200277.
101. GORETTA, L. A. and OTREMBA, R. R. (Nalco Chem. Co.), US Patent 4 196 272, 1 Apr. 1980.
102. MORAWETZ, H., *Macromolecules in Solution*, 2nd edn, 1975, Wiley–Interscience, New York, p. 469.
103. KLESPER, F., GRONSKI, W. and BARTH, V., *Makromol Chem.*, 1970, **139**, 1.
104. MCQUARRIE, D. A., MCTAGUE, J. P. and REISS, H., *Biopolymers*, 1965, **3**, 653.
105. PLATÉ, N. A., *Pure Appl. Chem.*, 1976, **46**, 49.
106. MERLE, Y. and MERLE-AUBRY, L., *Macromolecules*, 1983, **16**(6), 1009.
107. ARCUS, C. L., *J. Chem. Soc.*, 1949, 2732.

108. RATCHFORD, W. P. and FISHER, C. H., *J. Amer. Chem. Soc.*, 1949, **69**, 1911.
109. CONIX, A., SMETS, G. and MOENS, J., *Ric. Sci. 25 Supp.*, *Simp. Int. Chim. Macromol.*, *Milan–Turin*, 1954, p. 200; *Chem. Abstr.*, 5411545.
110. MARCUS, J. and SMETS, G., *J. Polym. Sci.*, 1957, **23**, 931.
111. KOROSTELEVA, E. A. and MISHCHENKO, K. P., *Zh. Prikl. Khim.* (*Leningrad*), 1980, **53**(8), 1921–3.
112. VOLK, H. and LAMPHERE, J. C. (Dow Chemical Co.), US Patent 4 151 140, 2 Nov. 1977.
113. CHANG, C. J. and STEVENS, T. E. (Rohm and Haas Co.), US Patent 4 423 199, 27 Dec. 1983.
114. ALMAEV, R. KH., GUBINA, A. V., RAKHIMKULOV, I. F. and BAZEKINA, L. V., *Tr. Bashk. Gos. Nauchno-Issled. Proektn. Inst. Neft. Prom-sti*, 1978, **53**, 44.
115. KHARIV, I. YU. and SIVETS, L. I., USSR 883 137, 23 Nov. 1981; Appl. 2 879 249, 1 Feb. 1980.
116. PHILLIPS, K. G. and BINGHAM, M. E. (Nalco Chemical Co.), US Patent 4 283 507, 11 Aug. 1981.
117. ZIL'BERMAN, E. N., STARKOV, A. A., DANOV, S. M. and ABRAMOVA, L. I., USSR 643 512, 25 Jan. 1979. From *Otkrytiya, Izobret.*, *Prom. Obraztsy, Touvarnye Znaki*, 1979, **3**, 89; *Chem. Abstr.*, 169561.
118. ZIL'BERMAN, E. N., STARKOV, A. A., EREMEEV, I. V., TRACHENKO, V. I. and KOLESNIKOV, V. A., *Vysokomol. Soedin.*, *Ser. B.*, 1979, **21**(1), 30.
119. KAWAKAMI, S., TAJIRI, S., TASHIRO, H. and ISAOKA, S. (Sumitomo Chemical Co. Ltd), Ger. Offen. 2 842 090, 5 Apr. 1979.
120. KURENKOV, V. F., VERIZHNIKOVA, A. S. and MYAGCHENKOV, V. A., *Vysokomol. Soedin.*, *Ser. A*, 1984, **26**(3), 535.
121. ZIL'BERMAN, E. N., ABRAMOVA, L. I., ALPHON'SHIN, G. N. and BRAZHKINA, S. N., *Vysokomol Soedin.*, *Ser. B*, 1980, **22**(10), 774.
122. Mitsubishi Chemical Industries Co. Ltd, Jpn. Kokai Tokkyo Koho 81 104 903, 21 Aug. 1981.
123. Mitsubishi Chemical Industries Co. Ltd, Jpn. Kokai Tokkyo Koho 81 104 904, 21 Aug. 1981.
124. Mitsubishi Chemical Industries Co. Ltd, Jpn. Kokai Tokkyo Koho 57 202 308, 11 Dec. 1982.
125. Marathon Oil Co., Jpn. Kokai Tokkyo Koho 58 04 091 (83 04 091), 11 Jan. 1983.
126. HAUGHTON, M. H. and MARTIN, C. G. (Albright and Wilson Ltd), Eur. Pat. Appl. 94 495, 4 Jan. 1984.
127. KANGAS, D. A., 'Sulphonic acids and sulphonate monomers' in *Functional Monomers*, R. A. Yocum and E. B. Nyquist, Eds, Vol. 1, 1973, Dekker, New York, p. 489.
128. Japan Exlan Co. Ltd, Jpn. Kokai Tokkyo Koho 58 15 512 (83 15 512), 28 Jan. 1983.
129. Japan Exlan Co. Ltd, Jpn. Kokai Tokkyo Koho 58 23 811 (83 23 811), 12 Feb. 1983.

130. HAYASHI, T., ODA, Y. and EMURA, N. (Toyo Soda Manuf. Co.), *Kenkyu Hokoku*, 1983, **27**(2), 81.
131. OKUBO, M., TACHIKA, H., ANDO, M., TANGE, T., YAMASHITA, S. and MATSUMOTO, T., *Nippon Setchaku Kyokaishi*, 1982, **18**(4), 153.
132. KURENKOV, V. F., VERIZHNIKOVA, A. S., SEVERINOV, A. V., KUZNETSOV, E. V. and MYAGCHENKOV, V. A., (USSR) Deposited Doc. SPSTL 13 Khp-D81, 1981, p. 116; *Chem. Abstr.*, 9772879.
133. Dainippon Ink and Chemicals Inc., Jpn. Kokai Tokkyo Koho 82 108 113, 6 Jul. 1982.
134. Sanyo Chemical Industries Ltd, Jpn. Kokai Tokkyo Koho 82 28 111, 15 Feb. 1982.
135. CHONDE, Y. and KRIEGER, I. M., *J. Appl. Polym. Sci.* 1981, **26**(6), 1819.
136. KURENKOV, V. F., VERIZHNIKOVA, A. S., KUZNETSOV, E. V. and MYAGCHENKOV, V. A., *Izv. Vyssh. Uchebn. Zaved., Khim. Khim. Tekhnol.*, 1982, **25**(2), 221.
137. VAN DER HOFF, B. M. E., *J. Polym. Sci.*, 1958, **33**, 487.
138. KOENIG, J., SUELING, C. and KORTE, S. (Bayer A–G), Ger. Offen, DE 3 225 521, 12 Jan. 1984.
139. HOPKINS, T. R. (Lubrizol Corp.), Brit. Pat. Appl. 2 102 817, 9 Feb. 1983.
140. GORETTA, L. A. and OTREMBA, R. R. (Nalco Chemical Co.), US Patent 4 331 792, 25 May 1982.
141. IZUMI, Z., KIUA, H., WATANABE, M. and UCHIYAMA, H., *J. Polym. Sci.*, 1965, **A3**, 272.
142. HENRICI-OLIVE, G. and OLIVE, S., *Makromol Chem.*, 1960, **37**, 71.
143. IZUMI, Z., KIUCHI, H. and WATANABE, M., *J. Polym. Sci., Part A1*, 1963, 705.
144. VAN DER HOFF, J. W., BRADFORD, E. B., TARKOWSKI, H. L., SHAFFER, J. B. and WILEY, R. M., 'Polymerization and polycondensation processes', *ACS Advan. Chem. Ser.*, 1962, **34**, 32.
145. KURENKOV, V. F. and MYAGCHENKOV, V. A., *Eur. Polym. J.*, 1979, **15**, 849.
146. KURENKOV, V. F., VAGAPOVA, A. K. and MYAGCHENKOV, V. A., *Eur. Polym. J.*, 1982, **18**(9), 763.
147. PONRATHNAM, S., MILAS, M. and BLUMSTEIN, A., *Macromolecules*, 1982, **15**(5), 1251.
148. BLUMSTEIN, A., PONRATHNAM, S. and BELLANTONI, E., *J. Polym. Sci., Polym. Lett. Ed.*, 1980, **18**(4), 299.
149. BLUMSTEIN, A., MILAS, M., OZCAYIR, Y. and BELLANTONI, E., *ACS Polym. Prepr., Div. Polym. Chem.*, 1983, **24**(1), 249.
150. REHAK, A., FOLDES-BEREZHNYKH, T., CZAJLIK, I. and TUDOS, F., *Magy, Kem. Foly.*, 1982, **88**(8), 340; *Chem. Abstr.*, 97145358.
151. KONSULOV, V., *God. Sofii, Univ., Khim. Fak.*, 1980, **71**(2), 95; *Chem. Abstr.*, 103944.
152. MCCORMICK, C. L. and CHEN, G. S., *J. Polym. Sci., Polym. Chem. Ed.*, 1982, **20**(3), 817.

153. SUNG, Y. K., LEE, S. Y. and JEJIK, O., *Pollimo*, 1983, **7**(6), 392; *Chem. Abstr.*, 10068807.
154. McCORMICK, C. L., PARK, L. S., CHEN, G. S. and NEIDLINGER, H. H., *ACS Polym. Prepr., Div. Polym. Chem.*, 1981, **22**(1), 137.
155. BRESLOW, D. S. and KUTNER, A., *J. Polym. Sci.*, 1958, **27**, 295.
156. KANGAS, D. A., *J. Polym. Sci. A-1*, 1970, **8**, 1813.
157. SHULZ, G. V., HENRICI-OLIVE, G. and OLIVE, S., *Z. Phys. Chem.*, 1980, **27**, 1.
158. ROTH, H. H., *Ind. Eng. Chem.*, 1957, **49**, 1820.
159. GOETHALS, E. J., *Synthesis and Properties of Polymers with Sulphur-containing Functional Groups.* (Moore, Ed.), 1973, D. Reidel Pub. Co., Dordrecht, p. 321.
160. PALYANICHKO, I. G., SIDORENKO, L. N., PLAKIDIN, V. L., KOZHUSHKOVA, I. and IZOTOV, V. YA., USSR 857 149, 23 Aug. 1981.
161. GIBSON, H. W. and BAILEY, F. C., *Macromolecules*, 1980, **13**(1), 34.
162. THALER, W. A. (Exxon Research and Engineering Co.), Eur. Pat. Appl. 71 347, 9 Feb. 1983.
163. THALER, W. A., *Macromolecules*, 1983, **16**(4), 623.
164. PERNICONI, A. C. and YOUNG, H. F., US Patent 3 730 900, 1 May 1973.
165. MARCINSKI, M., OLKOWSKA, J. and ATAMANCZUK, B., *Zesz. Nauk.—Wysza Szk. Pedagog. im. Powstancow Slask. Opolu, (Ser.): Chem.* 1981, **4**, 55; *Chem. Abstr.*, 9598381.
166. LAISAAR, S., *Eesti NSV Tead. Akad. Toim., Keem.* 1979, **28**(4), 266; *Chem. Abstr.*, 9259392.
167. Japan Synthetic Rubber Co. Ltd, Jpn. Kokai Tokkyo Koho 58 152 861 (83 152 861), 10 Sept. 1983.
168. VORONKOV, M. G., PLATONOVA, A. T., ANNENKOVA, V. Z., ANNENKOVA, V. M., KAZIMIROVSKAYA, V. B. and UGRYUMOVA, G. S., Fr. Demande 2 444 053, 11 Jul. 1980.
169. ROSE, J. B. (ICI Ltd), Eur. Pat. Appl. 41 780, 16 Dec. 1981; GB Appl. 80/18 915, 10 Jun. 1980.
170. FANTA, G. F., STOUT, E. I. and DOANE, W. MC., Brit. Patent 1 569 481, 18 Jun. 1980.
171. SANDLER, S. R. and KARO, W., *Polymer Syntheses*, Vol. 1, 1974, Academic Press, New York, p. 366.
172. Maruzen Oil Co. Ltd, Jpn. Kokai Tokkyo Koho 80 52 307, 16 Apr. 1980.
173. HARTMANN, M., HIPLER, U. C. and CARLSOHN, H., *Acta Polym.*, 1980, **31**(3), 165.
174. KHARDIN, A. P., KARGIN, YU. N., BAKHTINA, G. D., LENIN, A. S. and GAIDUKOV, V. A., *Vysokomol. Soedin., Ser. B*, 1981, **23**(9), 678.
175. FIELDHOUSE, J. W. and GRAVES, D. F., 'Phosphorus Chemistry', *ACS Symp. Ser.*, 1981, **171**, 315.
176. SIDORCHUK, I. I., ABBASOVA, B. G. and EFENDIEV, A. A., *Azerb. Khim. Zh.*, 1981, **3**, 73.
177. LEVIN, YA. A., FRIDMAN, G. B., GURSKAYA, V. SH., GAZIZOVA, L. KH., SHULYNDIN, S. V. and IVANOV, B. E., *Vysokomol, Soedin., Ser. A.*, 1982, **24**(3), 601.

178. FINKE, M. and RUPP, W. (Hoechst A–G), Ger. Offen. 3 210 775, 29 Sept. 1983.
179. KOBAYASHI, S., HUANG, M. Y. and SAEGUSA, T., *Polym. Bull.* (*Berlin*), 1982, **6**(7), 389.
180. CURRELL, B. R., GRZESKOWIAK, R. and LYNN, M. E., *Br. Polym. J.*, 1981, **13**(3), 122.
181. PLOCHOCKA, K., PENCZEK, S., SKARZYNSKI, J., BORECKA, B. and ZIELINSKA, J. (Instytut Chemii Przemyslowej), Pol. 112 479, 10 May 1982.
182. SHAPKIN, N. P. and SHCHEGOLIKHINA, N. A., *Zh. Obshch. Khim.*, 1981, **51**(1), 3.
183. GAZIZOV, T. KH., GOROKHOVSKAYA, I. V., KIBARDIN, A. M. and PUDOVIK, A. N., *Vysokomol. Soedin., Ser. B,* 1979, **21**(10), 777.
184. KOBAYASHI, S., OKAWA, M., NIWANO, M. and SAEGUSA, T., *Polym. Bull.* (*Berlin*), 1981, **5**(6), 331.
185. MORR, M. and KULA, M. R. (Gesellschaft fur Biotechnologische Forschung m.b.H.), PCT Int. Appl. 81 00 571, 5 Mar. 1981.
186. TOMALIA, D. A., 'Relative heterocyclic monomers' in *Functional Monomers* (R. H. Yocum and E. B. Nyquist, Eds.), Vol. 1, 1974, Marcel Dekker, New York.
187. LUSKIN, L. S., 'Basic monomers' in *Functional Monomers* (R. H. Yocum and E. B. Nyquist, Eds.), Vol. 1, 1974, Marcel Dekker, New York, p. 555.
188. LI, F., GU, Z., YE, W. and FENG, X., *Gaofenzi Tongxun* 1983, **2,** 122.
189. Rohm and Haas Co., Philadelphia, Penn., USA.
190. MATSEVA, L. V., *Uzb. Khim. Zh.*, 1969, **13,** 56.
191. SHIMOMURA, T., KUWABARA, Y., TSUCHIDA, E. and SHINOHARA, I., *Kogyo Kagaku Zasshi*, 1968, **71,** 283.
192. KETURKA, V. and MIKUCONIS, H., *Aktual. Probl. Razvil. Nauchn. Issled. Molodykh Uch. Spets. Vil'nyus, Gosuniv. im. V. Kapsukasa, Mater. Konf. Molodykh Uch. Spets. Estestv. Khim. Fak.*, 1980, p. 98; *Chem. Abstr.*, 94122040.
193. EHRLICH, G. and DOTY, P., *J. Amer. Chem. Soc.*, 1954, **76,** 3764.
194. SUZUKI, N., WADA, Y. and FURUNO, A. (Nitto Electric Industrial Co. Ltd, Mitsubishi Rayon Co. Ltd), Ger. 2 749 295, 6 Sept. 1979.
195. AMAYA, H., HANDA, R. and OSIMA, I. (Nitto Chemical Industry Co. Ltd, Mitsubishi Rayon Co. Ltd), Jpn. Kokai Tokkyo Koho 80 16 010, 4 Feb. 1980.
196. EGOYAN, R. V., AKOPYAN, F. T. and BEILERYAN, N. M., *Arm. Khim. Zh.*, 1979, **32**(9), 704; *Chem. Abstr.*, 92147303.
197. DZHUMADILOV, T. K., BAKAUOVA, Z. KH. and BEKTUROV, E. A. *Izv. Akad. Nauk Kaz. SSR, Ser. Khim.*, 1981, **2**, 31; *Chem. Abstr.*, 9543798.
198. Mitsui Toatsu Chemicals, Inc., Jpn. Kokai Tokkyo Koho 57 121 007 (82 121 007), 28 Jul. 1982.
199. Japan Atomic Energy Research Institute, Taki Chemical Co. Ltd, Jpn. Kokai Tokkyo Koho 81 135 505, 23 Oct. 1981.
200. Nitto Chemical Industry Co. Ltd, Mitsubishi Rayon Co. Ltd, Diafloc Co. Ltd, Jpn. Kokai Tokkyo Koho 57 177 008 (82 177 008), 30 Oct. 1982.

201. KEGGENHOFF, B. and ROSENKRANZ, H. J. (Bayer A–G), Ger. Offen. 2 926 103, 8 Jan. 1981.
202. CHERNENKOVA, YU. P., ZIL'BERMAN, E. N., SHVAREVA, G. N. and KRASAVINA, L. B. (USSR), Zh. Prikl, Khim. (Leningrad), 1980, 53(2), 378; Chem. Abstr., 92198830.
203. CHERNENKOVA, YU. P., ZIL'BERMAN, E. N. and SHAVAREVA, G. N., Vysokomol. Soedin., Ser. B, 1982, 24(2), 119.
204. DIMA, M., OPREA, S. and DUMITRIU, E., Rev. Roum. Chim., 1981, 26(1), 131.
205. BILA, J. and HRABAK, F., Eur. Polym., J., 1984, 20(1), 69.
206. LI, F., CUI, Q. and FENG, X., Gaofenzi Tongxun, 1983, 5, 396.
207. HRABAK, F., BILA, J. and HYNKOVA, V., Eur. Polym. J., 1982, 8(10), 927.
208. YUKAWA, J., KATAOKA, K. and TSURUTA, T., Polym. J., 1979 11(11), 895.
209. FRECHET, J. M. J., ACS Polym. Prepr., 1983, 24(2), 340.
210. ONYON, P. F., Trans. Faraday Soc., 1955, 51, 400.
211. HOWARD, G. J. and LAI, S., J. Polym. Sci., Polym. Chem. Ed., 1979, 17(10), 3273.
212. SVESHNIKOVA, T. G., SMIRNOVA, L. A., SEMCHIKOV, YU. D., KAMYSHENKOVA, L. I. and REVUNOVA, I. V. (USSR), Deposited Doc. SPSTL 994 Khp-DS1, 1982; Chem. Abstr., 98161228.
213. SUTYAGIN, V. M. and LOPATINSKII, V. P., Vysolomol. Soedin., Ser. B, 1983, 25(2), 74.
214. AYYAKUMOVA, N. I., BUTOVETSKAYA, V. L. and NEKHOROSHIKH, L. M. (USSR), Deposited Doc. SPSTL 13 Khp-D81 (1981), 124; Chem. Abstr., 9792876.
215. NATANSOHN, A., MAXIM, S. and FELDMAN, D., Polymer, 1979, 20(5), 629.
216. TRUBITSYNA, S. N. and ASKAROV, M. A., Izv. Vyssh. Uchebn. Zaved., Khim. Khim. Tekhnol., 1982, 25(3), 361.
217. GOGOI, A. K. and DASS, N. N., J. Polym. Sci., Polym. Chem. Ed., 1983, 21(12), 3517.
218. TANG, L. C. and FRANCOIS, B., Eur. Polym. J., 1983, 19(8), 707, 715.
219. LINDSELL, W. E., ROBERTSON, F. C. and SOUTAR, I., Eur. Polym. J., 1983, 19(2), 115.
220. RAYNAL, S., LECOLIER, S., NDEBEKA, G. and CAUBERE, P., J. Polym. Sci., Polym. Lett. Ed., 1980, 18(1), 13.
221. SOUM, A. and FONTANILLE, M., 'Anionic polymers', ACS Symp. Ser., 1981, 166, 239.
222. CARLINI, C., J. Polym. Sci., Polym. Chem. Ed., 1980, 18(3), 799.
223. OHSHIMA, I. and NAKASHIMA, Y. (Nitto Chemical Industry Co. Ltd), Fr. Demande 2 472 558, 3 Jul. 1981.
224. RUZIEV, R. R., KRAPIVIN, A. M., DZHALILOV, A. T., TOPCHIEV, D. A., ASKAROV, M. A. and KABANOV, V. A., Dokl. Akad., Nauk. Uzb. SSR, 1981, 1, 35; Chem. Abstr., 98126659.
225. Nippon Oils and Fats Co. Ltd, Jpn. Kokai Tokkyo Koho 57 131 226 (82 131 226), 14 Aug. 1982.

226. Kyowa Gas Chemical Industry Co. Ltd, Jpn. Kokai Tokkyo Koho 82 120 560, 27 Jul. 1982.
227. ARNDT, P. J., LOWITZ, J. and WENZEL, F. (Roehm G.m.b.H.), Ger. 2 848 627, 7 Feb. 1980.
228. KUESTER, E., DAHMEN, K. and BARTHELL, E. (Chemische Fabrik Stockhausen und Cie), Ger. Offen. 2 857 432, 3 Jul. 1980.
229. GIPSON, R. M., HOTCHKISS, P. and NIEH, E. C. Y. (Texaco Development Corp.), US Patent 4 212 820, 15 Jul. 1980.
230. MOSS, P. H. and NIEH, E. C. Y. (Texaco Development Corp.), US Patent 4 180 643, 25 Dec. 1979.
231. Mitsubishi Gas Chemical Co. Inc., Jpn. Kokai Tokkyo Koho 81 103 210, 18 Aug. 1981.
232. Nitto Chemical Industry Co. Ltd, Mitsubishi Rayon Co. Ltd, Diafloc Co. Ltd., Jpn. Kokai Tokkyo Koho 80 09 630, 23 Jan. 1980.
233. Nitto Chemical Industry Co. Ltd, Mitsubishi Rayon Co. Ltd, Diafloc Co. Ltd, Jpn. Kokai Tokkyo Koho 80 09 628, 23 Jan. 1980.
234. Nitto Chemical Industry Co. Ltd, Mitsubishi Rayon Co. Ltd, Diafloc Co. Ltd, Jpn. Kokai Tokkyo Koho 80 09 629, 23 Jan. 1980.
235. CHULPANOV, K. A., ISMAILOV, I., RAKHMATULLAEV, KH., DZHALILOV, A. T. and ASKAROV, M. A., Vysokomol. Soedin., Ser. B, 1983, 25(3), 147; Chem. Abstr., 98179984.
236. ISMAILOV, I., CHULPANOV, K. A., RAKHMATULLAEV, KH., DZHALILOV, A. T. and ASKAROV, M. V., Uzb. Khim. Zh., 1982, 2, 26; Chem. Abstr., 97128123.
237. ISHIGAKI, I., OKADA, T., SASUGA, T., TAKEHISA, M. and MACHI, S., J. App. Polym. Sci., 1981, 26(2), 741.
238. Sumitomo Chemical Co. Ltd, Jpn. Kokai Tokkyo Koho 57 121 009 (82 121 009), 28 Jul. 1982.
239. Sumitomo Chemical Co. Ltd, Jpn. Kokai Tokkyo Koho 58 27 711 (83 27 711), 18 Feb. 1983.
240. Sumitomo Chemical Co. Ltd, Jpn. Kokai Tokkyo Koho 57 121 008 (82 121 008), 28 Jul. 1982.
241. ISHIGAKI, I., FUKUZAKI, H., OKADA, TOSHIMI, OKADA, TOSHIO, OKAMOTO, J. and MACHI, S., J. Appl. Polym. Sci. 1981, 26(5), 1585.
242. Osaka Yuki Kagaku Kogyo Co. Ltd, Jpn. Kokai Tokkyo Koho 58 129 010 (83 129 010), 1 Aug. 1983.
243. Nippon Kayaku Co. Ltd, Jpn. Kokai Tokkyo Koho 82 12 011, 21 Jan. 1982.
244. Mitsubishi Chemical Industries Co. Ltd, Jpn. Kokai Tokkyo Koho 80 07 826, 21 Jan. 1980.
245. ARNDT, P. J., ROSS, K. and WENZEL, F. (Roehm GmbH), Fr. Demande 2 491 476, 9 Apr. 1982.
246. Dainippon Ink and Chemicals Inc., Kawamura Physical and Chemical Research Inst., Jpn. Kokai Tokkyo Koho 57 158 208 (82 158 208), 30 Sept. 1982.
247. Nippon Kayaku Co. Ltd, Jpn. Kokai Tokkyo Koho 81 135 501, 23 Oct. 1981.

248. Mitsubishi Chemical Industries Co. Ltd, Jpn. Kokai Tokkyo Koho 81 143 205, 7 Nov. 1981.
249. CABESTANY, J., TROUVE, C. and DEPERNET, D. (Societé Française Hoechst SA), US Patent 4 396 752, 2 Aug. 1983.
250. Societé Française Hoechst SA, Jpn. Kokai Tokkyo Koho 80 68 911, 20 May 1980.
251. Sanyo Chemical Industries Ltd, Jpn. Kokai Tokkyo Koho 80 89 308, 5 Jul. 1980.
252. HONDA, T., ITOH, H., SAITO, J. and MITSUISHI, T. (Mitsui Toatsu Chemicals Inc), Fr. Demande 2 489 338, 5 Mar. 1982.
253. Mitsubishi Chemical Industries Co. Ltd, Jpn. Kokai Tokkyo Koho 81 81 315, 3 Jul. 1981.
254. VALLIN, D., HUGEUT, J. and VERT, M., Polymer Amines Ammonium Salts, Invited Lect. Contrib. Pap. Int. Symp., 1979, (Pub. 1980), E. J. Goethals (Ed.), Pergamon, Oxford.
255. BUTLER, G. B., Polymer Amines Ammonium Salts, Invited Lect. Contrib. Pap. Int. Symp., 1979, (Pub. 1980), E. J. Goethals (Ed.), Pergamon, Oxford.
256. OTTENBRITE, R. M. and RYAN, W. S., Jr, Ind. Eng. Chem. Prod. Res. Dev., 1980, 19(4), 528.
257. BUTLER, G. B. and ANGELO, R. J., J. Amer. Chem. Soc., 1957, 79, 3128.
258. LANCASTER, J. E., BACCEI, L. and PANZER, H. P., J. Polym. Sci., 1976, B14, 549.
259. WYROBA, A., Polimery (Warsaw), 1981, 26(4), 139; Chem. Abstr., 95220422.
260. KABANOV, V. A., TOPCHIEV, D. A. and NAZHMETDINOVA, G. T., Vysokomol. Soedin., Ser. B, 1984, 26(1), 51.
261. WANDREY, C., JAEGER, W. and REINISCH, G., Acta Polym., 1981, 32(4), 197.
262. DUMITRIU, E., OPREA, S. and DIMA, M., Mater. Plast. (Bucharest), 1981, 18(4), 202; Chem. Abstr., 96163281.
263. WANDREY, C., JAEGER, W. and REINISCH, G., Acta Polym., 1981, 32(5), 257.
264. REINISCH, G., JAEGER, W., HAHN, M. and WANDREY, C., Proc. IUPAC Macromol. Symp., 28th, 1982, p. 83.
265. WYROBA, A., Przem. Chem., 1981, 60(11–12), 531; Chem. Abstr., 9739481.
266. HAHN, M., JAEGER, W. and REINISCH, G., Acta Polym., 1983, 34(6), 322.
267. TOPCHIEV, D. A., BIKASHEVA, G. T., MARTYNENKO, A. I., KAPTSOV, N. N., GUDKOVA, L. A. and KABANOV, V. A., Vysokomol. Soedin., Ser. B, 1980, 22(4), 269.
268. TOPCHIEV, D. A., NAZHMETDINOVA, G. T., KARTASHENSKI, A. I., NECHAEVA, A. V. and KABANOV, V. A., Izv. Akad. Nauk SSSR, Ser. Khim., 1983, 10, 2232; Chem. Abstr., 10034887.
269. TOPCHIEV, D. A. and NAZHMETDINOVA, G. T., Vysokomol. Soedin., Ser. A., 1983, 25(3), 636.

270. HAHN, M., JAEGER, W., SEEHAUS, F., WANDREY, C. and REINISCH, G., Ger. (East) 156 979, 6 Oct. 1982.
271. BALLSCHUH, D., OHME, R. and RUSCHE, J., Ger. (East) 141 029, 9 Apr. 1980.
272. OHME, R., BALLSCHUH, D. and RUSCHE, J., Ger. (East) 141 028, 9 Apr. 1980.
273. Nippon Kayaku Co. Ltd, Jpn. Kokai Tokkyo Koho, 81 16 448, 17 Feb. 1981.
274. BERGTHALLER, P. (Agfa-Gevaert AG), Ger. Offen. 2 946 550, 27 May 1981.
275. HARADA, S. and HASEGAWA, S., *Makromol. Chem. Rapid Commun.*, 1984, **5**(1), 27.
276. EGOROV, V. V., BATRAKOVA, E. V., TITKOVA, L. V., DEMIN, V. V., ZUBOV, V. P. and BARNAKOV, A. N., *Vysokomol. Soedin., Ser. B,* 1982, **24**(5), 370.
277. MIELKE, I. and RINGSDORF, H., *J. Polym. Sci., Part C,* 1970, **31**, 107; *Polym. Letters,* 1971, **9**, 1; *Makromol. Chem.,* 1971, **142**, 319.
278. SALAMONE, J. L., SNIDER, B. and FITCH, W. C., *J. Polym. Sci., A1,* 1971, **9**, 1493.
279. KABANOV, V. A., ALIEV, K. V., PATRIKEEVA, T. I., KAVGINA, O. V. and KAVGIN, V. A., *J. Polym. Sci.,* 1967, **C16**, 1079.
280. KABANOV, V. A., ALIEV, K. V. and KAVGIN, V. A., *Polym. Sci., USSR,* 1968, **10**, 1873.
281. SALAMONE, J. C., SNIDER, B. and FITCH, W. L., *J. Polym. Sci.,* 1971, **B9**, 13.
282. SALAMONE, J. C. and ELLIS, E. J., *ACS Polym. Prop., Div. Polym. Chem.,* 1972, **32**, 294.
283. SALAMONE, J. C., SNIDER, B. and FITCH, W. L., *J. Polym. Sci.,* 1971, **B9**, 13.
284. SALAMONE, J. C., MAHMUD, M. U., WATTERSON, A. C., OLSON, A. P. and ELLIS, E. J., *J. Polym. Sci., Polym. Chem. Ed.,* 1982, **20**(5), 1153.
285. EGOROV, V. V., KOSTROMIN, S. G., SIMAKOVA, G. A., TITKOVA, L. V., DREVAL, V. E., GOLUBEV, V. B., ZUBOV, V. P. and KABANOV, V. A., Deposited Doc., 1979, VINITI 551; *Chem. Abstr.,* 92164340.
286. EGOROV, V. V., ZUBOV, V. P., GOLUBEV, V. E., SHAPIRO, YU. E., DREVAL, V. E., TITKOVA, L. V. and KABANOV, V. A., *Vysokomol. Soedin., Ser. B,* 1981, **23**(11), 803.
287. EGOROV, V. V., ZUBOV, V. P., LACHINOV, M. B., KHACHATURYAN, O. B. and GOLUBEV, V. B., *Vysokomol. Soedin., Ser. A,* 1981, **23**(4), 848.
288. OHTSUKA, Y., KAWAGUCHI, H. and HAYASHI, S., *Polym.,* 1981, **22**(5), 658.
289. LIU, L. J. and KRIEGER, I. M., *J. Polym. Sci., Polym. Chem. Ed.,* 1981, **19**(11), 3013.
290. BRONSTEIN-BONTE, I. Y. and LINDHOLM, E. P. (Polaroid Corp.) US Patent 4 340 522, 20 Jul. 1982.
291. SOKOLOV, V. P., KORABLEVA, G. R., CHIKUNOVA, L. A. and SEMENOV, K. F., USSR, 952 856, 23 Aug. 1982. From *Otkrytiya, Izobret., Prom. Obraztsy, Tovarnye Znaki,* 1982, **31**, 128.

292. Kyoritzu Yuki Co. Ltd, Jpn. Kokai Tokkyo Koho 81 04 606, 19 Jan. 1981.
293. Kyoritsu Yuki Co. Ltd, Jpn. Kokai Tokkyo Koho 82 98 505, 18 Jun. 1982.
294. OTANI, H., FURUNO, A. and INUKAI, K. (Nitto Chemical Industry Co. Ltd, Mitsubishi Rayon Co. Ltd), Jpn. Kokai Tokkyo Koho 79 16 593, 7 Feb. 1979.
295. Kyoritsu Yuki Co. Ltd, Jpn. Kokai Tokkyo Koho 82 31 904, 20 Feb. 1982.
296. ABE, G., Jpn. Kokai Tokkyo Koho 82 14 603, 25 Jan. 1982.
297. Kyoritsu Yuki Co. Ltd, Jpn. Kokai Tokkyo Koho 82 31 902, 20 Feb. 1982.
298. KREBS, R. F., MAREK, P. J. and PHILLIPS, K. G. (Nalco Chem. Co.), US Patent 4 405 728, 20 Sept. 1983.
299. DRAGAN, S., BARBOIU, V., PETRARIU, I. and DIMA, M., *J. Polym. Sci. Polym. Chem. Ed.*, 1981, **19**(11), 2869.
300. TANOKA, H., *J. Polym. Sci.*, 1978, **B16,** 87.
301. DRAGAN, S., PETRARIU, I. and DIMA, M., *J. Polym. Sci., Polym. Chem. Ed.*, 1980, **18**(7), 2333.
302. KAWABE, H., *Bull. Chem. Soc. Jpn.*, 1981, **54**(7), 1914.
303. LUCA, C., PETRARIU, I. and DIMA, M., *J. Polym. Sci., Polym. Chem. Ed.*, 1979, **17**(12), 3879.
304. PETRARIU, I., ROTARU, M. and DRAGAN, S., *Rev. Roum. Chim.*, 1980, **25**(1), 145; *Chem. Abstr.*, 938742.
305. MEUNIER, G., BOIVIN, S., HEMERY, P., BOILEAU, S. and SENET, J. P., *Polym.*, 1982, **23**(6), 861.
306. MEUNIER, G., BOIVIN, S., HEMERY, P., SENET, J. P. and BOILEAU, S. *Polymer Sci. Technol.* (Plenum), 1983, **21** (Modif. Polymer), 293.
307. SHAIKHOVA, M. K., MUMINOV, K. M., DZHALILOV, A. T. and ASKAROV, M. A., *Dokl. Akad. Nauk Uzb. SSR*, 1980, **8,** 47; *Chem. Abstr.*, 9525715.
308. DANICHER, L., LAMBIA, M. and LEISING, F., *Bull. Soc. Chim. Fr.*, 1979, **9–10**(2), 544.
309. GHESQUIERE, D., MORCELLET-SAUVAGE, J. and LOUCHEUX, C., *Macromol. Synth.*, 1982, **8,** 79.
310. BOUCHER, E. A. and MOLLETT, C. C., *J. Chem. Soc., Faraday Trans. 1,* 1982, **78**(1), 75.
311. BOUCHER, E. A. and KHOSRAVI-BABADI, E., *J. Chem. Soc., Faraday Trans. 1,* 1983, **79**(8), 1951.
312. RINAUDO, M. and DESBRIERES, J., *Macromol. Chem. Phys.*, 1981, **182**(2), 373.
313. GHESQUIERE, D., CAZE, C. and LOUCHEUX, C., *Polym. Bull. (Berlin),* 1983, **10**(5-6), 282.
314. TAGHIZADEH, T., CAZE, C. and LOUCHEUX, C., (a) *Polym. Bull. (Berlin),* 1980, **3**(11), 593; (b) *Eur. Polym. J.,* 1982, **18**(10), 907.
315. LUCA, C., BARBOIU, V., PETRARIU, I. and DIMA, M., *J. Polym. Sci., Polym. Chem. Ed.,* 1980, **18**(7), 2347.
316. SERITA, H., OHTANI, N. and KIMURA, C., *Yukagaku*, 1981, **30**(2), 96; *Chem. Abstr.*, 94157436.

317. Dainippon Ink & Chemicals Inc., Kawamura Physical & Chemical Research Inst., Jpn. Kokai Tokkyo Koho 82 108 103, 6 Jul. 1982.
318. HAMBRECHT, J., NAARMANN, H., RICHTER, K., REICHEL, F. and HOEHR, L. (BASF AG), Ger. Offen. 3 047 688, 22 Jul. 1982.
319. MORISHIMA, Y., HASHIMOTO, T., ITOH, Y., KAMACHI, M. and NOZAKURA, S., *J. Polym. Sci., Polym. Chem. Ed.*, 1982, **20**(2), 299.
320. SALAMONE, J. C., TSAI, C. C., OLSON, A. P. and WATTERSON, A. C., 'Ion Polymers', *ACS Adv. Chem. Ser.*, 1980, **187**, 337.
321. MERLE, Y., MERLE-AUBRY, L. and SELEGNY, E., Polymer Amines Ammonium Salts, Invited Lect. Contrib. Pap. Int. Symp. 1979 (Pub. 1980), p. 113, E. J. Goethals (Ed.), Pergamon, Oxford.
322. Seiko Chemical Industry Co. Ltd, Jpn. Kokai Tokkyo Koho 58 164 633 (83 164 633), 29 Sept. 1983.
323. KAWAGUCHI, H., HOSHINO, H., AMAGASA, H. and OHTSUKA, Y., *J. Colloid Interface Sci.*, 1984, **97**(2), 465.
324. KAWAGUCHI, H., HOSHINO, H. and OHTSUKA, Y., *Proc. IUPAC Macromol. Symp. 28th*, 1982, p. 609.

Chapter 2

SMECTITIC CLAYS AS IONIC
MACROMOLECULES

GERHARD LAGALY

Institut für anorganische Chemie der Universität Kiel, Federal Republic of Germany

1 INTRODUCTION

Smectitic clay minerals represent a special type of two-dimensional ionic macromolecule (Figs 1 and 3). Macromolecular silicate layers carry negative charges that are balanced by gegen-ions. One of the differences from ionic polymers proper lies in the stiffness of the layers, which are about 1 nm thick. These layers rarely occur as isolated individual units, but aggregate to form crystalline structures. This structural order gives these materials several advantages over true ionic polymers.

(a) Many types of charged and uncharged organic compounds, including polymers, interact with the external and internal silicate surfaces.
(b) Surfaces are planar and the reactions are not complicated by conformational changes of the polymer backbone.
(c) Much information can be gained about the arrangement of the interacting molecules on the internal surfaces from X-ray studies.
(d) The charges are distributed in distinct patterns that are not affected by conformational changes.
(e) Layers with different charge densities are available.
(f) Comprehensive studies on charge pattern interactions can be undertaken.

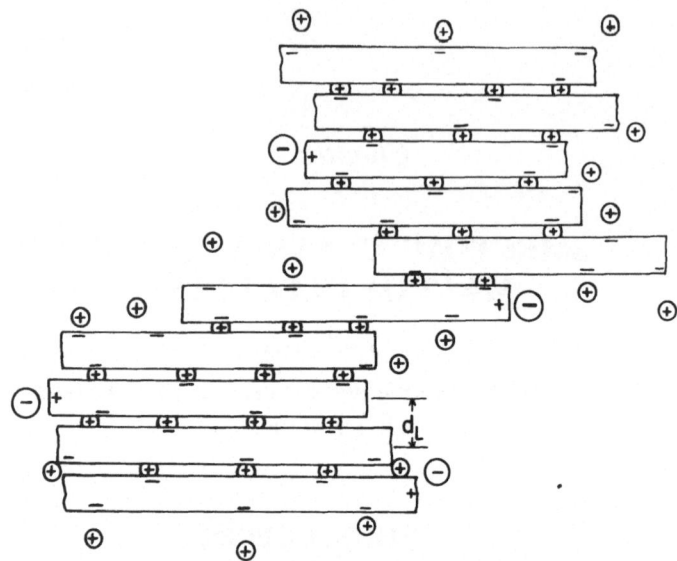

FIG. 1. Structure of smectitic clays (2:1 layer silicates) made up of macromolecular silicate anions; d_L = basal spacing.

2   CLAY MINERAL STRUCTURE

2.1   Principles

The crystal structure of clay minerals[1] consists of silicate layers that
contain sheets of $[SiO_4]$ tetrahedra and sheets of $[Me(O, OH)_6]$
octahedra, where Me = Metal, (Fig. 2).

In 2:1 type minerals (Fig. 2a) the silicate layers consist of a central
octahedral sheet sandwiched between two tetrahedral sheets, so that
the oxygen ions of the $[Me(O, OH)_6]$ sheet also belong to the
tetrahedral sheets. Common cations are Me = Al^{3+}, Fe^{3+}, Mg^{2+}, Fe^{2+}.
Minerals with Me = Al^{3+} or Fe^{3+} are classified as dioctahedral, those
with Me = Mg^{2+} or Fe^{2+} as trioctahedral.

The structural unit contains two 'chemical units' Si_4O_{10} or
$(Si, Al)_4O_{10}$, of which the general chemical formula may be written:

Dioctahedral: $\{Al_2(OH)_2Si_4O_{10}\}$ pyrophyllites
Trioctahedral: $\{Mg_3(OH)_2Si_4O_{10}\}$ talcs

The layers in talc and pyrophyllite are electrostatically neutral.
A large variety of minerals results from isomorphous substitutions

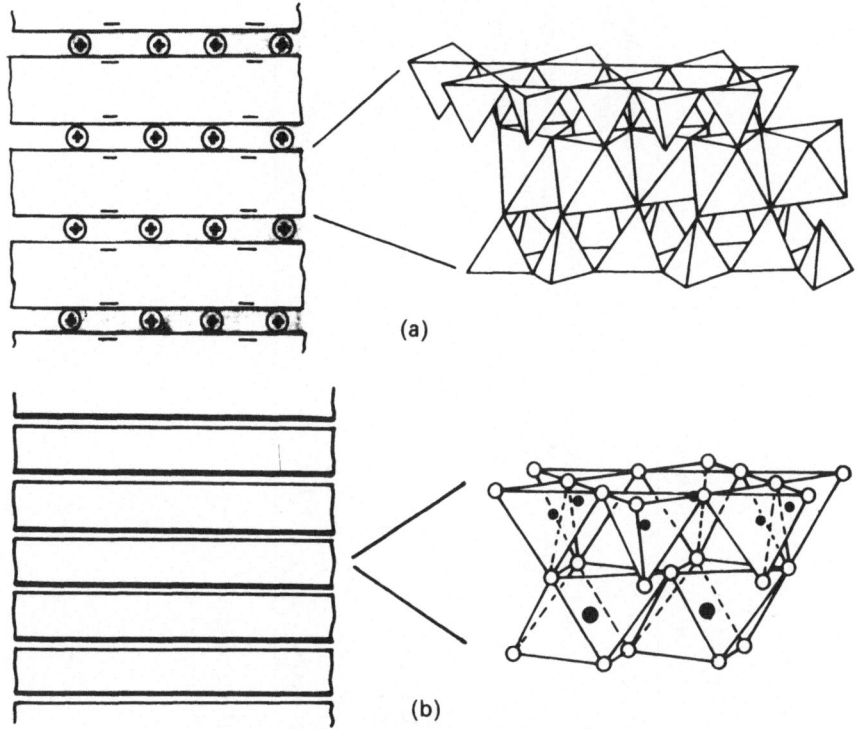

FIG. 2. Crystal structure of (a) 2:1 clay minerals: micas, vermiculites, smectites; (b) 1:1 clay minerals: kaolinite.

and negative charging of the layers (Table 1). The layer charge arises by some combination of four mechanisms:

(1) substitution of Me^{3+} (frequently Al^{3+}) for Si^{4+} in the tetrahedral sheets;

(2) substitution of Me^{2+} (frequently Mg^{2+}) for Al^{3+} in the octahedral sheet (in hectorite and the synthetic laponite: Li^+ for Mg^{2+});

(3) existence of vacancies in octahedral sites;

(4) existence of proton defects (O^{2-} instead of OH^--ion).*

---

* Cross-substitutions, $Fe^{3+} + O^{2-} \rightleftharpoons Fe^{2+} + (OH)^-$, are observed during redox reactions but do not change the layer charge.

TABLE 1

CLASSIFICATION OF CLAY MINERALS

Layer type	Group	Subgroup	Species[a]	Layer charge ξ $(eq/(Si, Al)_4O_{10})$	Intracrystalline swelling	Cation exchange capacity (meq/100 g) Inorganic ions	Organic ions
1:1	Serpentine—kaolin	Kaolins	Kaolinite	0	—	up to 10	up to 10[e]
2:1	Talc-pyrophyllite	Talcs	Talc	0	—	—	—
		Pyrophyllites	Pyrophyllite				
	Smectites	Saponites	Saponite, hectorite	0.2–0.6	+ + +	60–120	60–120
		Montmorillonites	Montmorillonite, beidellite				
	Vermiculites		Vermiculite	0.6–0.8	+ +[b]	100–170	100–170
	Mica	Micas	Muscovite, biotite	~1	(+)[c]	up to 35[d]	up to 200

[a] Only a few examples are given.
[b] Can be blocked in presence of K^+, NH_4^+, Rb^+, Cs^+ as interlayer cations.
[c] Swelling only after exchange of the potassium interlayer cations.
[d] Depending on particle size (see ref. 2).
[e] See refs 3, 4.

Neglecting proton defects, the general formula is then given by:

$$M_{x+y/v}^{v+}(H_2O)_n\{(Al, Mg)_{2-3}^{(6-y)+}(OH)_2(Si_{4-x}Al_xO_{10})^{x-}\}^{(x+y)-}$$

where $(x + y) = \bar{\xi}$ is the layer charge (in eq/(Si, Al)$_4$O$_{10}$). The inter-layer cations M^+ balance the negative layer charges and are exchangeable by other cations. Commonly, water molecules are held between the layers of vermiculites and smectites.

Besides the 2:1 layer clay minerals there is the group of 1:1 layer silicates (serpentine–kaolin group). The silicate layers consist of a tetrahedral and an octahedral sheet (Fig. 2b). The idealized compositions are

Dioctahedral: $\{Al_2(OH)_4Si_2O_5\}$ kaolin minerals.

Trioctahedral: $\{Mg_3(OH)_4Si_2O_5\}$ serpentine minerals.

The layers are uncharged and only minute amounts of cations as defects may be bound between the layers. The cohesion between the layers mainly arises from hydrogen bonds, dipole–dipole interactions and van der Waals forces.[1]

In halloysite the dioctahedral 1:1 layers are separated by water monolayers. The structure generally incorporates considerable inter-layer randomness.

2.2 The Layer Charge of Smectites

Most studies described in this chapter refer to smectites. The layer charge $\bar{\xi}$ is about 0.3 eq/(Si, Al)$_4$O$_{10}$. The layers making up a crystal do not possess exactly the same charge density and charge density distribution.[5] One may easily imagine that the extent and type of substitutions and defects vary as a result of the natural origin. Variations within the layers are also likely, for instance from the centre to the edges, or areas with higher charge density may alternate with lower-charged parts.

An extreme case occurs with silicate layers with pronounced unsymmetrical charge density distribution. Rectorite and rectorite-like minerals (Fig. 3) consist of layers that exhibit different charge densities at both sides.[1,6]

The alkylammonium method[7] provides a simple means for determining the layer charge and the charge distribution. Charge distribution is known to vary widely and ranges from 'uniform' distribution (less common) to mixed-layer structures (very common).[5,6] The

FIG. 3. Electron micrograph showing the aggregated macromolecular silicate anions (almost regularly interstratified specimen from Honami mine, Japan,[6] after exchange of octadecylammonium ions, by courtesy of Dr H. Vali (Thesis, Technische Universität, Munich, 1983) and Professor Dr H. Köster, Technische Universität, Munich, FRG.

mixed-layer charge distribution of smectites is frequently observed and very likely reflects the cooperative nature (Section 6) of some processes during alteration of the precursor minerals into smectites.

2.3 Structure at the Edges
At the crystal edges, an accumulation of charges is attenuated by the adsorption of protons by O^{2-} or OH^- ions (Fig. 4). Generally, the compensation is not complete so that cations or anions, as gegen ions, balance the edge charges. In solution, the edge charge depends on the pH. At pH < pzc (point of zero charge) an excess of protons charges the edges positively; when pH > pzc the edges obtain a negative charge. The pzc of the edge surface of smectites most probably lies between pH 4 and 5. The pH-dependent edge charges bring about an additional ion-exchange capacity—anion exchange at pH < pzc, cation

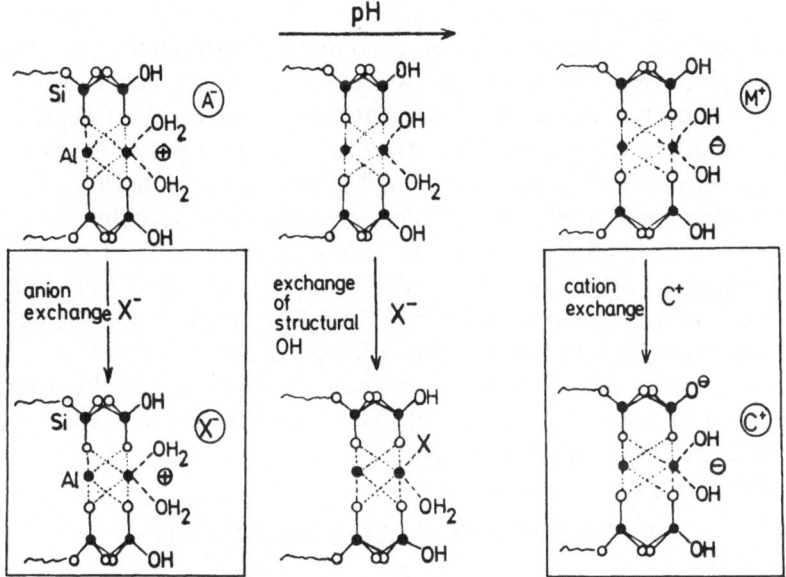

FIG. 4. Reactions on the edge faces of the silicate layers. (With kind permission of the Royal Society.[8])

exchange at pH > pzc. The exchange of structural OH-groups seems to be promoted at pH-values around the pzc.

The interaction of positive edge charges with negative surface charges favours the aggregation of the clay plates. These aggregations form gel-like structures which decisively influence rheological properties.[9,10]

2.4 Occurrence, Size and Shape

Natural samples are commonly used in clay mineral studies. They are formed by weathering or hydrothermal alteration of precursor minerals, or they are newly formed under hydrothermal conditions. The varying properties of smectites are, at least partially, caused by their different origins and history.

The main clay mineral in bentonites is montmorillonite (typically 20–95%). Admixed are quartz (sometimes cristobalite), micas, chlorites, feldspars, carbonates (including calcite, dolomite and siderite) and small amounts ($\leq 1\%$) of mixed-layer minerals, iron oxides and organic materials (for examples see ref. 11). A few bentonites contain a small percentage of kaolinite.

Iron oxides and organic materials are removed by chemical reactions[1] and carbonates can be decomposed with dilute acids. Since a complete separation of the smectites from all other admixed materials cannot be achieved, a bentonite is usually fractionated by sedimentation techniques. The fraction $<2\,\mu m$ frequently contains almost pure smectite and only a small percentage of impurities.

Electron micrographs of smectites generally show very thin undulating sheets; individual particles are barely discernible, so that reliable dimensions of the individual crystals cannot be obtained from micrographs. Sedimentation experiments yield equivalent Stokes diameters that do not represent the true crystal dimensions.

As discussed in more detail in Section 7.3, sodium smectite crystals dispersed in water disintegrate into very thin platelets of a few layers, or even into the individual silicate layers.

3 INTRACRYSTALLINE REACTIVITY

3.1 Intracrystalline Reactions

The best known compounds among layer structures with intracrystalline reactivity are 2:1 type clay minerals. Niobates, titanates, molybdates, uranates and phosphates may be mentioned as reacting-layer structures of different chemical natures.[12–14] Their interlayer spaces are accessible to ions or neutral molecules. The spacing (basal spacing d_L, Fig. 1) between the layers is variable and follows the changes of the interlayer structure. It can easily be obtained from the d-values of the basal reflections that are generally easily discernible on X-ray powder patterns.

The following specification does not imply that a distinct layer compound participates in all types of reaction. The reactivity of the 2:1 layer clay minerals is also very variable. Micas permit only restricted intracrystalline reactions, whereas smectites are highly reactive towards all types of reactants.

The intracrystalline reactions may be classified as follows:

(1) Exchange of inorganic and organic cations for the interlayer cations.
(2) Reversible hydration or solvation of the interlayer space.
(3) Displacement of interlayer molecules (water, neutral organic molecules) by other neutral compounds.
(4) Complexing of the interlayer cations.

(5) Conformational changes of interlayer ions or molecules.

(6) Redox reactions.

The intracrystalline reactions of neutral layer compounds (1:1 type clay minerals) are less manifold. They include

(1) Intercalation of neutral molecules.

(2) Displacement reactions.

Some examples illustrating the different reactions are discussed in the following sections.

3.2 Cation Exchange

Smectites and vermiculites easily exchange their interlayer cations (in natural samples these are commonly Ca^{2+}, Mg^{2+} and Na^+). Among inorganic ions, the preference for potassium ions to sodium ions increases with the layer charge and is particularly strong for highly charged vermiculites and micas. The potassium ions are so tightly fixed between the mica layers that the hydration shells are displaced, and the cation exchange requires special conditions (see Section 6.4). The fixation of potassium ions is explained by an optimal fit into the hexagons of the surface oxygen atoms.

Smectites bind a large variety of organic cations. The basal spacing, after appropriate drying, is determined by the dimensions of the cations and is nearly independent of their packing density, as long as the cations are arranged in monolayers. If the monolayer capacity is exceeded and bilayers form, the spacing increases by about the diameter of the cation. Frequently, intermediate spacings are observed for two particular reasons:

(1) the cations retain a high degree of flexibility in the interlayer space (orientation or conformation can change with packing density) and

(2) the number of interlayer cations varies randomly (within certain limits) from interlayer space to interlayer space.

As a consequence of the heterogeneous charge distribution, the effect of the randomness of the cation density distribution predominates for smectites. A regular variation of the interlayer cation density can lead to superstructures with largely increased spacings.[5]

The cation exchange capacity (CEC, usually in meq/100 g) is related to the layer charge, ξ. For most smectites the molar mass of the formula unit is

$$360 + \xi \cdot MM_{cation} = 372 \text{ g mol}^{-1}$$

for $\xi = 0 \cdot 3$ and calcium ions (molar mass $MM_{cation} = 40 \text{ g mol}^{-1}$). Thus

$$CEC = \frac{\xi \times 1000}{372} \times 100 = 269\xi \text{ (meq/100 g)}$$

A smectite with $\xi = 0 \cdot 3$ (eq/(Si, Al)$_4$O$_{10}$) has an interlayer cation exchange capacity of 81 meq/100 g.

The experimental CEC is generally higher because the cations at the crystal edges contribute to the CEC. For many smectites the interlayer CEC is about 80% of the total CEC (measured at pH ~ 6).[7]

The highly charged 2:1 clay minerals (illites, micas) do not exchange their interlayer cations (Table 1); instead, the observed CEC results from external surface cations. With organic cations a complete exchange may be achieved but requires very long reaction times, e.g. up to several months for micas.

3.3 Interlayer Hydration

The interlayer space of smectites and vermiculites contains variable amounts of water, depending on the relative humidity or the salt concentration in salt solutions. The driving force for the uptake of water is the hydration of the interlayer cations. Uncoordinated water molecules between the hydrated cations complete the filling of the interlayer space.

Recently, much information about the interlayer structure, including coordination of the water around the cations, positions of the interlayer cations and their mobility, has been gained.[15-22] The properties of the interlayer water differ from the bulk water properties; one particular difference is the increased acidity of the interlayer water.[23,24]

The state of hydration is commonly expressed as the number of water layers, i.e. one, two or four corresponding to spacings of about $1 \cdot 2$, $1 \cdot 5$ and $2 \cdot 0$ nm respectively. However, these are not to be considered as strongly organized layers. The extent and type of hydration, cation positions, cation density, orientation of the water molecules on the silicate surface and water packing density are affected by the type of the cations in the smectite, the layer charge and layer charge distribution. States of hydration ascribed to one, two and four water layers can thus be found within spacings of 1·2–1·3 nm, 1·5–1·6 nm and 1·9–2·0 nm.

The different states of hydration are attained by varying the water vapour pressure or by immersing the smectite ' in salt solutions. Spacings intermediate between the plateaux (Fig. 5) are produced by a

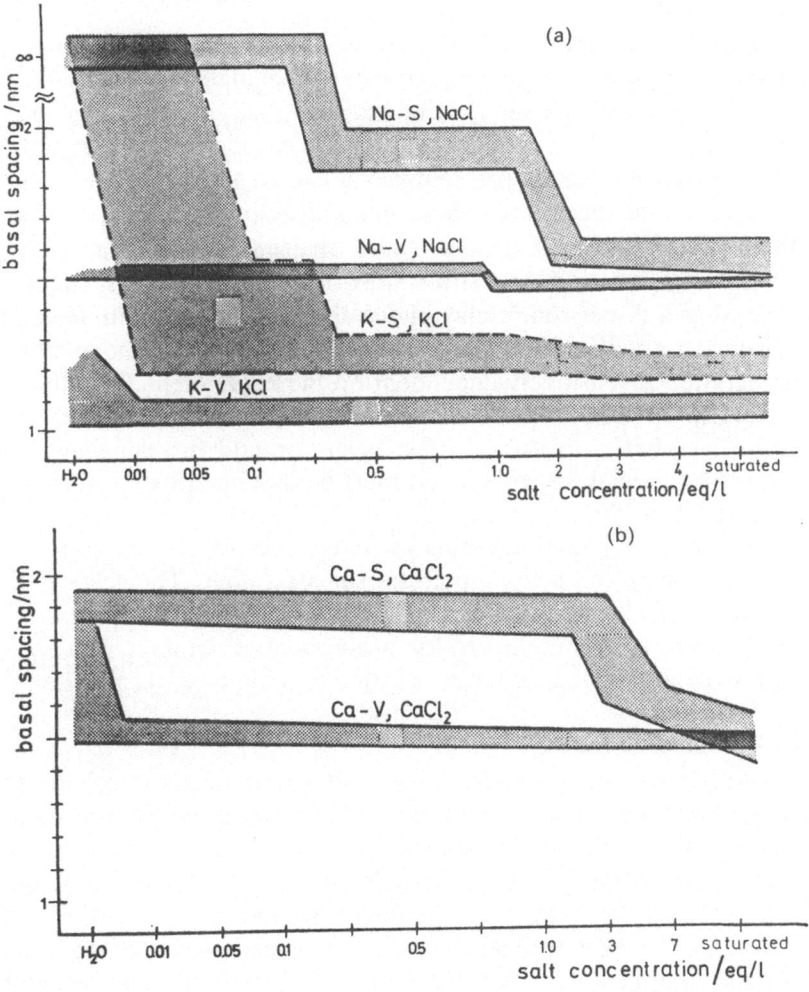

FIG. 5. Basal spacings of smectites (S) and vermiculites (V) in contact with aqueous salt solutions (data from about 25 different smectites and about five different vermiculites).

random distribution of interlayer spaces in different states of hydration, which arise mainly as a consequence of charge heterogeneity. The different hydration states of sodium, potassium and calcium smectites is evident and the particular behaviour of potassium ions is clearly demonstrated. Even at low potassium salt concentrations, the spacing can be reduced to $1 \cdot 1$–$1 \cdot 3$ nm. More highly charged potassium vermiculites collapse to $1 \cdot 0$–$1 \cdot 1$ nm at KCl concentrations as low as $0 \cdot 01$ M or less.

Particular attention should be directed to the behaviour of sodium, potassium and lithium smectites in water or diluted salt solutions. At 'indefinite swelling' ($d_L \to \infty$), the spacings exceed 2 nm. The basal reflections in the X-ray pattern are very broadened or completely absent and no regular interlayer spacing can be measured. According to Norrish,[25] the spacing of sodium montmorillonite increases linearly with $1/c^{1/2}$ (c = salt concentration). This 'spacing' is certainly not a true spacing but a mean value. The state of 'indefinite swelling' may be described as a dispersion of clay plates disintegrated into thin lamellae consisting of a few, sometimes only one, silicate layers. The extent of disintegration is sensitively dependent on the experimental conditions. The interlayer cations no longer reside at the silicate surface, but become distributed in diffuse ionic layers around the lamellae. The system is a colloidal dispersion and must be described by the theory of colloid stability.[26–31]

In presence of limited amounts of water, gels are formed consisting of frameworks of the macromolecular silicate anions. The gels contain isolated silicate layers—commonly about 20–30%—and aggregates of several layers (e.g. five layers for a Na-montmorillonite from Wyoming), depending on the interlayer cation and the type of smectite.[22,32]

3.4 Interlayer Solvation

Like water molecules, neutral organic polar molecules can solvate the interlayer cations. The structural model is in principle the same: solvated cations and solvent molecules between the solvation shells. The extent of cation solvation depends on the cation density, the type of smectite and the strength of the liquid structure.

Polar liquids exhibit distinct liquid structures with strong intermolecular interactions. The intermolecular forces have to compete with the solvation force of the cations and the field arising from the surface charges. Strong solvation forces and the low 'self-preservation tendency' of the liquid promote extensive solvation shells. An instructive example has been reported by Annabi-Bergaya et al.[33] Methanol molecules in the solid state are associated in zigzag chains. Similar associates are assumed to exist in liquid methanol. This type of association is maintained on the surface of 'Li$^+$-smectite, but is completely disturbed on Ca^{2+}-smectite surfaces by the stronger electrical field of the Ca^{2+} ions.

A question related to the interlayer solvation is the location of the interlayer cations. The cations may reside on the silicate surface, with some of their co-ordination sites occupied by surface oxygen ions, or they may be completely surrounded by solvent molecules. The results of electron spin resonance measurements[34] support the latter concept of interlayer solvation. For strong cation–solvent interactions the cations are completely solvated; otherwise they remain attached to the surface.

3.5 Displacement Reactions

Solvated smectites and vermiculites are frequently prepared from air-dried samples, an instance being the treatment with ethylene glycol and glycerol. This reaction is widely used for identifying smectites and vermiculites.[1,35] The solvent molecules, usually present in excess, have to displace the water molecules from the interlayer cations. With 'weak' solvents, some water molecules may remain co-ordinated to the cations and a complete substitution of the interlamellar water by the organic liquids is not achieved. In particular, multivalent cations retain water molecules in the presence of solvents with low Gutmann donor numbers.[34]

Liquids that cannot displace the hydration water, e.g. nitriles, may be introduced by exposing the dried samples to the liquid or its vapour. The number of suitable liquids is restricted because the opening of the interlayer spaces of dried samples is often kinetically impeded (see Sections 6.2, 6.3).

3.6 Complex Formation

As is the case for a cation in solution, the hydration or solvation of an interlayer cation may also be considered as complex formation. A distinction between hydrated or solvated interlayer cations and true complex formation may easily be found. Thus, in the case of true cation complexes all interlayer molecules are co-ordinated to the cations, whereas in solvated interlamellar regions solvent molecules are present between the solvated cations. It should be noted that the term complex is loosely used, (glycol 'complexes', alkanol 'complexes') and does not imply true complex formation.

The first true interlamellar complex was prepared by Weiss and Hofmann,[36] who reacted Ni^{2+}-batavite with diacetyl dioxime to form an interlamellar Ni^{2+}-complex. Recently, several types of complexes have become of interest for different applications. For example, interfacial systems are proposed for the photosensitized splitting of water. A useful complex colloidal system was found containing colloidally dispersed smectites with bipyridine Ru(II) complexes. The importance of the architecture of the system for light-induced electron transfer and catalytic reactions has been particularly emphasized.[37]

In quest of selective heterogeneous catalysts, rhodium phosphonium interlayer complexes were developed which, for instance, catalyze the hydrogenation of terminal olefins.[38,39] Further new intercalated clay catalysts are pillared clays in which the macromolecular silicate anions are prised apart by pillars of polycondensed Al^{3+}-ions or hydroxyzirconium cations. Pillars of other inorganic materials are obtained by hydrolysis of interlayer silicon or iron acetylacetonate cations, or halogeno cluster cations such as $Nb_6Cl_{12}^{2+}$ or $Ta_6Cl_{12}^{2+}$ which are hydrolyzed to the corresponding oxides.[39-41]

Resolution of optically active compounds by stereoselective adsorption has often been attempted. Some success was achieved by Yamagishi,[42] who was able to resolve, at least partially, several Co-III chelates into two isomers using an optical active Ni-complex on montmorillonite. Recently, Yamagishi and Fujita[43] reported different adsorption properties of the enantiomeric and the racemic [bis(chelated) cobalt(III)]$^+$ complexes by colloidally dispersed sodium montmorillonite. Figure 6 illustrates that the complex added as pure enantiomer ($+$ or $-$) occupies the cation exchange sites. From racemic mixtures, racemic pairs are adsorbed and stacked stereoregularly on the surface, so that twice the amount of complex is adsorbed (charge compensation of the additonal complex ions by adsorption of anions occurs). In the case of enantiomeric ions, the dense packing is impeded by steric interference between the ligands of neighbouring chelates.

Mortland and Pinnavaia in 1971[44] detected a special group of interlamellar complexes when Cu^{2+}-montmorillonite was reacted with arenes. In the yellow type I benzene complexes, the interlayer water is partially displaced from Cu^{2+} and the benzene molecules interact by their π-electrons with Cu^{2+}. After complete removal of the water red

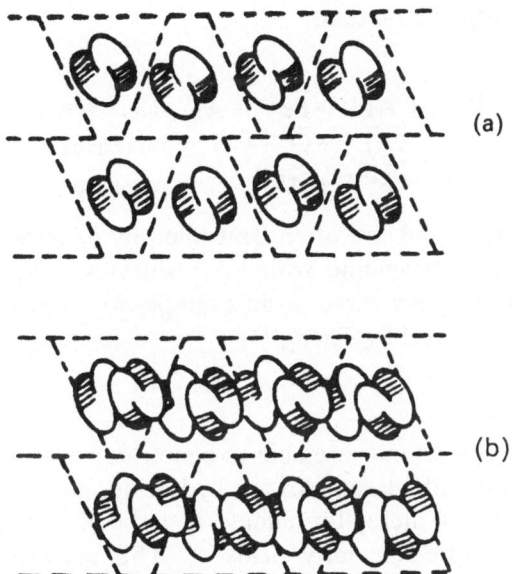

FIG. 6. Schematic representation of 'racemic adsorption' of the cobalt-III complex $Co(PAN)_2^+$, where PAN = 1-(2-pyridylazo)-2-naphthol.[43] (a) Adsorption of the complex cations in monolayers from solutions of the pure enantiomers. (b) Adsorption in bilayers from solutions of racemic complex ions. (With kind permission of Academic Press.)

type II complexes form in which the aromaticity of the ligands is lost[45,46] (see also Table 2 in ref. 8).

3.7 Further Reactions

The best known conformational change of intercalated molecules is the rotational isomerization of long-chain compounds described in Section 6.5.

Many redox reactions proceed by electron transfer between interlayer species and structural Fe^{2+} or Fe^{3+} ions. Octahedral Fe^{2+} ions in micas oxidize Ag^+ ions penetrating from a $AgNO_3$ solution into the expanded interlayers of micas (muscovite and phlogopite, in Ba^{2+} form).[47] The colloidally dispersed metallic silver develops an intense blue colour. The loss of negative layer charges is partially balanced by deprotonation of structural hydroxyl ions and probably by ejection of octahedral cations; structural ions are indicated by a bar in the following descriptive equations:

$$\begin{aligned}
&\qquad\qquad\qquad\qquad\qquad\qquad \textit{Layer charge}\\
&\overline{Fe}^{2+} + Ag^+ \rightarrow \overline{Fe}^{3+} + Ag^0 \quad \text{decreases}\\
&\overline{OH}^- \rightarrow \overline{O}^{2-} + H^+ \quad \text{increases}\\
&\overline{M}^{n+} \xrightarrow{?} M^{n+} \qquad\qquad \text{increases}
\end{aligned}$$

Colour clay reactions are often controlled by electron transfer. The colour reaction of benzidine with layer silicates is the most widely studied system and may serve as an example for such reactions. This field has been surveyed by Theng.[48]

3.8 Intercalation

Several neutral compounds are able to penetrate between the layers of 1:1 type clay minerals. Examples are hydrazine, dimethyl sulphoxide, urea, formamide, N-methylformamide and the K^+, NH_4^+, Rb^+, and Cs^+ salts of short-chain fatty acids.[8,49] The maximum degree of reaction depends on the type of kaolin. Based on these differences Weiss et al.[50] developed a method for a more detailed classification of kaolins (see also refs 4 and 7). Numerous organic compounds are intercalated by displacement reactions.[49] Intercalated guest molecules, for instance ammonium acetate, are displaced by compounds that are not directly intercalated.

Fully hydrated halloysites react with a large number of guest molecules[51] which more easily displace the interlayer water than is possible with the closed interlayer spaces of kaolins. With partially or completely dehydrated samples, reactivity more strongly depends on the mineralogical factors, e.g. size, crystallinity and iron content.

A third type of reaction is the 'Schleppreaktion' (entraining reaction[49]). Reactive guest molecules can entrain non-reacting organic molecules between the layers.

4   INTERACTION WITH CATIONIC SURFACE-ACTIVE AGENTS

4.1   Binding of Cationic Surfactants

The macromolecular silicate anions strongly bind cationic surface-active agents. Smectites and more lowly-charged vermiculites react with different types of surfactants (Table 2); highly charged vermicul-

TABLE 2

COMMON CATIONIC SURFACE-ACTIVE AGENTS WITH HIGH AFFINITY FOR
MACROMOLECULAR SILICATE ANIONS

commonly used for applications

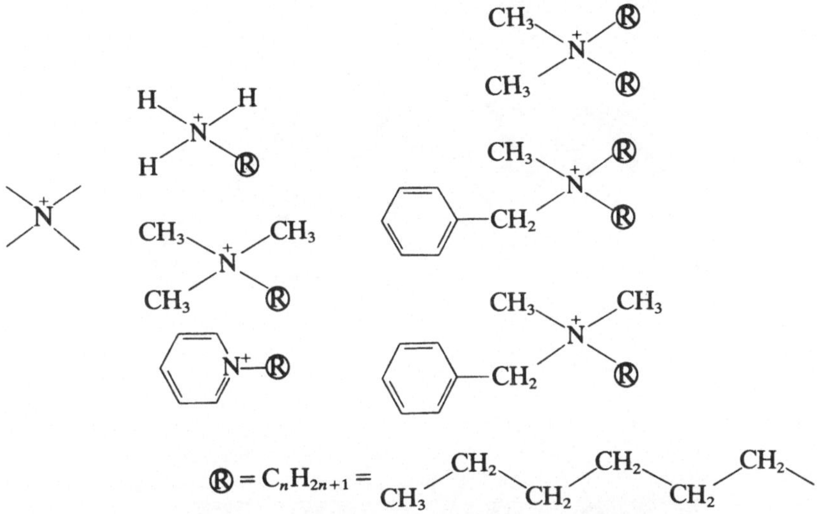

$$\circledR = C_nH_{2n+1} =$$

ites and micas exchange only primary alkylammonium ions for the interlayer cations. The affinity of the macroanions towards the surfactants increases with increasing chain length.[52]

During adsorption from aqueous solutions, all interlayer cations are exchanged at saturation, but the concentration of interlayer alkylammonium ions generally exceeds the CEC. Van der Waals interactions and the formation of distinct bimolecular interlayer structures promote an additional uptake of alkylammonium ions together with their gegen-ions and some water molecules. The additional surfactant cations and their anions are removed by washing. After drying, pure alkylammonium derivatives are obtained, with all interlayer cations replaced by surfactant cations.

4.2 Orientation of the Surfactants

The binding of surfactants between the silicate layers does not destroy the crystalline order of the structure. Basal spacing measurements reveal many details about the surfactant orientation of two-dimensional macromolecular systems. The main orientations are

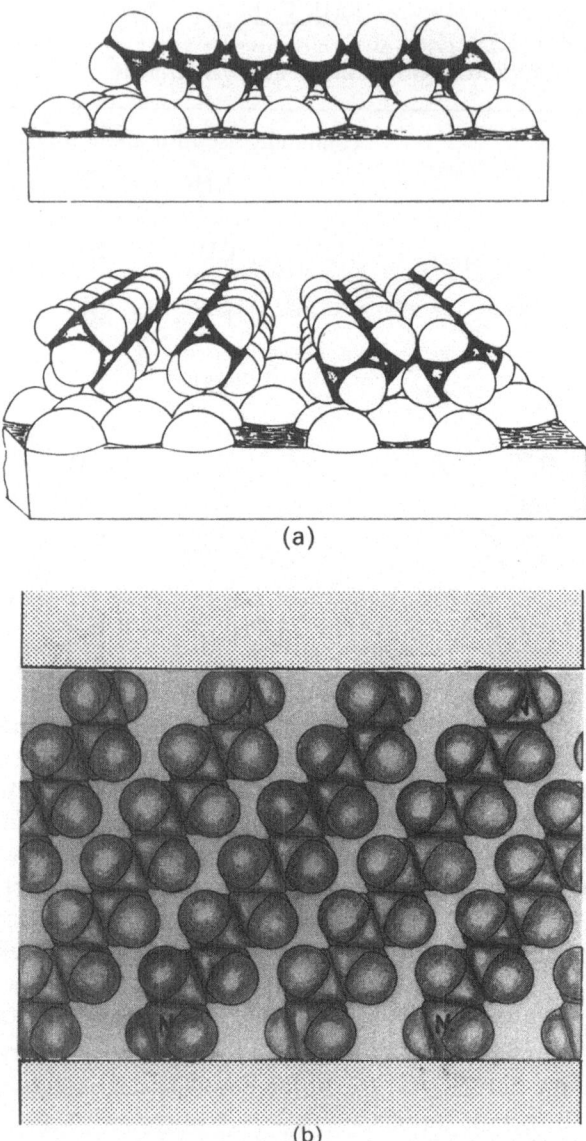

(a)

(b)

FIG. 7. Alkyl chains of surfactants on the silicate anions: (a) flat-lying chains; (b) paraffin-type structures. (Fig. 7b with kind permission of Steinkopff-Verlag.[53])

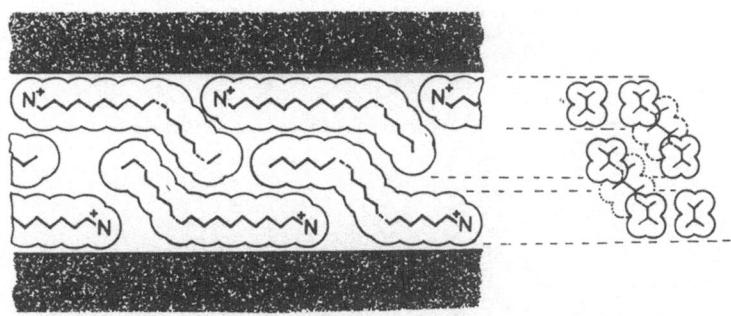

FIG. 8. The pseudotrimolecular interlayer structure. (With kind permission of the Clay Minerals Society.[53])

classified as monolayers, bilayers, pseudotrimolecular layers,* all with flat-lying chains, and paraffin-type structures (Figs 7 and 8). The orientation depends on the geometrical conditions.

The surface charge density is in the range $8-34\,\mu C\,cm^{-2}$ for a layer charge $\bar{\xi}$ of $0\cdot2-1\cdot0\,eq/(Si, Al)_4O_{10}$ (Table 3). The equivalent area A_e (area per interlayer cation) is

$$A_e = a_0b_0/2\bar{\xi}$$

(The structural unit with the basal area a_0b_0 contains two chemical units with the layer charge $\bar{\xi}$.)

The equivalent area has to be compared with the molecular area of the surfactant. For instance, primary n-alkylammonium ions $C_nH_{2n+1}NH_3^+$, which lie flat on the surface, require an area, A_c, expressed by the equation:[54]

$$A_c = 0\cdot127 \times 0\cdot45n + 0\cdot14 = 0\cdot057n + 0\cdot14 \quad (nm^2)$$

A typical smectite ($\bar{\xi} = 0\cdot30$) has an equivalent area of $0\cdot78\,nm^2$. Thus, decylammonium ions ($A_c = 0\cdot71\,nm^2$) lie flat on the surface and are aggregated to monolayers. Tetradecylammonium ions ($A_c = 0\cdot94\,nm^2$) are too bulky and form bilayers. With increasing chain length monolayer structures are followed by bilayer and finally

* The electrostatic repulsion between the cationic groups impedes a true trimolecular layer arrangement. Essentially, the structure is bimolecular with some chain ends shifted above one another so that the interlayer separation corresponds to the thickness of three alkyl chains. The chains can easily assume the required conformations by kinking (Section 6.5).

TABLE 3

SURFACE CHARGE DENSITY OF THE MACROMOLECULAR SILICATE ANIONS OF 2:1 CLAY MINERALS

Species	Lattice dimensions[a]			Equivalent area[b] A_e (nm²)	Layer charge[a] $\xi (eq/(Si, Al)_4 O_{10})$	Surface charge $(eq/(Si, Al)_4 O_{10})$	Surface charge density σ_0 ($\mu C cm^{-2}$)
	a_0 (nm)	b_0 (nm)	$a_0 b_0$ (nm²)				
Mica	0.521	0.900[d]	0.469	0.230	1.0	0.5	34.2
Vermiculites	0.531	0.920[e]	0.489	0.245	1.0	0.5	32.8
	0.534	0.925	0.494	0.309	0.8	0.4	26.0
				0.412	0.6	0.3	19.4
Beidellite	0.518	0.899	0.466	0.530	0.44	0.22	15.2
Montmorillonites	0.521	0.902	0.470	0.470	0.5	0.25	17.0
				0.588	0.4	0.2	13.6
				0.783	0.3	0.15	10.2
Hectorite	0.525	0.918	0.482	1.004	0.24	0.12	8.0

[a] Two units $(Si, Al)_4 O_{10}$ per unit cell $(a_0 b_0)$.
[b] $A_e = a_0 b_0 / 2\xi$.
[c] $\sigma_0 = 1.6 \times 10^{-19} \, \xi / a_0 b_0 \times 10^{-14}$.
[d] Muscovite.
[e] Biotite.

pseudotrimolecular arrangements. Paraffin-type structures with the chains pointing away from the surface were only observed for higher charge densities ($\geqslant 0.6$) found in vermiculites and micas.

The basal spacing as a function of n should increase in steps from 1·36 nm (monolayers) to 1·77 nm (bilayers) and 2·1–2·2 nm (pseudotrimolecular layers). Plateaux at these spacings are generally observed but the transitions are frequently broadened with spacings intermediate between 1·36, 1·77 and 2·1–2·2 nm. These intermediate spacings belong to non-integral (00l)-reflections and indicate the interlayer separations to be non-identical. The cause is the random variation of the charge density within certain limits. A useful method could be developed to ascertain this charge heterogeneity from the basal spacings of the alkylammonium derivatives.[7]

The formation of close-packed alkyl chain mono- and bi-layers raises several noteworthy points. As was first shown by Brindley and Hoffmann,[55] alkyl chains are keyed into the surface structure (Fig. 7) so that the methyl groups make close contact with the surface oxygen atoms. The resulting basal spacings are 1·32–1·34 nm. The compatibility of the organic and silicate dimensions make possible a dense aggregation of alkyl chains but not a close packing. When close-packed, the chains will be shifted out of the keying positions and the spacings expanded slightly. In fact, the shortest experimental spacings are 1·32 nm which increases to 1·36 nm for longer alkylammonium ions.

At the crystal edges, the alkyl chains are partially expelled from the interlayer space and the interlamellar area per alkylammonium ion becomes somewhat larger than A_e; the difference depends on the particle diameter. For layer charge determinations a correction must be introduced which becomes effective for diameters <40 nm.

The distribution of the charges in the silicate layer certainly differs from the charge pattern displaced by the N^+-centres of close-packed alkylammonium ions. Maximum electrostatic interactions and optimal packing density cannot be achieved simultaneously. The experimental results indicate that the van der Waals attractions, which increase with packing density, do overcome the electrostatic interactions (see also Section 7.4).

At charge densities $\geqslant 9\,\mu C\,cm^{-2}$ ($\bar{\xi} \geqslant 0.6\,eq/(Si, Al)_4 O_{10}$) the alkyl chains point away from the surface (paraffin-type structures). The chains may be assumed to tilt to occupy the entire space.[53] The experimental results on vermiculites establish that the required low tilting angles (angles between chain axis and surface) are not realized but adjust to about 55°. This renders possible optimal binding of the NH_3^+ group through three hydrogen bonds to surface oxygen ions.

The preference of angles of 50–60° also results from an additional geometrical constraint.[56] The optimal lateral packing of the tilted chains assumes that a CH_2-group of one chain lies between the CH_2-groups of neighbouring chains. This condition is met excellently for the alkylammonium derivatives of $KNiAsO_4$,[12] whereas in case of vermiculites the geometrical dimensions do not completely fit, so that the chain orientation will be less regular.

Paraffin-type arrangements are thus distinguished by a noticeable preference for tilting angles of 50–60°, even if the charge density is increased still further, up to $1\cdot0$ eq/$(Si, Al)_4O_{10}$.

4.3 Interlamellar Liquid Structures

The binding of surfactants does not abolish the general expansibility of the layer structure. The organic derivatives take up a large variety of organic liquids and even water. This may be demonstrated by two examples. Figure 9a reports the spacings of alkylammonium vermiculites under several organic liquids and water. The large increase of the basal spacing is evident. The spacings in water do not reveal any particular features compared with the organic liquids. In contrast, the sorption capacity of alkylammonium smectites (Fig. 9b) is more strongly dependent on the kind of organic liquids, and the interlamellar sorption of water is generally more restricted.

The fundamental process is the displacement of the alkyl chains from the surface, so that only the polar end groups remain in contact with it. The voids between the chains are occupied by liquid molecules. The interlayer space remains expansible and the chains possess enough flexibility to accommodate to the associated liquid molecules. The mutual adaptation between the alkyl chains and the liquid structure appears to be essential and distinguishes the interlamellar liquid uptake from pure pore filling in zeolites and other porous materials.

The concept[57] of liquid molecules associated in 'clusters' in the interlamellar space and between the chains is illustrated in Fig. 10. In the case of inorganic interlayer cations, the electrical field arising from the cations and the surface charges breaks up the liquid structure and the liquid molecules form solvation shells around the cations. Inasmuch as the cationic centres of the alkylammonium ions are fixed near

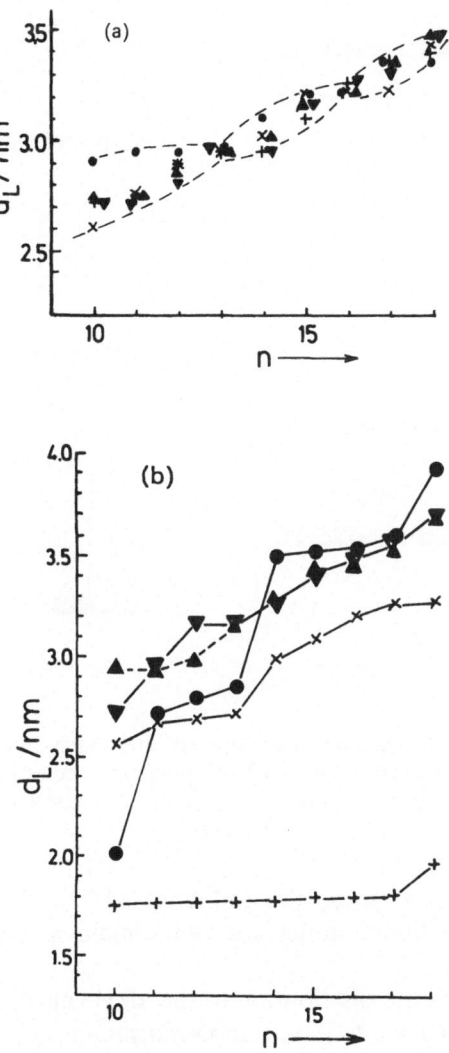

FIG. 9. Basal spacings d_L of alkylammonium vermiculite (a) and montmorillo-nite (b) under different liquids: H_2O (+), ethanol (×), formamide (●), dimethylformamide (▲), and dimethyl sulphoxide (▼); n = chain length of the alkylammonium ions $C_nH_{2n+1}NH_3^+$; vermiculite from Young River, Australia; montmorillonite from Wyoming, USA ('Greenbond').

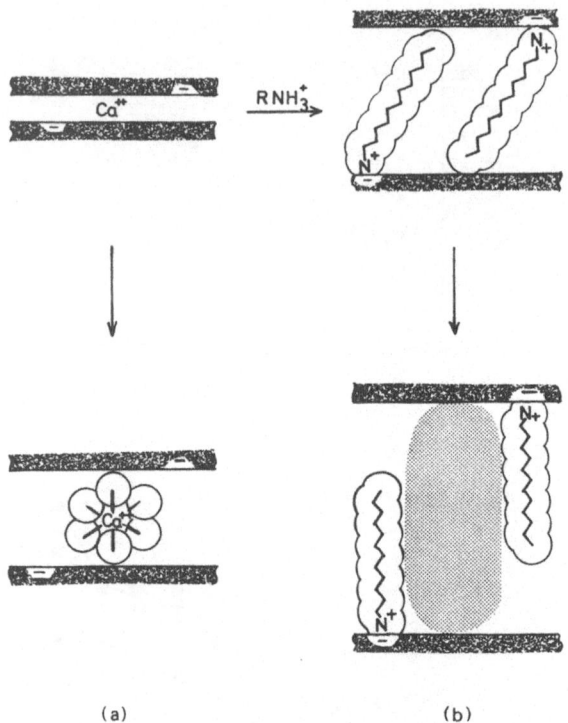

(a)   (b)

FIG. 10. Polar molecules between the silicate macroanions. (a) Inorganic interlayer cations: solvation shells of the polar molecules around the cations. (b) Organic interlayer cations: polar molecules associated to liquid-like 'clusters'.

the negative surface sites, the influence of the charges is strongly reduced and the liquid molecules can remain associated in distinct clusters.

Figure 11 shows the distribution of the alkyl chains in the $a_0 b_0$-plane for vermiculite ($\bar{\xi} = 0.67$) and montmorillonite ($\bar{\xi} = 0.33$). In vermiculites, the perpendicular alkylammonium ions of the monolayer occupy two-thirds of the possible sites and enclose cavities of about $0.6–0.7$ nm diameter. Neighbouring cavities are separated by the shells of alkylammonium ions. Though some interconnections are created by displaced alkyl chains (upper left of Fig. 11a), isolated cavities are considered as being typical. In montmorillonites with about half of the alkylammonium ions, the cavities are no longer isolated and a

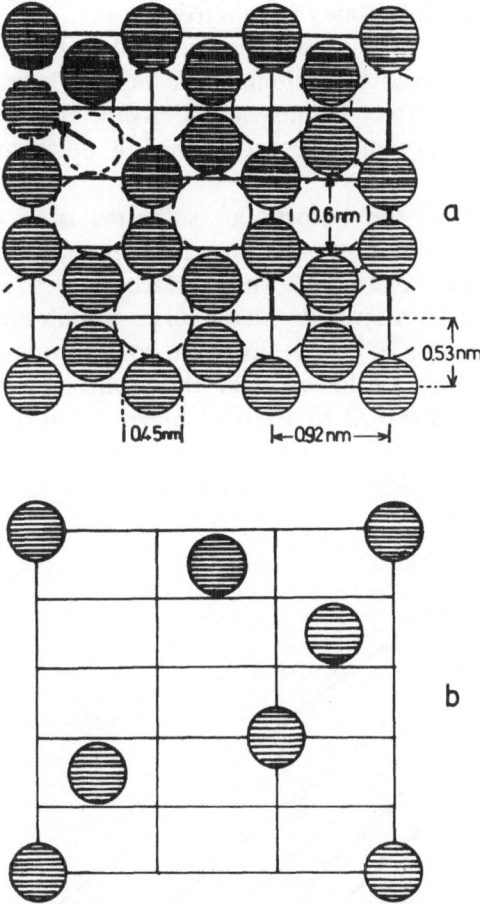

FIG. 11. The cavities between perpendicular alkyl chains (chains hatched). (a) Vermiculites with alkyl chain monolayers, about four chains on the surface area of three structural units. (b) Montmorillonites with alkyl chain bilayers, about one chain on the same area.

continuous network of liquid molecules develops between the chains (Fig. 11b). Network formation requires a large volume and enforces an additional increase of the interlayer separation. The chains assume a bimolecular arrangement with tilting angles of about 56° with chains possibly somewhat shortened by *gauche*-bonds. The conformations of the alkyl chains are more multifarious than in vermiculites, and the spacings depend more strongly on the type of liquids.

4.4 Adsorption from Binary Liquid Mixtures

The surface becomes hydrophobic when the inorganic cations are progressively displaced by cationic surfactants. The changes in selectivity towards organic liquids are derived from the adsorption isotherms for binary mixtures containing a polar and a non-polar liquid.[58,59]

Figure 12 shows the adsorption isotherms of a montmorillonite (with different amounts of hexadecylpyridinium ions) from methanol–benzene mixtures. The surface excess $n_1^{\sigma(n)}$ is plotted against the mole fraction x_1 of methanol in equilibrium. The azeotropic point $(x_1 = x_1^a$ for $n_1^{\sigma(n)} = 0)$ gives the composition of the adsorbed phase in the range of linearity of the isotherms. With increasing hydrophobicity, $(X = 0.22$ to $1.00)$ x_1^a moves from 0.8 to 0.3: at low

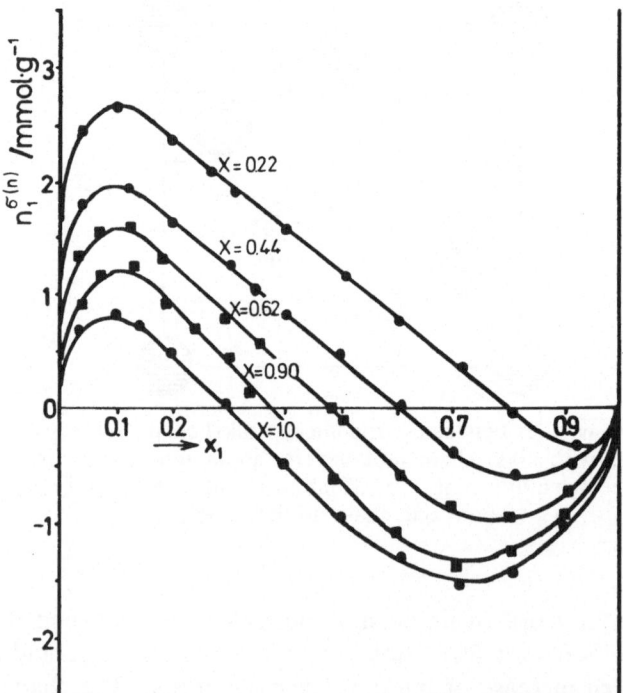

FIG. 12. Excess adsorption isotherms from methanol/benzene mixtures:[59] montmorillonite (Mád, Hungary) with increasing amounts of hexadecylpyridinium ions; X = mole fraction of interlayer hexadecylpyridinium ions; $n_1^{\sigma(n)}$ = surface excess; x_1 = mole fraction of methanol.

coverage by hexadecylpyridinium ions the interlamellar ratio methanol/benzene is $0 \cdot 8/0 \cdot 2 = 4/1$, but it is shifted to $0 \cdot 3/0 \cdot 7 = 0 \cdot 43/1$ at high coverage.

From the surface excess isotherms (composite isotherms), the volume, V^s, of the adsorbed phase (per g silicate or per unit $(Si, Al)_4O_{10}$) is obtained. Comparison with the calculated free interlayer volume, V_f (volume of the interlayer space as calculated from the basal spacing minus volume of the interlayer cations), gives an interesting insight into the interlamellar liquid structure.[59] The free volume V_f is appreciably larger than V^s. Only a part of the free interlayer volume can strictly be assigned to the adsorption phase that has a composition different from the bulk. The remaining volume, $V_f - V^s$, must be occupied by methanol and benzene molecules in a ratio similar to that in the bulk phase.

A different behaviour was observed for alkylammonium vermiculites in contact with a binary mixture of two polar liquids $(DMSO-H_2O)$.[57] In this case, V^s (from the adsorption isotherm) was larger than V_f (from basal spacings). Due to the stronger interactions between the highly polar molecules, the liquid structure between the chains on the *external* surfaces impose structuring effects on the adjacent liquid phase that change the composition near the surface from that of the bulk. The DMSO ratio in these 'lyospheres' around the particles is nearly independent of the bulk composition (for $x_{DMSO} = 0 \cdot 1$ to $0 \cdot 9$), but depends on the alkyl chain length $(DMSO/H_2O = 3 \cdot 7$ and $5 \cdot 4$ for $n = 12$ and $n = 18$, respectively).

4.5 Hydrophobic Surfaces

It is often assumed that hydrophobicity of a surface simply requires a complete covering by surfactants. The silicate surfaces available with different charge densities are excellent models to verify this concept. The simple but instructive experiments of Weiss[60] established that the critical chain length of alkylammonium ions decreases with increasing chain packing density but a complete coverage of the surface by alkyl chains is not required. The above model of interlamellar liquid structures suggests the following explanation (Fig. 13).[57] The surface is hydrophobic when the distance between the chains remains below a critical value, so that all water molecules between the chains are strongly organized into clusters. If the chains are further apart from each other, the bulk water phase penetrates between the water clusters around the chains and makes the surface hydrophilic.

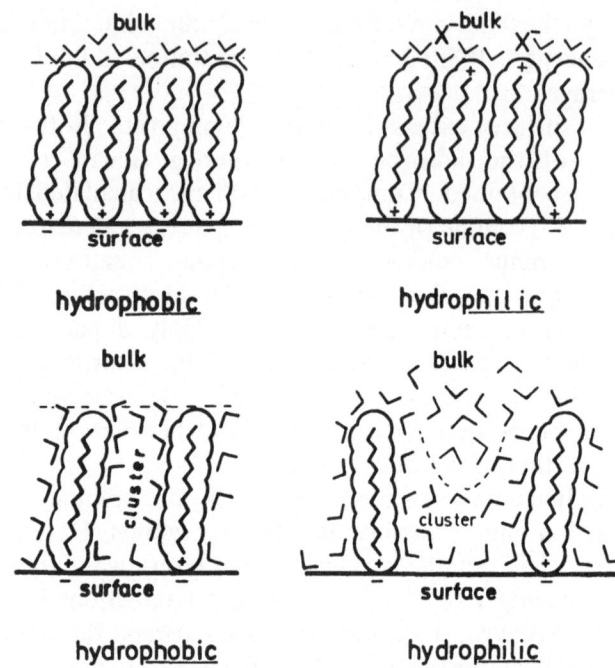

FIG. 13. Hydrophobic and hydrophilic surfaces.

It may be added that the excess of cationic surfactants that is adsorbed, together with the gegen-ions, generally makes the surface hydrophilic (Fig. 13).

5 INTERACTIONS WITH POLYMERS

5.1 General Aspects

The study of interactions between macromolecules and solid surfaces is of immediate interest in interface chemistry, biochemistry, biophysics and medical science. For this reason, if for no other, the use of clay minerals as solid adsorbents should be advanced. Many macromolecules penetrate between the layers of smectitic clays. This offers possibilities of gathering information about the orientation of polymers on solid surfaces which cannot be obtained in other systems.

On the other hand, the interactions of clay minerals with organic polymers have received a considerable amount of attention because of their great agricultural and industrial application (see Section 8).

The amount of adsorbed polymer generally depends on the levels of polymer concentration, pH ionic strength, and temperature. The macromolecules are adsorbed in trains, loops and tails. The proportion of trains, loops and tails is a fundamental parameter for describing polymer adsorption but its determination is one of the problems encountered in polymer chemistry. The interaction of polymers with solid surfaces is usually very strong, as indicated by the very steep increase of the adsorption isotherms at low equilibrium concentrations. The adsorption of poly(ethylene glycol)s on Ca-montmorillonite, showing increasing affinity with increasing molecular weights,[61-63] may serve as an example.

Polymer adsorption occurs sometimes very fast, sometimes very slowly. Generally, the polymer is not desorbed from the surface by diluting the equilibrium solution or changing pH and ionic strength (cf. ref. 61). This is a direct consequence of the large number of contacts between the surface and the polymer.

The macromolecule attached to the surface at many separate sites is not able to desorb simultaneously from all sites before some detached segments re-adsorb. If, however, the vacated sites could be occupied by competing agents, some of the polymer may be removed from the surface. So it is possible to desorb, at least partially, polyacrylamide from kaolinite by means of sodium tripolyphosphate and sodium metaphosphate.[64]

The apparent irreversibility of polymer adsorption has two important consequences: the system remains in a non-equilibrium state, and, frequently, the kind of products obtained and the amount of polymer adsorbed are sensitive to the methods of preparation[65] and can also depend on the solid-to-liquid ratio.[66] Subtle differences in the way the reactants are brought together might be significant. The result of these disparities can be larger than the effect of the molecular weights or the special kinds of polymer samples used.

5.2 Adsorption of Non-Ionic Polymers

Clay minerals offer three types of surfaces to the macromolecules: the external basal plane and edge surfaces and the internal surfaces. The adsorption on the external surfaces is comparable with that on other solid surfaces. The interlamellar adsorption exhibits some special features because the macromolecules have to penetrate between the layers. In general, macromolecules are adsorbed in trains and the required lattice expansion is relatively modest (Table 4).

TABLE 4

INTERCALATION OF NEUTRAL POLYMERS

| Polymer | Interlayer cation | Basal spacing[a] (nm) | | Reference |
		In solution	Dried	
Poly(vinyl	Ca²⁺	1·9–2·0	1·48	68
alcohol)	Na⁺	diffuse	1·36	69
Poly(ethylene	Ca²⁺	1·7–2·2	1·3–1·7	63
glycol)	Na⁺	1·3–1·6	1·33	
Poly(vinyl	Ca²⁺	1·7–1·9	1·6	70
pyrrolidone)	Na⁺	1·7–3·4	1·5	
Dextran	Ca²⁺	1·4–1·6	1·4	71
	Na⁺	1·6	1·3	
	Na⁺	1·4–1·7[b]	1·4–1·6	72

[a] Maximum and minimum values under different conditions or for polymers with different molar masses.
[b] 52% R.H.

The spacings after drying may represent approximately the layer separation required by the macromolecule, but may not be fundamentally related to the exact dimensions of the macromolecule. They may be too high because of an incomplete desorption of interlayer water, or too low because of an interstratification of polymer-free and polymer-bearing interlayers.

Adsorption of poly(vinyl alcohol)[62] (PVA) was first investigated by Emerson[67] and Greenland.[68] It is, however, not as uncomplicated as might be expected at first sight.

With calcium ions as the interlayer cations, the poly(vinyl alcohol) molecules penetrate between the layers but the layer separation in aqueous solution is mainly determined by the hydration of the interlayer cations (basal spacing about 2 nm). In the presence of sodium ions the amount of PVA adsorbed is about twice that on Ca^{2+}-smectite.[69] If all PVA molecules were adsorbed in trains, the internal and external surfaces would be completely covered with a close-packed monolayer of PVA molecules and the interlayers should contain bilayers of PVA with basal spacings of 1·8 nm. However, in aqueous solution or after air-drying, the PVA sodium smectite gives very diffuse basal reflections. The silicate layers are no longer parallel but are embedded in a gel of PVA and water (Fig. 14). After drying *in*

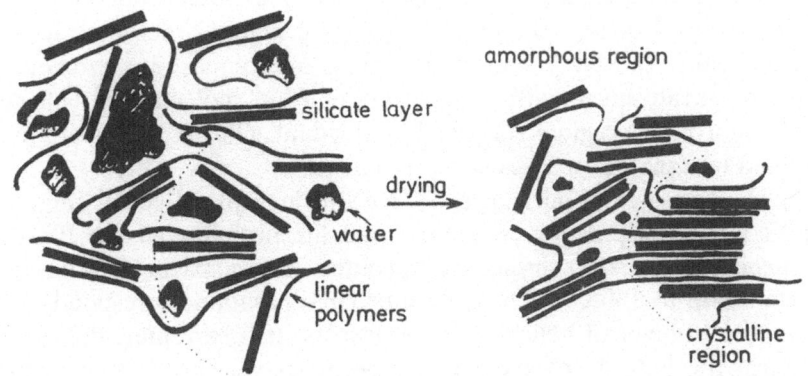

FIG. 14. Sodium smectites and poly(vinyl alcohol) (PVA): silicate lamellae embedded in a matrix of PVA, and PVA molecules preventing aggregation of layers to crystalline structures (interlayer sodium ions not shown).

vacuo, the basal spacing is reduced to 1·36 nm and no other spacings are observed. Only 50% of the PVA adsorbed can be incorporated into smectites, with spacings of 1·36 nm. During the removal of the water only a few of the silicate layers can aggregate. A large fraction remains in a disordered state, since PVA molecules bridging the layers

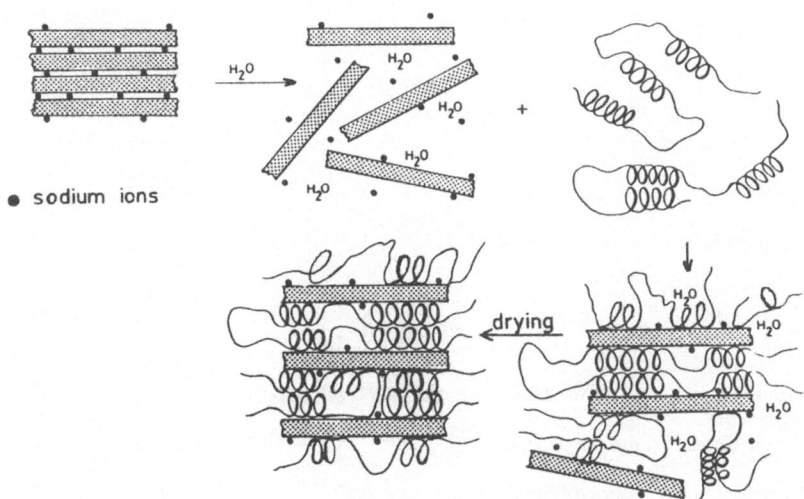

FIG. 15. Adsorption of PVA by sodium beidellite in the presence of boric acid (boric acid promotes the helical structure of PVA).

inhibit a rearrangement to crystalline order. The observation of a basal reflection of 1·36 nm, therefore, suggests a highly ordered structure, but in reality the PVA sodium smectites are composed of packets of parallel silicate layers with monolayers of PVA molecules in between and amorphous regions in which individual silicate layers are embedded in a matrix of PVA.

In the presence of boric acid, PVA forms helical molecules (Fig. 15) and colloidal dispersions of sodium smectite probably adsorb PVA in its helical form. After drying the smectites show a spacing of 4 nm that is stable up to 130°C.[69] The interlamellar separation corresponds to a bimolecular layer of helical PVA molecules. It is extremely difficult to confirm the helical arrangement in the interlayer space by any other method.

5.3 Polymerization in the Interlayer Space

Instead of adsorbing polymers between the layers, one may try to bring about adsorption of the corresponding monomers which are then polymerized in the interlayer space (Table 5). For instance, acrylonitrile directly adsorbed by alkali- and alkaline-earth-montmorillonite is polymerized in the interlayer space after initiation by benzoyl peroxide at 50°C.[73] The monomer 6-amino caproic acid can be polycondensed to nylon by heating at 240–250°C. The polycondensa-

TABLE 5

INTERLAMELLAR POLYMERIZATION

Monomer	Polymer	Interlayer cation	Basal spacing[a] (nm)	Reference
Acrylonitrile	polyacrylonitrile	Li^+	1·6–1·7	73
		Na^+, K^+	1·4–1·5	
		Ca^{2+}, Ba^{2+}	1·3–1·4	
Aminocaproic acid	nylon	Na^+	1·42–1·57	74
		Mg^{2+}, Ca^{2+}	1·55–2·28	
		Cu^{2+}, Co^{2+}	1·48–2·28	
Styrene Spontaneous	polystyrene	$(CH_3)_3N^+(C_{18}H_{37})$	3·3	75
Diazomethane	polymethylene	$H^+, Mg^{2+}, Mn^{2+}, Ni^{2+}, Cu^{2+}$	1·4–1·5	76
4-Vinylpyridine	poly(4-vinyl pyridine)	Na^+	1·47	77

[a] Varying with the interlamellar amounts of polymer.

tion increases the spacing from 1·7 nm (aminocaproic acid in Mg- and Ca-montmorillonite) up to 2·3 nm.[74]

In some cases, the montmorillonite has to be made hydrophobic by alkylammonium ions. The hydrophobicity and the interlayer expansion promote the uptake of monomers. Exchange of trimethyl octadecylammonium ions for the inorganic interlayer cations enables the accumulation of styrene and the polymerization to polystyrene (basal spacing 3·3 nm).[75]

Usually, the polymerization has to be initiated by suitable initiator molecules (e.g. peroxides). In some cases the monomers spontaneously polymerize in the interlayer space. The polymerization can be catalyzed by the increased acidity of hydration water around the interlayer cations, in particular the Al^{3+} ion. Probably still more active potential acidic initiation sites are hydrated lattice Al^{3+} ions at layer edges.[78] Examples are the spontaneous polymerization of diazomethane (to polymethylene),[76] and of vinylpyridine.[77] Further potential initiation sites are Lewis acid centres and redox centres, in particular iron ions.[79] The polymerization of benzene to poly(p-phenylene) on Cu^{2+}-smectites[80,81] is promoted by the interactions between benzene and Cu^{2+} ions (Section 3.6).

5.4 Interactions with Cationic Polymers

Among ionic polymers only cationic macromolecules are able to penetrate between the layers. Their positive centres can displace interlayer cations, but complete displacement is rarely attained. The polycation has to enter the interlayer space (Fig. 16), and as penetration proceeds the increasing number of contacts reduces the mobility of the polyion. The polyions do not occupy the whole interlayer space but accumulate in more or less broad zones near the crystal edges.

Some of the cationic charges are not balanced by surface charges and retain their gegen-ions which can be exchanged by other anions (cf. Figs 16, 29, 30). A distinct anion exchange capacity (AEC) was observed after adsorption of a cationic polysulphone.[82] CEC decreases and AEC increases with increasing loading with polymers (Fig. 17). The sum of CEC plus number of adsorbed segments is constant. This indicates that the exchange of the inorganic interlayer cations is not impeded by the accumulation of polymers at the crystal edges.

The anion exchange capacity corresponds to the number of segments that are not balanced by negative surface charges. The segments

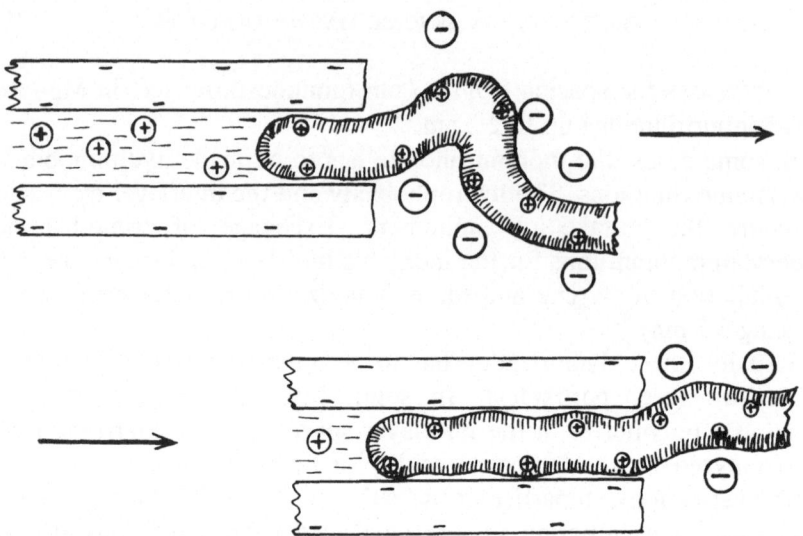

FIG. 16. Penetration of polycations between the silicate macroanions: decreasing mobility with increasing penetration.

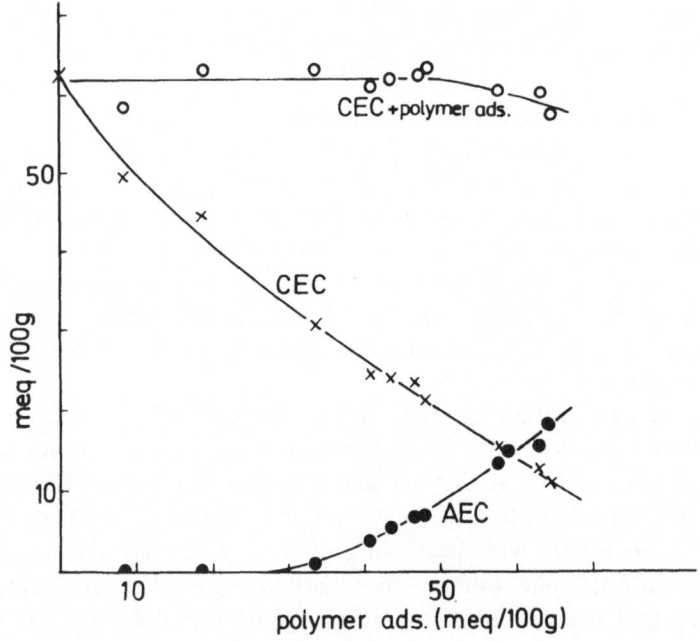

FIG. 17. Cation (CEC) and anion (AEC) exchange capacity of montmorillonite loaded with increasing amounts of polycations (diallyldimethylammonium–SO$_2$ copolymer[82]).

may be those protruding into the solution in loops and tails. Thus, the number of segments in loops and tails may be estimated from the AEC. At low polymer loading (AEC ~ 0), the cationic polymer is adsorbed almost completely in trains; at high loading, about 75% of the segments are in direct contact with the clay surface and 25% are protruding into the solution. Similar results were obtained for the adsorption of ionenes (Section 7.4). In general, a proportion of 20–30% of 'free' segments will be a realistic figure for montmorillonites.

It should be emphasized that the possibility of close contacts between the macromolecules and the surface atoms is of decisive importance and can overcome the opposing effects of electrostatic interactions. This is exemplified for the adsorption of ionenes in Section 7.4.

5.5 Protein Adsorption

Papers on protein adsorption multiplied in the years 1950–1960. Recently, the number of papers on protein–clay interactions has decreased but studies of protein–solid interactions in general has advanced.

Protein adsorption is pH-dependent. The adsorption often goes through a maximum near the isoelectric point of the protein (Fig. 18).[83,84] The adsorption maximum demonstrates that the protonation of the protein is not the only effect of pH changes, which also influence the conformation of the protein molecules. For example, progressive unfolding with increasing distance from the isoelectric point should decrease the adsorption (see also Section 7.4).

A strange effect may be noted which has to be considered in protein adsorption studies.[84] Protein molecules with low affinity for the substrate may be in close-packed arrangements. If the molecules are attached on the surface by stronger interactions that change the conformation and the orientation, the amount of material adsorbed per unit surface area can be reduced. One arrives at the paradoxical conclusion that stronger interactions can reduce the amounts adsorbed.

The fundamental question arising with clay–protein complexes is that of localization of the protein. Proteins of minor complexity (e.g. salmine), that can unfold, penetrate between the layers.[52,65,85] High molar masses, large sizes, high complexity and inability to unfold prevent most proteins from entering the interlamellar spaces to any

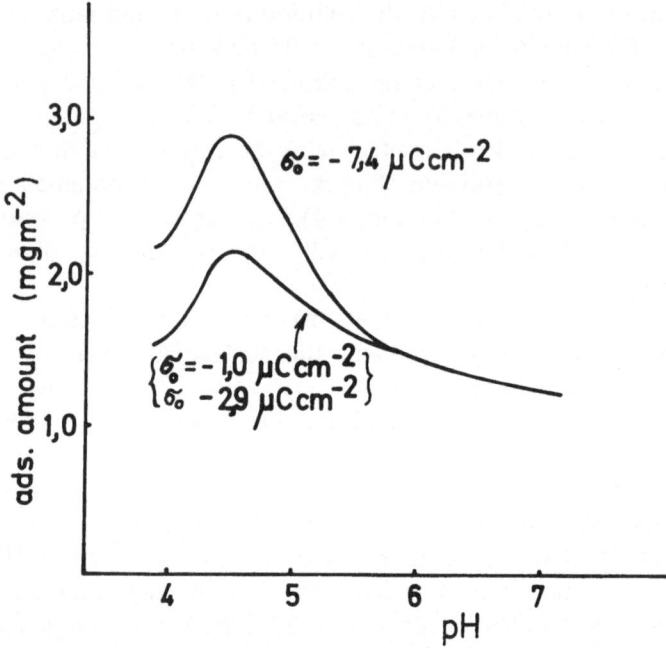

FIG. 18. pH-dependent adsorption of human plasma albumin on negatively charged polystyrene (σ_0 = surface charge density of polystyrene, isoelectric point (iep) of albumin: pH = 4·2–5·0.[83,84])

measurable degree. At most, they penetrate between the layers at the edges to some extent, so that they are anchored to the clay particles very tightly. As a consequence, these proteins can only be partially removed and the branches between the layers are protected from chemical attack and often cause an increased thermal resistance.[86,87]

If a dispersion of alkali smectites comes into contact with proteins, the highly dispersed silicate layers become embedded in the protein matrix. The basal reflections of such materials are broadened and make it difficult to identify the clay mineral.

A very instructive system is reported by Larsson and Siffert.[88] In the presence of sodium ions the smectite adsorbs lysozyme between the layers by a disaggregation–reaggregation mechanism (Fig. 19). Lysozyme is bound between silicate layers during the dispersion process without any need for diffusion in lateral directions. This explains how complete saturation is achieved which very probably would not be attained by lateral diffusion.

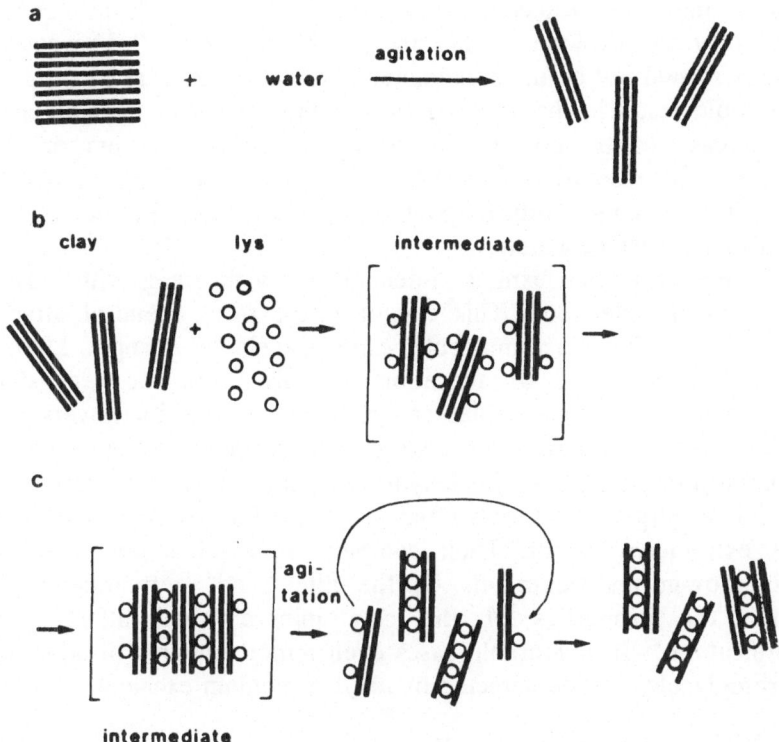

FIG. 19. Interlayer adsorption of lysozyme by a disaggregation–reaggregation mechanism. (Larsson and Siffert,[88] with kind permission of Academic Press, Inc.)

Both types of protein complexes—with and without interlamellar penetration—have some properties in common, namely that they decrease the cation exchange capacity and modify the properties of the clays. Clementz[89] described the effect of adsorption of asphalts and resins on the properties of the clays. Heavy fractions may be adsorbed onto clays in petroleum reservoirs and lead to alterations in rock properties. This may be one of the reasons why similar clay minerals often behave entirely differently in different rocks.

Early investigations on clay–protein complexes had already included studies on the interactions on these complexes with enzymes.[90] Later, interest in such studies ceased; however, more recently studies on the effects of clays on microbial reactions have revived. Particulate materials inhibit some host–parasite interactions. For example, *E. coli* bacteria may be protected from bacteriophage attack in aquatic systems by an envelope of clay around the bacterium.[91] The consequences should not be underestimated. Protection of fecal bacteria by clays could result in the accumulation of the organisms in sediments. The release of bacteria by welling-up of bottom waters or the desorption in estuarine systems due to reduction of electrolyte concentration could return potentially dangerous enteric microorganisms to surface waters.

Soil smectites often exhibit ill-defined X-ray diagrams with mostly diffuse basal reflections. This cannot result from chemical attack. Proteins and other macromolecules can cause similar changes. During the prolonged period of alteration in soils, even the very slow penetration of macromolecules can expand smectitic interlayers and produce more or less disordered structures. Besides proteins and their decomposition products, carbohydrates and derivatives should be mentioned. Organic materials often make a definitive identification of clays extremely difficult. Their complete removal is not generally attained by normal treatments. On the other hand, pretreatments can change the properties of the clay minerals and affect their identification.[92] In favourable cases even minor amounts of adhering macromolecules can be detected by alkylammonium exchange.[93]

5.6 Interactions with Anionic Polymers

The negative charge of anionic polymers does not prevent the macromolecules from being adsorbed by clay minerals. Polyanions do not penetrate into the interlayer spaces, but considerable amounts can be adsorbed on the external surfaces. For instance, polyacrylamides exhibit a strong affinity towards clay minerals.[66,94] Essentially non-ionic polyacrylamide is adsorbed at up to 10 mg per g by Georgia kaolin.[66] The adsorption is nearly pH-independent as long as the pH remains below 8. Above pH 8, the amounts adsorbed are decreased.

Commonly, polyacrylamides contain carboxylic groups (Fig. 20). From aqueous solutions (pH ~ 7·5), up to 3 mg polyamide/g montmorillonite are adsorbed. The adsorption increases with increasing proportion of carboxylic groups. As these groups (and dissociation constant $pK_a = 6·8$) are ionized at pH ~ 7·5, the strange effect is observed that the adsorption increases with increasing negative charge density of the polymer. Siffert and Espinasse[95] have proposed that the carboxylic groups are bound by exchanging structural OH-groups on the crystal edges (Figs 4, 20); the chelate effect promotes this reaction.

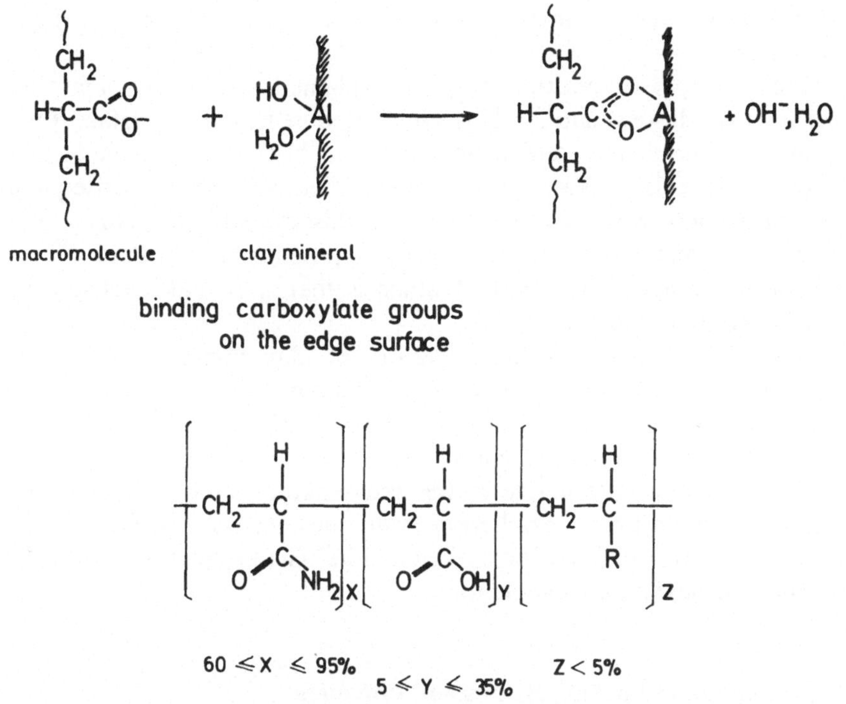

macromolecule   clay mineral

binding carboxylate groups
on the edge surface

$60 \leqslant X \leqslant 95\%$ $Z < 5\%$
$5 \leqslant Y \leqslant 35\%$

FIG. 20. Binding of carboxyl groups of hydrolyzed polyacrylamide on the surfaces.[95] (With kind permission of the Royal Society, London.[8])

Even sodium polyacrylate is adsorbed. The adsorption reaches a maximum (up to 1 mg/g) at $pH \approx pK_a$ of the polyacrylic acid.[95]

The adsorption of charged polyacrylamides can be increased considerably in the presence of sodium chloride, but electrolyte effects are absent for the adsorption of uncharged polyacrylamide on kaolinite.[66,94] The simplest explanation is that the addition of salts reduces excessively high surface potentials of the clay particles which oppose polyanion adsorption. If polyacrylamides are used for improving oil recovery, significant losses due to adsorption on the clay minerals of surrounding rocks are expected, in particular when the salinity of the medium is high.

5.7 Grafting Reactions

The use of clays in practical applications often requires a stable attachment of polymers onto the surface by covalent bonds. Grafting is a highly effective method for reinforcing natural and synthetic polymers.

The principle of grafting may be explained by the reaction with hexamethyl diisocyanate.[96] This compound reacts with surface OH-groups of vermiculite and forms urethane bridges. Some molecules retain a free isocyanate group which then initiates a subsequent polymerization with diols and further diisocyanate molecules. The reaction is analogous to the industrial preparation of Perlon®. A polyurethane matrix is obtained which is thermally and mechanically reinforced by vermiculite.

A second example is the silylation of clay minerals with vinyl chlorosilanes, followed by a polycondensation step which covers the clay particles with polysiloxane layers. Remaining double bonds initiate further chemical reactions.[97]

An interesting alternative is the crosslinking of polyethylene and other thermoplastics by irradiation techniques onto clay surfaces which are primed with organic 'polyethylene-like' chains such as poly(vinyl alcohol) or hexamethylenediamine.

5.8 Colloidal Clay Dispersions and Polymers

Polymers enjoy a broad technical application as sensitizing or stabilizing agents for colloidal clay dispersions. Addition of small amounts of polymer causes the colloidal dispersion to flocculate, whereas larger amounts stabilize the colloidal state. The presence of polymers also greatly influences the sensitivity of clay dispersions towards salts. Generally, very small amounts of polymer make the dispersion very sensitive (sensitizing action). At higher polymer loadings the dispersions tolerate large amounts of salts without any coagulation.

Acting as a sensitizing agent, the macromolecules are adsorbed on several particles at the same time. The polymer bridges promote particle linking and flocculation. Often the particle bridging must be attained by addition of a certain amount of salt that reduces excessively high negative surface potentials.

The most effective flocculation agents for clays are polyanions. A few segments are attached to the clay particles and the majority form bridges between the particles. Polycations, which are strongly adsorbed in trains on the clay mineral surface, are less effective because of the limited number of segments available for interparticle bridging. Optimal flocculation then requires very distinct conditions, depending on the concentrations of the clay and the salt.[98] Uncharged polymers are generally not very effective as flocculation agents.[99]

In the protection (stabilization) range, the polymers envelop the particles. The theory of steric stabilization by adsorbed polymers is very complex. It is based on (1) the steric hindrance of the adsorbed macromolecules as the surfaces approach causing volume restriction (entropic stabilization) and (2) osmotic phenomena, which are often more dominant.[100-102] Generally, the polymer coat reduces the attraction between the particles to the level of polymer–polymer interactions in solution, and make it so low that only under theta-conditions does the system become unstable.[103]

Typical polymers acting mainly by osmotic stabilization are poly(ethylene oxide)s, modified starches and carboxymethylcellulose in water. The protective action of these materials is a matter of common industrial practice.

6 COOPERATIVE PHENOMENA

6.1 Cooperative Reactions

A linear macromolecule may be described as a sequence of appropriately chosen subunits, $\cdots$ AAAA $\cdots$. The subunits should exist in two states A or B. Transitions between A and B may be produced by changing external conditions:

$$\cdots\cdot \text{AAAAAA} \cdots\cdot \underset{k_1'}{\overset{k_1}{\rightleftharpoons}} \cdots\cdot \text{AAA}\boxed{\text{B}}\text{AAA} \cdots\cdot$$

$$\underset{k_2'}{\overset{k_2}{\rightleftharpoons}} \cdots\cdot \text{A}\boxed{\text{BBBBB}}\text{A} \cdots\cdot$$

The process is called cooperative if the elementary transition of A into B is affected by the properties of the other parts in the chain. At positive cooperativity a subunit in state B favours the transition of the neighbouring A into B. The nucleation rate (k_1, step I) is then slow compared with the propagation rate (k_2, step II).[104]

The principle can easily be illustrated for the well-known helix–coil transition. The helix is stabilized by hydrogen bonds $\rangle$C=O $\cdots$ H—N$\langle$. A transition of the coil into the helix requires that C=O and N—H groups move into suitable positions which promote the formation of the hydrogen bonds. If a nucleus, i.e. a hydrogen bond in the correct position, is formed, neighbouring C═O and N—H groups are held in positions which enable further hydrogen bonds to form rapidly. The probability of forming a hydrogen bond depends on the position of the neighbouring groups.

Cooperative reactions are not restricted to transitions A ⇌ B. The binding of monomeric molecules, for instance surfactants, on linear macromolecules can also exhibit cooperativity. B and A are then defined as occupied and vacant binding sites, respectively.

Cooperative reactions related to linear macromolecules, especially biopolymers, are now fairly well understood. There is no theory for two-dimensional cooperativity, which is expected to occur during intracrystalline reactions. Some phenomenological observations are described in the following sections.

6.2 Intercalation

The problem is illustrated in Fig. 21. In opening the interlayer spaces to the guest molecules the layers cannot simply move apart perpendicularly to the plane of layers, because all interlayer bonds would have to be broken simultaneously. Instead, at any arbitrary crystal edge the layer must curl. A broad transition zone, whose size depends on the elastic properties of the layer, is created that moves to the centre of the layer. This mechanism was first proposed by Weiss on the basis of numerous experiments with kaolin.[49,50,105–110]

After nucleation at N (Fig. 21) the layer 'rolls up' and promotes the interpenetration of guest molecules at the neighbouring sites along edge 1. If a few molecules succeed in rolling up the layer, then a large number of guest molecules can rapidly penetrate between the layers, and so the reaction front parallel to 1 moves to the centre of the layer. Because of the anisotropy of the layer, only those nuclei N will survive which initiate the most favourable elastic deformation of the layer.

The reaction front parallel to 1 exerts a geometrical constraint to possible reaction fronts at other crystal edges. The simultaneous development of reaction fronts at 2 and 6 is rendered more difficult. More likely is a nucleation at edge 3 if the distance to the starting front 1 exceeds a critical value. For small distances the radius of curvature would be too small and forces required for bending of the layers would be too great. A small crystal, thus, opens its interlayer at only one crystal edge, for instance at 1, and one reaction front moves from 1 to 4. In larger crystals two or more reaction fronts can be initiated simultaneously and the reaction proceeds faster than is the case with

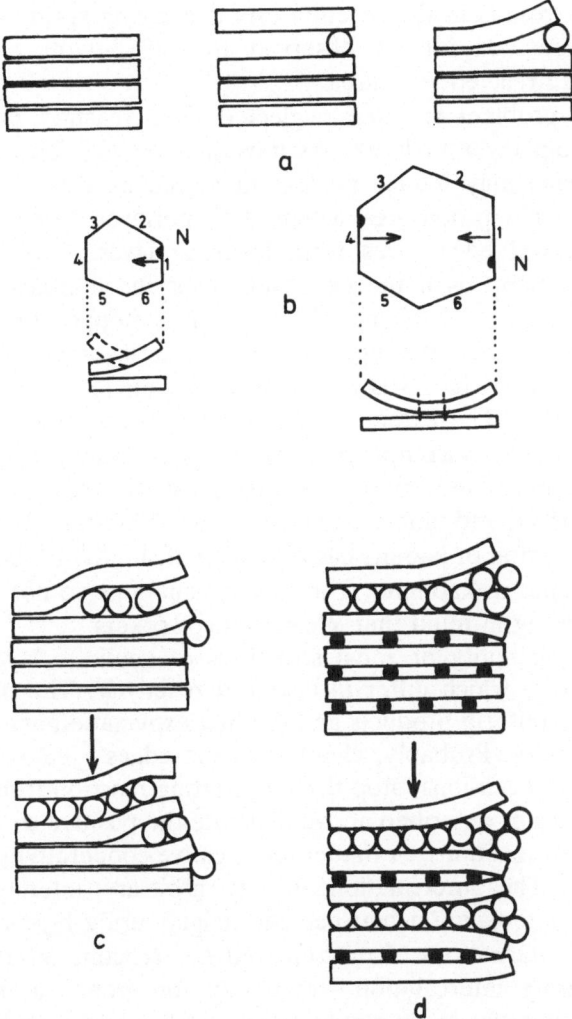

FIG. 21. Problems related to the opening of the interlayer spaces of layer structures. (a) Penetration of molecules or ions requires elastic deformation of the layers. (b) Nucleation (N) and formation of a single reaction front at small crystals and two or more reaction fronts at large crystals. (c) Propagation of the reaction from interlayer to interlayer along the crystal edge faces; reaction at the first interlayer nucleates the reaction at the next interlayer. (d) Reaction at an interlayer prevents the reaction at the adjacent interlayers.

small particles. In still larger particles the diffusion paths become too long and this results in the reaction rate decreasing again. A reaction rate maximum was indeed observed for $3 \cdot 8 – 5 \cdot 0 \, \mu m$ particles of kaolinite when reacted with urea.[106]

A further problem is often neglected. The reaction has also to propagate from layer to layer. As illustrated in Fig. 21c, a reacting interlayer space may induce nucleation at the adjacent layers. The reaction not only propagates within the interlayer spaces, but also along the crystal edge faces from layer to layer. One can easily imagine what happens if the expansibility of the interlayer space is blocked by a structural defect. The reaction is stopped or drastically slowed down because the opening of an interlayer space within a packet of unreacted layers is much more difficult to nucleate than at terminal surfaces. The extent of reaction reaches an almost constant level below 100%. Incomplete reactions have been frequently observed for kaolins and, in a few cases, for zirconium phosphate, $Zr(HPO_4)_2 \cdot 2H_2O$, and niobium oxyphosphate, $NbOPO_4 \cdot 3H_2O$.[111] The incomplete reaction of kaolins has also been attributed to the presence of different types of kaolinite. The two explanations do not conflict in principle, bearing in mind that Weiss and coworkers[50,108,110] could not exclude the possibility of zonal structures of kaolinite layers aggregated to packets which differ in chemical reactivity. The phosphates are synthetic, uniform products and the first explanation appears to be more appropriate. Probably, charges on the edges ($\geqslant$P—O$^-$ groups and cations as gegen-ions) stop the propagating reaction fronts.

The mechanisms described above illustrate the positive cooperativity of intercalation reactions. A model for negative cooperativity is shown in Fig. 21d. The intercalation in any arbitrary interlayer space obstructs the intercalation between the neighbouring layers. In some cases regular alternations of reacting and non-reacting interlayers are produced during intercalation—frequently for graphites, rarely for phosphates but never for kaolinite and crystalline silicic acids.

The 'self-preservation tendency' of the liquid structure (Section 3.4) also affects the reaction rates. A maximum appears at a very low ratio of dimethylsulphoxide (DMSO) to kaolinite that almost corresponds to a monomolecular covering of the external surface by DMSO. At higher amounts of DMSO, the liquid-like association of DMSO molecules retards the intercalation. The nucleation mechanism recently proposed by Weiss[109] may also explain this fact.

Based on neutron scattering experiments, it is assumed that the adsorption of DMSO molecules at the terminal surfaces causes a reorientation of OH-groups or a migration of protons which initiate the elastic deformation of the layers and the opening of the interlayer space. The molecules then enter the interlayer space and the reaction rate obeys the equation of two-dimensional phase boundary reactions (Avrami–Erofeev equation[105]). It eventually changes to a two-dimensional diffusion-controlled reaction. When a monolayer of DMSO covers the particles, the molecules may possess an optimal orientation for initiating the hydroxyl reorientation or proton migration that is lost at higher coverages.

6.3 'Propping Open' Mechanisms

Displacement reactions generally proceed more easily than intercalations because the layers are already separated. They can be kinetically impeded if the exchange of small molecules by larger ones requires a large layer separation. A stepwise expansion frequently overcomes this difficulty. The 'propping open' mechanism procedure, firstly introduced by Brindley and Ray,[112] enables the interlamellar sorption of bulky molecules. Brindley and Ray prepared a series of Ca-montmorillonite primary alcohol complexes starting from the ethanol derivative which was simply prepared by exposing the dried montmorillonite to the liquid or its vapour. The ethanol complex was then placed in contact with hexanol, which easily displaced the ethanol. To prepare complexes with still higher alcohols, hexanol was displaced in a similar manner.

6.4 Cation Exchange

Cation exchange reactions of smectites generally occur in hydrated or solvated interlayer spaces and the effect of cooperative forces is attenuated or even eliminated. Cooperative effects are observed for the exchange of potassium ions in micas, because the interlayers of K^+-micas are closed and the potassium ions are strongly fixed between the surface oxygen atoms (basal spacing ~ 1 nm). Graf von Reichenbach and Rich[113] exchanged Ba^{2+} ions for the K^+ ions by several five-day treatments with $0 \cdot 1 \text{N} \, BaCl_2$ solutions at 120°C. The amount of K^+-ion release increases with the particle size from $<0 \cdot 08 \, \mu$m particles up to 2–$5 \, \mu$m particles and decreases with further increasing size (5–$20 \, \mu$m). Similar observations were reported by Mortland and Lawton.[114] The K^+-ion release also increases with the particle thickness.[115] The maximal reactivity of particles of intermediate size needs an explanation similar to that for kaolinite intercalations, since in both systems the reactants have to enter collapsed interlayer spaces.[116] The basal spacing of the Ba^{2+} micas is increased to 1·21 nm because of the uptake of a monolayer of water.

The release of K^+-ions from muscovite by Ba^{2+}-ions is initiated at certain layers preferentially and then spreads to the adjacent layers so that zonal structures form. Partially reacted crystals thus contain packets of reacted and non-reacted layers. Any superlattice structure throughout the process of K^+-ion exchange could not be observed.[117]

Probably the latter mechanism (Fig. 21d) is more likely for reactions which start by collapsing the layers. At least in vermiculites, K^+- or Cs^+-ion sorption can cause collapse of alternate layers producing regularly interstratified structures.[118]

Vermiculite from Palabory (Transvaal) prepared from an inter-stratified mica–vermiculite mineral exhibits a pronounced tendency for 1:1 regular interstratification during cation exchange reactions $Ca^{2+} \rightleftharpoons K^+$, $Na^+ \rightleftharpoons K^+$ and $Na^+ \rightleftharpoons C_nH_{2n+1}NH_3^+$ ($n = 6$, 8).[119] These super-structures are more probably caused by an unsymmetrical charge distribution than by cooperativity of the reactions.

The cooperativity influences the selectivity of the cation exchange. For instance, if expanded micas are re-exchanged with K^+ ions, the potassium selectivity decreases with decreasing particle size.[115]

6.5 Conformational Changes
A different type of cooperative reaction was studied on interlamellar bimolecular films. During the sorption of long-chain alkanols by

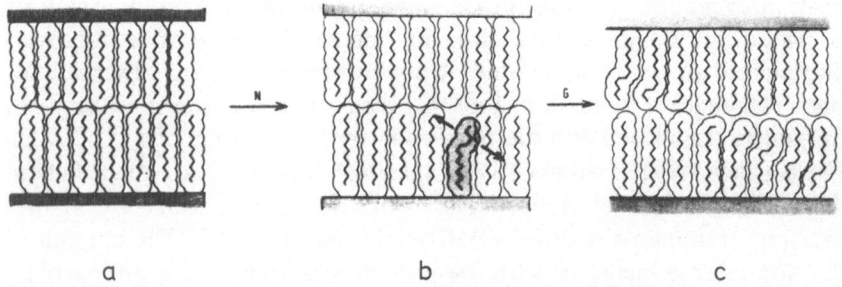

a   b   c

FIG. 22. Bilayers of long-chain compounds (e.g. alkylammonium ions and alkanol molecules) between silicate layers: (a) most of the chains in all-*trans* conformation; (b) nucleation (N) of a kink as a defect; (c) growth (G) of kink-blocks around the defect. (With kind permission of Verlag Chemie.[120])

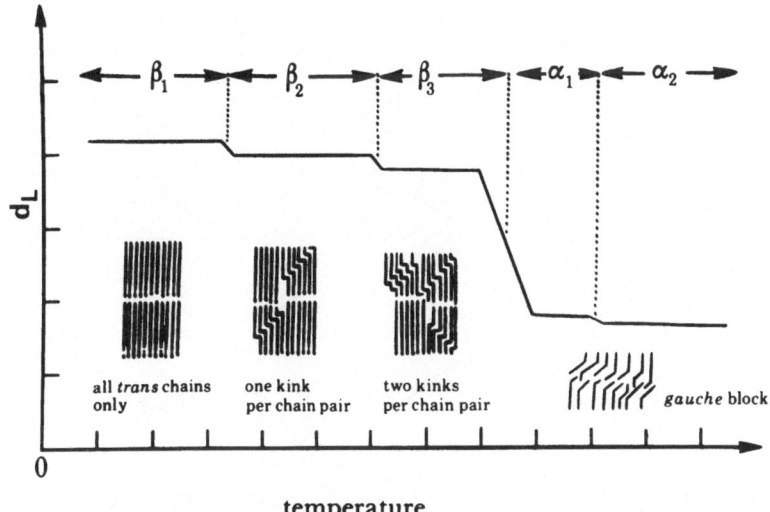

FIG. 23. Temperature-dependent phase transitions from all-*trans* blocks (β_1) to kink-blocks (β_2, β_3) and *gauche*-blocks (α_1, α_2) and related basal spacing changes. (Schematic; with kind permission from The Royal Society, London.[8])

alkylammonium smectites and vermiculites the alkylammonium ions are arranged so that all chains are directed perpendicular to the layers (Fig. 22). With increasing temperature the interlayer structure undergoes a series of changes that are brought out by stepwise decrease of spacings (Fig. 23) and discontinuities of the specific heats.[121,122] The changes are explained by the onset of rotational isomerization of the alkyl chains which results in the formation of different types of kink and *gauche*-blocks.[8,120]

Formation of kinks is a fundamental conformational change of alkyl chains aggregated to mono- or bi-molecular films. Figure 24 illustrates the rotations of the C—C bonds into the sequence *gauche*(+)–*trans*– *gauche*(−) (gt$\bar{\text{g}}$ = 2gl-kink*). Because both chain sections remain parallel, kinked chains do not destroy mono- or bi-molecular aggregations.

The decrease of the basal spacing reflects directly the shortening of the chains by the formation of kinks. A 2gl-kink shortens the overall length by 0·127 nm, higher order kinks by multiples of that distance. If

* 2gl-kink: (2l–1) *trans*-bonds between two *gauche*-bonds (g, $\bar{\text{g}}$), so that the chain shortening is l × 0·127 nm.

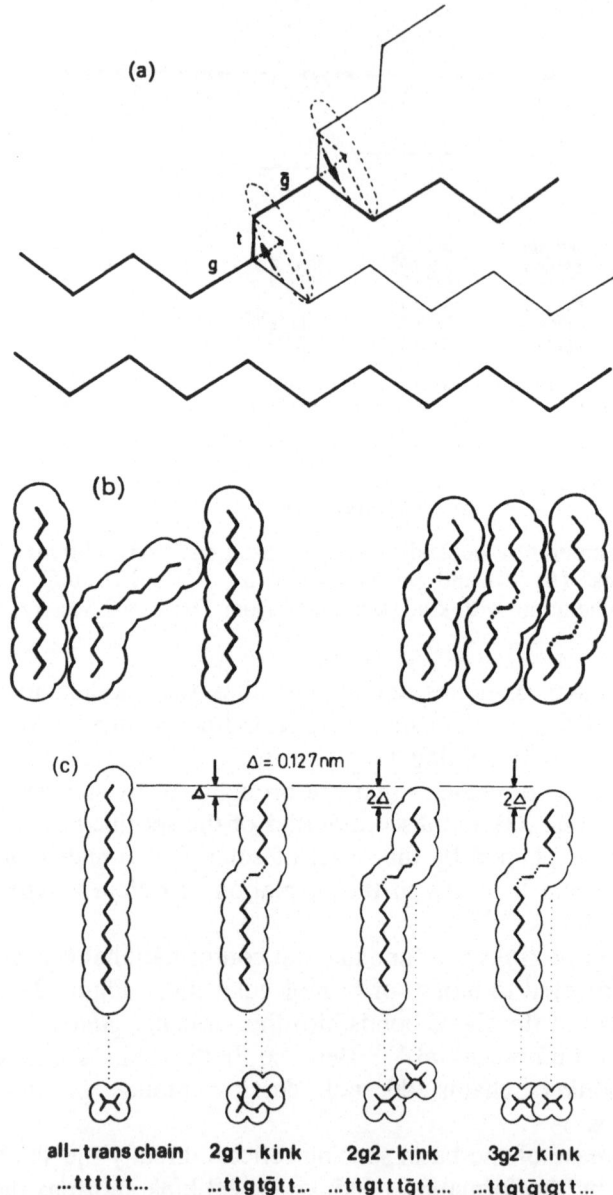

FIG. 24. Alkyl chains with kinks: (a) the rotations around C—C bonds (with kind permission of Verlag Chemie[120]); (b) a chain with an isolated *gauche*-bond between two all-*trans* chains and aggregation of kinked chains; (c) different types of kinks (*gauche*-bonds dotted).

spacings are decreased by 0·11–0·13 nm (or multiples thereof), then theory demands the formation of kinks in almost all chains (Fig. 22). It is unlikely that the kinks are randomly distributed because the van der Waals attraction between the loosely packed chains is reduced too strongly. Rather, kink blocks form where the regular displacement of the kinks in neighbouring chains guarantees a dense packing. This makes the process cooperative: in a lamella of alkyl chains the probability of kink formation in an alkyl chain depends on the state of the neighbouring chains (Figs 22, 24b). It is high if an adjacent chain already contains a kink. The cooperativity may be simply explained by steric interactions. The nucleation of a kink in a chain exerts a steric constraint on the adjacent chains and promotes the formation of kinks in neighbouring positions, so that by the displacement of the kinks the chains fit together.

In agreement with C_p-measurements,[121] the calculations of Pechhold et al.[123] and Baur[124] established the surprisingly low enthalpy of formation of kink-blocks with one kink per chain of about 3 kJ mol^{-1}, which is below the defect energy of a kink in an isolated chain (5·5 kJ mol^{-1}).[120] This result is explained tentatively by a partial compensation of the enthalpy demand by a gain in van der Waals energy due to an increase in packing density as the basal spacing decreases.

At higher temperatures the kink-blocks (β-phases) transform into gauche-blocks (α-phases). The reactions proceed by several intermediate stages. This is best demonstrated by the occurrence of a series of C_p-peaks in the transition range. The structural elements of gauche-blocks are isolated gauche-bonds and, more likely, gtg-conformations.[123]

The cooperativity is not strongly one-dimensional as suggested by Fig. 22c. The chain section displaced by the kink not only disturbs the chains in the same row but also those in neighbouring rows (Fig. 25). This makes kink-block formation highly cooperative and guarantees the sharp transitions.

The alkyl chains of the bimolecular films between uranium mica layers have about the same packing density, but the temperature-induced phase transitions are less sharp and the basal spacing decreases in irregular steps.[124] The square arrangement of the chains impedes a complete cooperative coupling, the cooperativity remaining one-dimensional. The growth steps take place within individual rows of chains, but each row requires its own nucleation (Fig. 25).

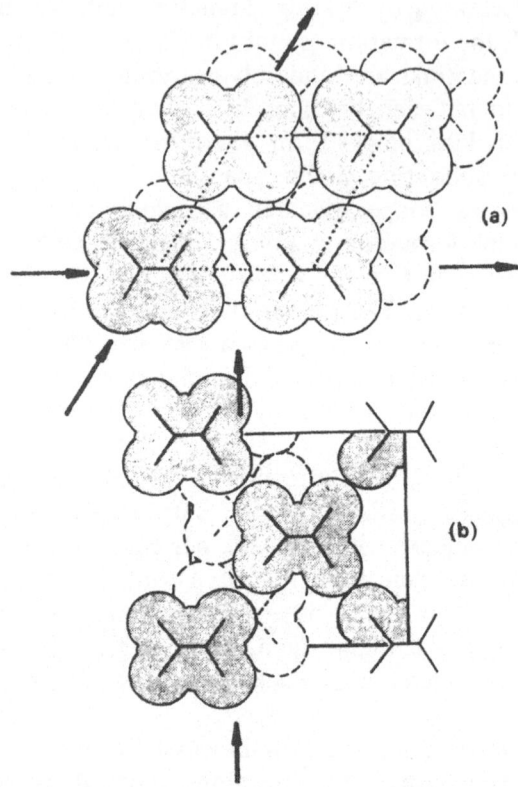

FIG. 25. Cooperativity of kink-block formation: (a) layer silicates: cooperative growth of the kink-blocks in two directions (arrows); (b) uranium micas: cooperative growth of the kink-blocks solely in one direction.

The looser packing density (about 0.24–$0.25\,nm^2$/chain) in vermiculites, smectites and uranium micas compared with that in paraffins ($0.19\,nm^2$/chain) promotes the nucleation of the kinks. In the interlayers of uvanite the chains are closely packed with a density of $0.21\,nm^2$/chain. The nucleation then requires such high temperatures that the structure directly transforms to *gauche*-blocks.[125]

Kinks and related structural defects were first postulated for crystalline, partially crystalline and melted polymers,[126] but the corresponding chain shortening could not be observed because of the random distribution of the defects (see also Fig. 9 in ref. 8).

In membranes with protein–lipid–protein organization, the alkyl

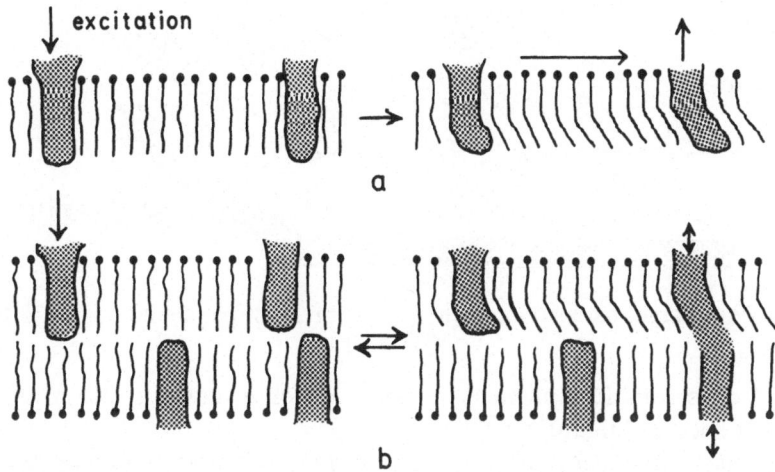

FIG. 26. A schematic model for information transfer within lipid layers (a) and formation of protein contacts (b) by cooperative transitions of alkyl chain conformations. (With kind permission of Springer Verlag.[13])

chains of the lipid layer are aggregated to bilayers. The alkanol complexes of alkylammonium silicates are certainly not models of biomembranes but they model the alkyl chain aggregations in the lipid layer. For instance, the effect of *cis*-unsaturated alkyl chains on the chain aggregation should be recognized.[8,127] A further hypothesis must be mentioned (Fig. 26). Information might be propagated within a lipid bilayer from one point of excitation to an acceptor site far away by conformational changes of the chains. Furthermore, the displacement of integral proteins in biomembranes by conformational changes of the chains might entail a switching effect, when two integral proteins come into contact.

7 CHARGE PATTERN INTERACTIONS

7.1 Charge Patterns

A charge pattern describes the distribution of charges of an ionic macromolecule. For ionic linear macromolecules this pattern is represented by the distance between the charged groups and varies with the conformation of the molecule. If electrostatic forces predominate, the interaction between ionic macromolecules should be governed by the fit or misfit of the charge patterns (Fig. 27). Such effects are claimed to

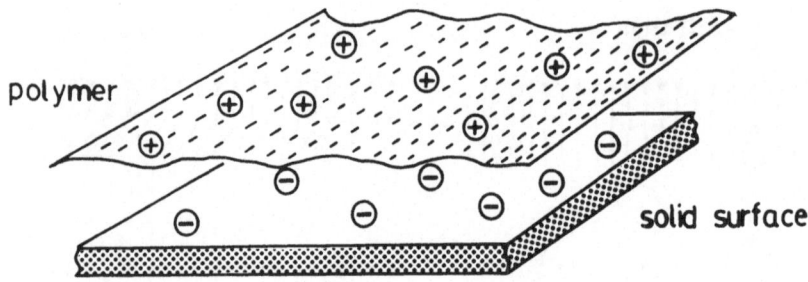

Fig. 27. Charge pattern interactions. (With kind permission of Springer Verlag.[13])

be of considerable biological importance, for instance in the recognition of specific binding sites.

Attempts to elucidate charge pattern interactions experimentally run into many difficulties. Most one-dimensional polymers are too flexible. Two-dimensional entities with distinct charge distributions would provide more suitable substrates but are rarely available. Initial insights into charge pattern interactions can be obtained using the silicate anions as rigid two-dimensional macromolecules. The surface charge densities are well known and samples with widely differing charge patterns are available.

7.2 Cation Exchange

As far as we know, the exchange with inorganic cations is hardly influenced by the charge pattern of the surface. Exchange depends mainly on the charge density. More familiar is the increased selectivity towards potassium ions at high layer charges.

The surface charge density can govern the stoichiometry of the exchange. Bivalent cations are exchanged for monovalent cations in equivalent ratios:

$$\frac{Na^+ \qquad Na^+}{\ominus \qquad \ominus} + Ca^{2+} \rightleftharpoons \frac{Ca^{2+}}{\ominus \qquad \ominus} + 2Na^+$$

If the distance between the surface charges becomes too large, the reaction proceeds between ions in equimolar proportions:

$$\frac{Na^+ \qquad Na^+}{\ominus \qquad \ominus} + 2Ca^{2+} + 2X^- \rightleftharpoons \frac{X^- \qquad X^-}{\underset{\ominus \qquad \ominus}{Ca^{2+} \quad Ca^{2+}}} + 2Na^+$$

The equimolar exchange was first observed on kaolinite.[128]

However, the binding of organic multivalent cations may be more sensitive to the surface charge patterns. One example was reported by Philen et al.[129] Competitive adsorption of diquat^{2+} (**I**) and paraquat^{2+} (**II**) cations by external surfaces of vermiculite and mica appeared to be related linearly to the surface charge density. Paraquat^{2+} cations

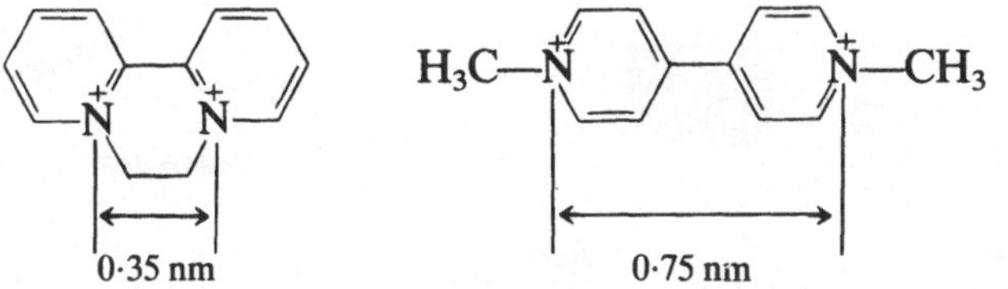

with larger distances between charges are preferentially bound on the lower charged surfaces.[130] The intercharge distances in paraquat are also compatible with the pattern at higher surface charge densities but steric hindrance between the bulky cations promotes the adsorption of diquat cations.

7.3 Selective Coagulation

When sodium smectites are added to water or dilute sodium salt solutions, they disintegrate into the individual silicate layers. A stable colloidal dispersion is formed that is coagulated by the addition of salts. If the critical coagulation concentration is exceeded, the silicate layers aggregate again to form particles.

These phenomena were observed in experiments conducted by Frey and Lagaly.[31] Colloidal dispersions were prepared from fractions $<0.1\,\mu$m and $>0.1\,\mu$m of sodium montmorillonite and sodium beidellite; these minerals were selected as examples of lowly and highly charged smectites. For both fractions of montmorillonite, the layer charge $\xi = 0.28$; for the $<0.1\,\mu$m fraction of beidellite, $\xi = 0.37$, and for the $>0.1\,\mu$m fraction, $\xi = 0.38$ eq/(Si, Al)$_4$O$_{10}$.

A $1:1$ mixture of both dispersions thus contains layers of montmorillonite and beidellite (Fig. 28). During the salt coagulation of the mixed dispersion, the layers can aggregate in different ways with the formation of four types of particles:

(1) particles made up either by highly charged or by lowly charged layers (separation);

(2) particles consisting of packets of highly and lowly charged layers (zonal mixed layers);

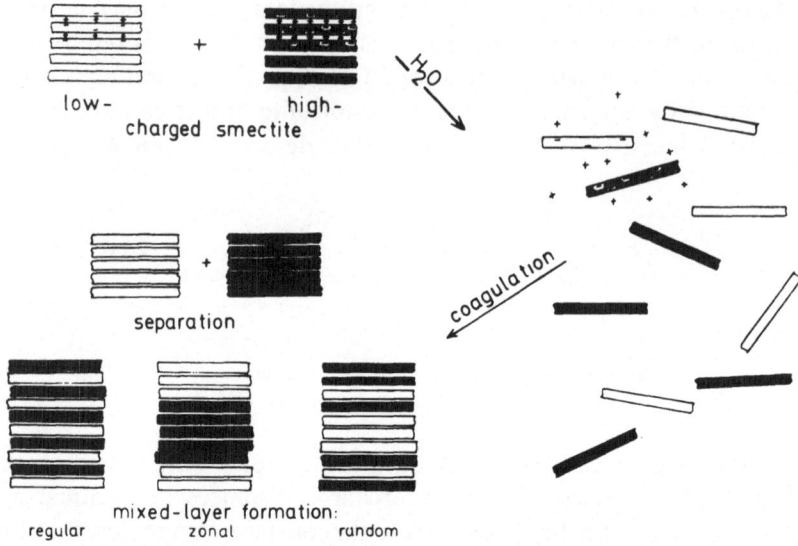

FIG. 28. Disaggregation of two sodium smectite crystals into individual silicate layers and different ways of reaggregation (coagulation by addition of sodium chloride; surface charges and interlayer cations not entirely shown). (With kind permission of Academic Press.[31])

   (3) particles with alternating highly and lowly charged layers (regular mixed-layers);
   (4) particles with a random sequence of highly and lowly charged layers (random mixed layers).

Investigation of the coagulates by alkylammonium exchange provides the only method that allows a distinction between the different type of particles. The results are summarized as follows.

   (i) The type of the aggregate depends on the particle size and is less influenced by the experimental conditions during coagulation.

  (ii) The mixed dispersion of fractions $> 0 \cdot 1 \, \mu$m leads to a separation into particles of highly charged layers and particles of lowly charged layers. (Types (1) and (2) cannot be distinguished by the alkylammonium method but the predominance of (1) could be proved indirectly.) The charge distribution of the highly charged crystals is similar to that in the original beidellite but is by no means identical.

(iii) Only the fraction $< 0.1\,\mu$m forms random mixed-layer particles that contain highly and lowly charged layers.

The experimental results clearly demonstrate that the 'self-aggregation' of colloidal silicate layers is governed by the charge patterns, only if the layers are large enough. The differences of the maximal repulsion energies (per plate) between differently charged layers at the critical coagulation points (DLVO theory: for details see Ref. 31) increase with the diameter of the layers. In the dispersions made from the larger particles, a high repulsive energy between the highly charged layers guarantees the separation into lowly and highly charged particles because the lowly charged layers aggregate first. For small particles, the difference between the repulsive energies is too small to enforce separation, and random mixing of the layers is promoted.

7.4 Interactions with Linear Polycations

Ionenes (Fig. 29) are very suitable macromolecules for studying charge pattern interactions. The distance between the positive centres can be varied by the number, n, of CH_2-groups. The ionenes penetrate between the layers of smectites, and interlayer cations are displaced by N^+-groups.

The experimental data—basal spacings, d_L, number of segments per unit, N, and number of cations displaced per segment, χ—depend on the segment length, n. The expected relations for loop and train adsorption are schematically illustrated in Fig. 29. For adsorption in loops between the layers, d_L increases with n, and N remains nearly constant, if every N^+ displaces one interlayer cation ($\chi = 1$). If the ionenes are adsorbed in trains, d_L is independent of n, but N decreases with n. The variation of χ with n is explained by Fig. 30. For small n, the short distances between the charges do not fit the surface charge pattern. Several N^+ groups do not displace interlayer cations and carry their gegen-ions. On average, a segment displaces less than one cation ($\chi < 1$). The conditions for displacing interlayer cations become optimal when, by increasing n, the distances between the positive centres fit those between the surface charges ($\chi = 1$). For still longer chains, each N^+ finds a negative surface charge ($\chi = 1$), if the chain possesses enough flexibility. However, the chains may lose their flexibility (1) because they are tightly fixed between the layers and (2) because of a mutual steric hindrance of neighbouring chains. Not all

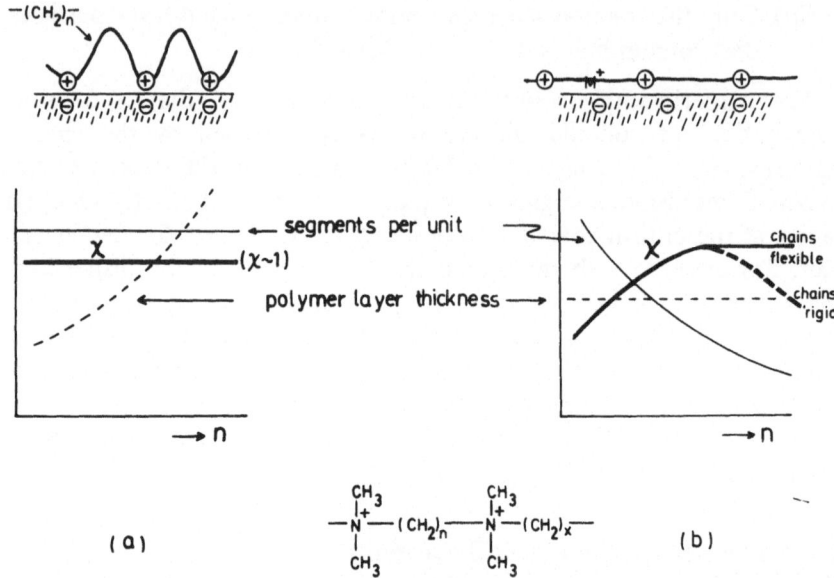

FIG. 29. Two models for ionene adsorption on silicate macroanions. (a) Adsorption in loops; each segment displaces an interlayer cation. (b) Adsorption in trains, χ as a function of n depends on the flexibility of the ionene chains on the surface (χ = interlayer cations displaced per segment).

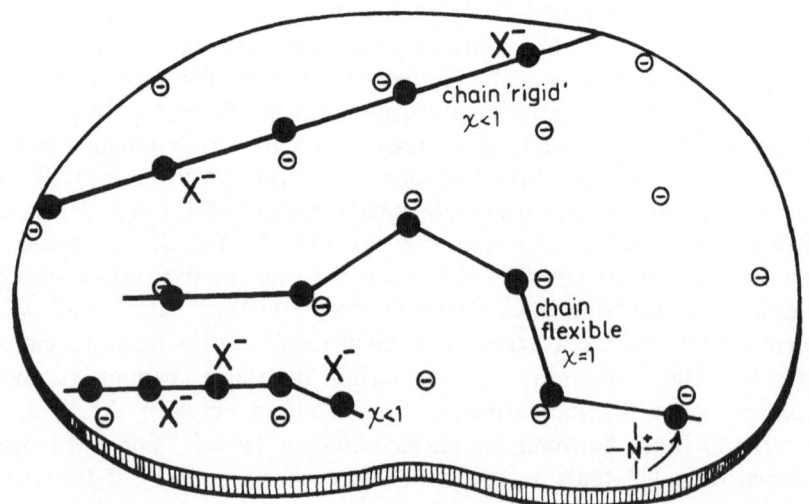

FIG. 30. Interaction of ionenes of different segment lengths n with the surface charge pattern.

N^+-groups are then balanced by surface charges and χ is expected to decrease again (dotted curve in Fig. 29b). Thus, adsorption and immobilization of chains demands that χ as a function of n increases to a maximum and then decreases.

One might first assume the fit to be optimal if the area per segment equals the equivalent area. This is true for beidellite (Fig. 31), but does not apply to other silicates.[69] Indeed, the equivalence is not to be expected because it requires that the dimensions of the equivalent area correspond to those of the segments—which is quite unlikely. It is also conspicuous that hectorite, beidellite and vermiculite have maxima at $n = 8$.

The segment lengths of ionenes with $n = 7$ $(l = 1{\cdot}02 \text{ nm})$ are

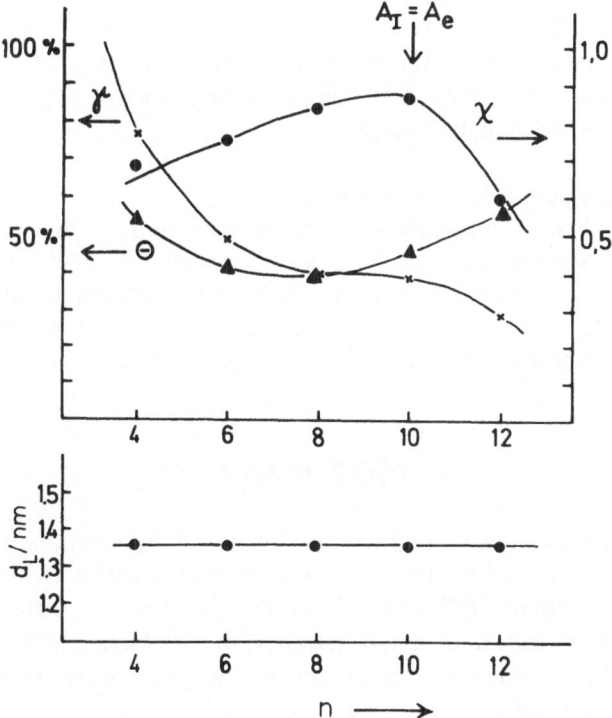

Fig. 31. Adsorption of ionenes on beidellite: $\theta =$ surface coverage, $\gamma =$ percentage of interlayer cations displaced, $\chi =$ interlayer cations displaced per segment, $d_L =$ basal spacings (after drying), $n =$ segment length. (With kind permission of Springer Verlag.[13])

compatible with the lattice dimensions ($2a_0 = 1.04$ nm) and the periodicity of the CH_2-groups in alkyl chains ($4 \times 0.254 = 1.02$ nm). The surface area $a_0 b_0$ of the unit cell of the smectites contains about 0.6 charges. Thus two unit cells possess about one charge, so that distances of $2a_0 = 1.04$ nm between surface charges will occur most frequently. This explains how an optimal fit is attained for ionenes with $n \approx 8$.

Ionene adsorption demonstrates the combined effects of charge pattern and steric interactions. The electrostatic interaction of the charges is important but is by no means exclusively decisive. The optimum steric fit is of great influence and can overcome the electrostatic interactions. Apparently this is a general principle which has often been overlooked. Protein adsorption, for instance, is certainly governed by these combined effects. Experiments of Lyklema and Norde[83,84] clearly illustrate that the charge interactions are not the most important factors (see Fig. 18 and Section 5.5). Protein adsorption has a maximum in the isoelectric region of the protein but it is most striking that, above this point, the negatively charged protein is still adsorbed by negative surfaces.

7.5 Replication in Inorganic Systems
The capability of replication is the basic requirement of biological systems. A fascinating hypothesis was developed by Weiss postulating that even inorganic systems are capable of replication. In studies with clays, it was demonstrated that the charge pattern of a silicate layer can be transmitted to freshly forming silicate layers during synthesis experiments.[131]

8 TECHNICAL USES

This section summarizes important technical applications.[2,9,132] The employment of kaolins and clays in ceramic industries is well known, but no more than 20% is used for this purpose. A further 20% is required as fillers in a large number of products (rubber, paints, plastics). More than 50% is applied in the paper industry as coating and filler materials.

Bentonites are used in their natural form or in activated forms by:

(1) soda activation: exchange of Ca^{2+} by Na^+ by treatment with soda;

(2) acid activation: reaction with mineral acids to produce leaching earths with partial or complete destruction of the structure;

(3) organic activation: exchange of the inorganic cations by cationic surfactants (Table 2).

Natural bentonites and leaching earths provide inexpensive adsorbents for the chemical and pharmaceutical industries, and for clarification of wines, beer, animal litter, and supports for pesticides. They are particularly suitable for pelletizing ores. Their use as catalysts, for example in the production of silicones, or catalyst supports has been recognized for a long time.

Sodium bentonites find common applications in foundry moulding sands, as drilling fluids for oil recovery and in construction engineering. These and many other applications profit by the peculiar rheological properties of the bentonite dispersions.

Organic-activated bentonites[133] are used purely as thickeners or to produce or improve thixotropic properties in organic systems such as paints, enamel paints, waxes and organic drilling fluids, and as additives for a large number of materials including tar, asphalts, fibreglass resins, mastics, lutes, putties, ointments and cosmetics, and in recently increasing amounts, in washing powders.

Clays enjoy a substantial application for filling and reinforcement of polymer systems, for instance natural and synthetic rubber, PVC and related thermoplastic materials. A highly effective method of reinforcing natural and synthetic elastomers is the grafting of the polymers to the clay surfaces.

Conditioning of soils by polymeric organic materials is related to the formation of water-stable aggregates of clay particles. Natural polysaccharides are particularly effective. Poly(vinyl alcohol) and 'Krilium' (hydrolyzed polyacrylonitrile) are applied as synthetic soil aggregate stabilizers.

Steric stabilization of clay dispersions by polyelectrolytes is common industrial practice. It is applied, for example, in the protective action of carboxymethylcellulose, modified starches and other materials towards clays in salt-water drilling fluids.

REFERENCES

1. BRINDLEY, G. W. and BROWN, G., *Crystal Structures of Clay Minerals and Their X-ray Identification*, 1980, Mineralogical Society, London.

2. WEISS, A., Z. Anorg. Allgem. Chem., 1958, **297**, 257; WEISS, A., MEHLER, H. and HOFMANN, U., Z. Naturforsch., 1956, **11b**, 435.
3. JEPSON, W. B., Phil. Trans. Roy. Soc. Lond., 1984, **A311**, 411.
4. WEISS, A., Z. anorg. Allgem. Chem., 1959, **299**, 92.
5. LAGALY, G. and WEISS, A., Proc. Internat. Clay. Conf. Mexico 1975, (S. W. Bailey Ed.) 1976, Applied Publishing, Wilmette, Ill., USA, p. 157.
6. LAGALY, G., Clays Clay Min., 1979, **27**, 1.
7. LAGALY, G., Clay Min., 1981, **16**, 1.
8. LAGALY, G., Phil. Trans. Roy. Soc. Lond., 1984, **A311**, 315.
9. VAN OLPHEN, H., Clay Colloid Chemistry, 1977, J. Wiley and Sons, New York.
10. LAGALY, G, in Zum Fliessverhalten von Stoffen und Stoffgemischen, (W.-M. Kulicke, Ed.), 1986, Hüthig u. Wepf Verlag, Basel–Heidelberg–New York, p. 147.
11. VOGT, K. and KÖSTER, H. M., Clay Min., 1978, **13**, 25.
12. BENEKE, K. and LAGALY, G., Z. Naturforsch., 1978, **33b**, 564; Am. Min., 1981, **66**, 432; Clay Min., 1982, **17**, 175.
13. LAGALY, G., Naturwiss., 1981, **68**, 82.
14. RÖSNER, C. and LAGALY, G., J. Solid State Chem., 1984, **53**, 92.
15. HOUGARDY, J., STONE, W. E. E. and FRIPIAT, J. J., J. Chem. Phys., 1976, **64**, 3840.
16. GIESE, R. F. and FRIPIAT, J. J., J. Colloid Interf. Sci., 1979, **71**, 441.
17. SUQUET, H., PROST, R. and PEZERAT, H., Clay Min., 1982, **17**, 231.
18. SPOSITO, G. and PROST, R., Chem. Reviews, 1982, **82**, 553.
19. ORMEROD, E. C. and NEWMANN, A. C., Clay Min., 1983, **18**, 289.
20. TUCK, J. J., HALL, P. L. and HAYES, M. H. B., J. Chem. Soc., Faraday Trans. I, 1984, **80**, 309.
21. FRIPIAT, J., CASES, J., FRANCOIS, M. and LETELLIER, M., J. Colloid Interf. Sci., 1982, **89**, 378; FRIPIAT, J. J., LETELLIER, M., and LEVITZ, P., Phil. Trans. Roy. Soc. London, 1984, **A311**, 287.
22. TSCHOUBAR, C. T., Phil. Trans. R. Soc. London, 1984, **A311**, 259.
23. MORTLAND, M. M., FRIPIAT, J. J., CHAUSSIDON, J. and UYTTERHOEVEN, J., J. Chem. Phys., 1963, **67**, 248.
24. CADY, S. S. and PINNAVAIA, T. J., Inorg. Chem., 1978, **17**, 1501.
25. NORRISH, K., Trans. Faraday Soc., 1954, **18**, 120.
26. CEBULA, D. J. and OTTEWILL, R. H., Clays Clay Min., 1981, **29**, 73; LUBETKIN, S. D., MIDDLETON, S. R. and OTTEWILL, R. H., Phil. Trans. Roy. Soc. London, 1984, **A311**, 353.
27. KLEIJN, W. B. and OSTER, J. D., Clays Clay Min., 1982, **30**, 383.
28. HELMY, A. K., NATALE, I. M. and MANDOLESI, M. E., Clays Clay Min., 1982, **30**, 49.
29. SCHRAMM, L. L. and KWAK, J. C. T., Clays Clay Min., 1982, **30**, 40.
30. VIANI, B. E., LOW, P. F. and ROTH, C. B., J. Colloid Interf. Sci., 1983, **96**, 229.
31. FREY, E. and LAGALY, G., Proc. Internat. Clay Conf., Oxford 1978, 1979, Elsevier Scientific, Amsterdam, p. 131; J. Colloid Interf. Sci., 1979, **70**, 46.

32. Pons, C. H., Rousseaux, F. and Tschoubar, D., *Clay Min.*, 1981, **16**, 23; 1982, **17**, 327.
33. Annabi-Bergaya, F., Cruz, M. I., Gatineau, L. and Fripiat, J. J., *Clay Min.*, 1981, **16**, 115.
34. Berkheiser, V. and Mortland, M. M., *Clays Clay Min.*, 1975, **23**, 404.
35. Brindley, G. W., *Clay Min.*, 1966, **6**, 237.
36. Weiss, A. and Hofmann, U., *Z. Naturforsch.*, 1951, **6b**, 405.
37. Nijs, H., van Damme, H., Bergaya, F., Habti, A. and Fripiat, J. J., *J. Molecular Catalysis*, 1983, **21**, 223.
38. Farzaneh, F. and Pinnavaia, T. J., *Inorg. Chem.*, 1983, **22**, 2216.
39. Pinnavaia, T. J., *Science*, 1983, **220**, 365.
40. Oades, J. M., *Clays Clay Min.*, 1984, **32**, 49–57.
41. Yamanaka, S., Doi, T., Sako, S. and Hattori, M., *Mater. Res. Bull.*, 1984, **19**, 161.
42. Yamagishi, A., *J. Chem. Soc., Dalton Trans.*, 1983, 679.
43. Yamagishi, A. and Fujita, N., *J. Colloid Interf. Sci.*, 1984, **100**, 136.
44. Mortland, M. M. and Pinnavaia, T. J., *Nature*, 1971, **229**, 75; Pinnavaia, T. J. and Mortland, M. M., *J. Phys. Chem.*, 1971, **75**, 3957.
45. Rupert, J. P., *J. Phys. Chem.* 1973, **77**, 784.
46. Pinnavaia, T. J., Hall, P. L., Cady, S. S. and Mortland, M. M., *J. Phys. Chem.*, 1974, **78**, 994.
47. Sayin, M., Beyme, B. and Graf von Reichenbach, H., *Proc. Internat. Clay Conf. Oxford 1978*, 1979, Elsevier Scientific, Amsterdam, Oxford, New York, p. 177.
48. Theng, B. K. G., The Chemistry of Clay–Organic Reactions, 1974, Adam Hilger, London.
49. Weiss, A., Thielepape, W. and Orth, H., *Proc. Internat. Clay. Conf. Jerusalem 1966*, 1966, Israel University Press, Jerusalem, p. 277.
50. Range, K. J., Range, A. and Weiss, A., *Proc. Internat. Clay Conf. Tokyo 1969*, Vol. 1, 1970, Israel University Press, Jerusalem, p. 3.
51. Churchman, G. J. and Theng, B. K. G., *Clay Min.*, 1984, **19**, 161.
52. Weiss, A., *Clays Clay Min.*, 1963, 191; *Angew. Chem.*, 1963, **75**, 113.
53. Lagaly, G. and Weiss, A., *Kolloid Z. Z. Polymere*, 1970, **238**, 485; Lagaly, G., *Clays Clay Min.*, 1982, **30**, 215.
54. Lagaly, G. and Weiss, A., *Kolloid Z. Z. Polymere*, 1971, **243**, 48.
55. Brindley, G. W. and Hoffmann, R. W., *Clays Clay Min.*, 1962, p. 546.
56. Brindley, G. W. and Moll, W. F., *Am. Miner.*, 1965, **50**, 1355.
57. Lagaly, G. and Witter, R., *Ber. Bunsenges. physik. Chemie*, 1982, **86**, 74; Lagaly, G., Witter, R. and Sander, H. in *Adsorption from Solution* (R. H. Ottewill, C. H. Rochester and A. L. Smith, Eds), 1983, Academic Press, London, p. 65.
58. Dékány, I., Szánto, F., Nagy, L. G. and Fóti, G., *J. Colloid Interf. Sci.*, 1975, **50**, 265.
59. Dékány, I., Szántó, F., Weiss, A. and Lagaly, G., *Ber. Bunsenges. physik. Chem.*, 1985, **89**, 62.

60. Weiss, A., *Kolloid Z. Z. Polymere,* 1966, **211,** 94.
61. Parfitt, R. L. and Greenland, D. J., *Clay Min.,* 1970, **8,** 305.
62. Burchill, S., Hall, P. L., Harrison, R., Hayes, M. H. B., Langford, J. I., Livingstone, W. R., Smedley, R. J., Ross, D. K. and Tuck, J. J., *Clay Min.,* 1983, **18,** 373.
63. Parfitt, R. L. and Greenland, D. J., *Clay Min.,* 1970, **8,** 317.
64. Dodson, P. J. and Somasundaran, P., *J. Colloid Interf. Sci.,* 1984, **97,** 481.
65. Armstrong, D. E. and Chesters, G., *J. Soil Sci.,* 1964, **98,** 39.
66. Hollander, A. F. and Somasundaran, P., in *Adsorption from Aqueous Solutions,* (P. H. Tewari, Ed.) 1981, Plenum Press, New York, London, p. 143.
67. Emerson, W. W., *J. Agr. Sci.,* 1956, **47,** 117.
68. Greenland, D. J., *J. Colloid Sci.,* 1963, **18,** 647.
69. Mesrogli, M., Thesis, University of Kiel, 1979.
70. Levy, R. and Francis, C. W., *J. Colloid Interf. Sci.,* 1975, **50,** 442; *Clays Clay Min.,* 1975, **23,** 85, 475.
71. Parfitt, R. L. and Greenland, D. J., *Soil Sci. Soc. Am. Proc.,* 1970, **34,** 862.
72. Olness, A. and Clapp, C. E., *Clays Clay Min.,* 1973, **21,** 289.
73. Kato, C., Kuroda, K. and Hasegawa, K., *Clay Min.,* 1979, **14,** 13.
74. Kato, C., Kuroda, K. and Misawa, K., *Clays Clay Min.,* 1979, **27,** 129.
75. Kato, C., Kuroda, K. and Takahara, H., *Clays Clay Min.,* 1981, **29,** 294.
76. Bart, J. C., Cariati, F., Erre, L., Gessa, C., Micera, G. and Piu, P., *Clays Clay Min.,* 1979, **27,** 429.
77. Friedlander, H. Z., *ACS Div. Polymer Chem. reprints,* 1963, **4,** 300.
78. Hawthorne, D. G. and Solomon, D. H., *Clays Clay Min.,* 1972, **20,** 75.
79. Solomon, D. H. and Loft, B. C., *J. Appl. Polym. Sci.,* 1968, **12,** 1253.
80. Stoessel, F., Guth, J. L. and Wey, R., *Clay Min.,* 1977, **12,** 255.
81. Soma, Y., Soma, M. and Harada, I., *J. Phys. Chem.,* 1984, **88,** 3034.
82. Ueda, T. and Harada, S., *J. Appl. Polym. Sci.,* 1968, **12,** 2395.
83. Koutsoukos, P. G., Mumme-Young, C. A., Norde, W. and Lyklema, J., *Colloids and Surfaces,* 1982, **5,** 93.
84. Lyklema, J. and Norde, W., *Croatica Chem. Acta,* 1973, **45,** 67; Norde, W., *Croatica Chem. Acta,* 1983, **56,** 705.
85. Talibudeen, O., *Trans. Farad. Soc.,* 1954, **51,** 582.
86. Schnitzer, M. and Kodama, H., *Clays Clay Min.,* 1972, **20,** 359.
87. Moinereau, J., *Clay Min.,* 1977, **12,** 75.
88. Larsson, N. and Siffert, B., *J. Colloid Interf. Sci.,* 1983, **93,** 424.
89. Clementz, D. M., *Clays Clay Min.,* 1976, **24,** 312.
90. Mortland, M. M. and Gieseking, J. E., *Soil Sci. Soc. Am. Proc.,* 1952, 10.
91. Roper, M. M. and Marshall, K. C., *Microbial Ecology,* 1978, **4,** 279.
92. Perez-Rodriguez, J. L. and Wilson, M. J., *Clay Min.,* 1969, **8,** 39.
93. Perez-Rodriguez, J. L., Weiss, A. and Lagaly, G., *Clays Clay Min.,* 1977, **25,** 243.

94. STUTZMAN, T. and SIFFERT, B., *Clays Clay Min.*, 1977, **25**, 392; ESPINASSE, P. and SIFFERT, B., *Clays Clay Min.*, 1979, **27**, 279.
95. SIFFERT, B. and ESPINASSE, P., *Clays Clay Min.*, 1980, **28**, 381.
96. SIFFERT, B. and BIABA, H., *Clays Clay Min.*, 1976, **24**, 303.
97. RUIZ-HITZKY, E. and FRIPIAT, J. J., *Clays Clay Min.*, 1976, **24**, 25.
98. KIM, H. S., LAMARCHE, C. and VERDIER, A., *Colloid Polym. Sci.*, 1983, **261**, 64.
99. THENG, B. K. G., *Clays Clay Min.*, 1982, **30**, 1.
100. OTTEWILL, R. H. and WALKER, T., *Kolloid Z. Z. Polym.*, 1968, **227**, 108.
101. OVERBEEK, J. T. G., *Adv. Colloid Interf. Sci.*, 1982, **16**, 17.
102. NAPPER, D. H. *Polymeric Stabilization of Colloidal Dispersions*, 1983, Academic Press, London.
103. SILBERBERG, A., *J. Macromol. Sci.—Phys.*, 1980, **B18**, 677.
104. SCHWARZ, G., *Biopolym.*, 1968, **6**, 873; *Eur. J. Biochem.*, 1970, **12**, 442; ENGEL, J. and SCHWARZ, G., *Angew. Chem.*, 1970, **82**, 468.
105. FENOLL, H.-A. P. and WEISS, A., *Quimica*, 1969, **65**, 769.
106. WEISS, A., BECKER, H. O., ORTH, H., MAI, G., LECHNER, H. and RANGE, K. J., *Proc. Internat. Clay Conf. Tokyo 1969*, Vol. 2, 1970, Israel University Press, Jerusalem, p. 180.
107. WEISS, A. and RANGE, K. J., *Proc. Internat. Clay Conf. Tokyo 1969*, Vol. 2, 1970, Israel University Press, Jerusalem, p. 185.
108. FERNANDEZ-GONZALES, M., WEISS, A. and LAGALY, G., *Keram. Z.*, 1976, **28**, 55.
109. WEISS, A., CHOY, J. H., MEYER, H. and BECKER, H. O., *Proc. Internat. Clay Conf., Bologna, Pavia 1981*, Abstracts, 1981, p. 331.
110. LECHNER, H., Thesis, University of Heidelberg, 1969; MAI, G., Thesis, University of Munich, 1969; ORTH, H., Thesis, University of Munich, 1970.
111. BENEKE, K. and LAGALY, G., *Inorg. Chem.*, 1983, **22**, 1503.
112. BRINDLEY, G. W. and RAY, S., *Am. Min.*, 1964, **49**, 106.
113. GRAF VON REICHENBACH, H. and RICH, C. I., *Clays. Clay Min.*, 1969, **17**, 23.
114. MORTLAND, M. M. and LAWTON, K., *Soil Sci. Soc. Am. Proc.*, 1961, **25**, 473.
115. ROSS, G. J. and RICH, C. I., *Clays Clay Min.*, 1973, **21**, 77.
116. GRAF VON REICHENBACH, H., *Proc. Internat. Clay Conf., Madrid 1972*, 1973, Division de Ciencias C.S.I.C., Madrid, p. 457.
117. KODAMA, H. and ROSS, G. J., *Proc. Internat. Clay Conf., Madrid 1972*, 1973, Division de Ciencias C.S.I.C., Madrid, p. 481.
118. SAWHNEY, B. L., *Clays Clay Min.*, 1972, **20**, 93.
119. GRUNER, L., LE DRED, R. and WEY, R., *C. R. Acad. Sci. Paris*, 1979, **288**, 661; LE DRED, R., GRUNER, L. and WEY, R., *C. R. Acad. Sci. Paris*, 1979, **288**, 1247; SAEHR, D., LE DRED, R. and WEY, R., *Proc. Internat. Clay Conf., Bologna, Pavia 1981*, 1982, Elsevier Scientific, Amsterdam, p. 133.
120. LAGALY, G., *Angew. Chem. Int. Ed. Engl.*, 1976, **15**, 575.
121. LAGALY, G., FITZ, S. and WEISS, A., *Angew. Chem. Int. Ed.*, 1973, **12**, 850.

122. BAUR, H., *Progr. Colloid Polym. Sci.*, 1975, **58,** 1.
123. PECHHOLD, W., LISKA, E., GROSSMANN, H. P. and HÄGELE, P. C., *Pure Appl. Chem.*, 1976, **46,** 127.
124. FERNANDEZ-GONZALES, M., WEISS, A., BENEKE, K. and LAGALY, G., *Z. Naturforsch.*, 1976, **31b,** 1205.
125. ERTEM, G. and LAGALY, G., *J. Colloid Interf. Sci.*, 1978, **66,** 12.
126. BLASENBREY, S. and PECHHOLD, W., *Rheolog. Acta*, 1967, **6,** 174; PECHHOLD, W., *Kolloid Z. Z. Polymere*, 1968, **228,** 1; PECHHOLD, W., *Colloid Polym. Sci.*, 1980, **258,** 269.
127. LAGALY, G., WEISS, A. and STUKE, E., *Biochem. Biophys. Acta*, 1977, **470,** 331.
128. WEISS, A., *Kolloid Z. Z. Polymere*, 1958, **158,** 22.
129. PHILEN, O. D., WEED, S. B. and WEBER, J. B., *Soil Sci. Soc. Am. Proc.*, 1970, **34,** 527.
130. NAIRN, D. R. and GUY, R. D., *Clays Clay Min.*, 1981, **29,** 205.
131. WEISS, A., *Angew. Chem. Int. Ed. Engl.*, 1981, **20,** 850.
132. ODOM, I. E., *Phil. Trans. Roy. Soc. London*, 1984, **A311,** 391.
133. JONES, T., *Clay Min.*, 1983, **18,** 394.

Chapter 3

STRUCTURE AND PHYSICAL PROPERTIES OF SOME CARBOXYLATED ELASTOMERS

M. Pineri

Commissariat à l'Energie Atomique, Institut de Recherches Fondamentales, Centre d'Etudes Nucléaires de Grenoble, France

1 INTRODUCTION

The properties and uses of carboxylated elastomers have been extensively reviewed by Jenkins and Duck.[1] The aim of this chapter is to concentrate on the structural studies which have been performed mainly on diene rubbers containing carboxylic acids either statistically dispersed along the chain or as terminal groups at both ends of low molar mass chains. Formation of carboxylic dimers produces inter-chain bonding, resulting in a moderate increase of tensile strength. Strong ionic bonds are obtained when the H^+ ion is exchanged by metal ions. Thermal lability of the ionic crosslinks has important implications, especially when considering the possibility of recovery. During the last few years the availability of sophisticated physical experiments like small-angle X-ray scattering using high fluxes from synchrotron machines, Mössbauer spectroscopy and extended X-ray absorption for fine structure, in addition to more conventional experiments using different ions as probes, has enabled us to gain a better knowledge of the structure of these materials.

The first part of this chapter is concerned with the neutralisation processes of the carboxylic elastomers and general considerations in the formation of ionic crosslinks. In the second part some physical results relating to the structure of some statistical carboxylated elastomers are considered. The last part of the chapter will be concerned only with carboxyl terminated elastomers and the main

141

results which have been obtained concerning both the structural determination and the physical properties will be reviewed.

2 NEUTRALISATION OF CARBOXYLIC RUBBERS

The metal carboxylate salts of carboxylic rubbers have been obtained by different synthetic routes. The earlier work described in the literature, in effect, corresponds to the addition of metal oxides.[2] Hydroxides of alkaline metals and acetates or acetylacetonates of divalent metals have been reacted in a Brabender plastograph at moderate temperatures.[3] Solution reactions have also been used by titrating the starting copolymer with solutions of methoxides.[3] This neutralisation technique has been extensively developed in the case of telechelic polymers (*telos* means end, and *chele* means claw, in Greek).

The simplest way to effect ionic crosslinking of carboxylic rubbers is by heating the acid copolymer in the presence of metal oxides. Divalent zinc and magnesium oxides seem to give materials with the best physical properties. Comparison of this efficacy has been made for different metal oxides and hydroxides.[1] Generally, it has been found that the best mechanical properties are obtained by addition of salts corresponding to twice the stoichiometric ratio.[2] Such a result implies that the crosslinking reaction does not involve a simple formation of $-COO^- \, -X^{2+} \, -^-OOC$ but rather suggests the presence of both partially neutralised $-COO^-M-OH$ complexes and residual oxide particles. There may be physical crosslinking on the surface of these particles. Up to now it has been difficult to make a quantitative analysis of the mechanical properties versus ionic crosslinking. Local structures of the ionic bonds which can be obtained by spectroscopic techniques such as electron spin resonance (ESR), Mössbauer or extended X-ray absorption for fine structure (EXAFS), have not yet been elucidated in these carboxylated elastomers.

A similar situation has been observed in neutralisation with acetates or acetylacetonates of divalent cations.[3] A Brabender plastograph heated at 150°C was used to mix the salts with the butadiene–methacrylic copolymers, allowing the carboxylic groups to ionise. The secondary reaction product (acetic acid or acetylacetone) was volatilised. Here again, an excess of salt is necessary to get a maximum resistance torque as measured in the plastograph. The increase of the viscosity is a direct indication of crosslinking even at 150°C. When analysing the structure of these materials in terms of possible ionic clustering, the question of this excess salt must be raised. Is this excess in the form of microparticles which act as filler, or is it mixed with the ionic clusters formed by the polymer carboxylates? Oxidative degradation has also been shown to occur during this high-temperature neutralisation step.

Solution reactions have been employed to improve the exchange reactions between the H^+ ions and the cations. Benzene–methanol (80:20) solutions of butadiene–methacrylic copolymer have been titrated with 0·1N sodium methoxide under a nitrogen atmosphere.[5] An equilibrium state was obtained by allowing the solutions to stand for several weeks at room temperature. Carboxylated telechelic polybutadiene and polyisoprene were dissolved in benzene and mixed at 55–60°C with methoxides in methanol solution.[6] A detailed analysis of this neutralisation process has been made by Jerome et al.[6] with these carboxylated telechelic polymers. Decahydronaphthalene has been used as the polymer solvent. Metal trimethylacetates, selected for their solubility, have been added in stoichiometric amounts. A continuous distillation extracts the trimethylacetic acid and shifts the neutralisation reaction until stoichiometry is obtained. The authors have followed the reaction by successive infrared and viscosity measurements. Removal of the secondary reaction product has been shown to be as essential for a stoichiometric reaction as the use of anhydrous alcohol.

3 GENERAL CONSIDERATIONS CONCERNING THE FORMATION OF THE IONIC CROSSLINKS

The thermally labile character of the ionic associations results in mechanical properties of the neutralised materials, corresponding to those of a crosslinked polymer with a fluidity at high temperature.

Condensation of charges by electric interaction gives the multiplets a structure first proposed by Eisenberg,[7] and results in strong crosslinking. The electrostatic energy of interaction between two dipoles represents an important part of the total energy of a basic pair (100 kcal for a monovalent carboxylate whose anion and cation are 3 Å apart). Such energies facilitate deformation of polymer chains, required to form these associations. In a recent article, Dreyfus[8] has shown that, in the case of monovalent carboxylates, these multiplets are very probably formed by the association of two carboxylate dipoles. Clustering of multiplets results from the existence of a residual energy of interaction. This energy has been calculated[8] and found to be $1.5\,kT$ for two multiplets, each one formed by two COO^--X^+ groups with an anion–cation distance of $3\,\text{Å}$ and a $8\,\text{Å}$ distance between each multiplet.

The basic ideas in the Dreyfus model for clustering can be summarised as follows.

(1) The residual electric energy between multiplets is much larger than kT when the multiplets are close.
(2) The polymer backbone fills the space between multiplets in a compact way involving hard core repulsion between monomers.
(3) The entropic energies needed to deform the chains from their natural configuration are negligible compared with the residual electric and hard core energies, (1) and (2) above.

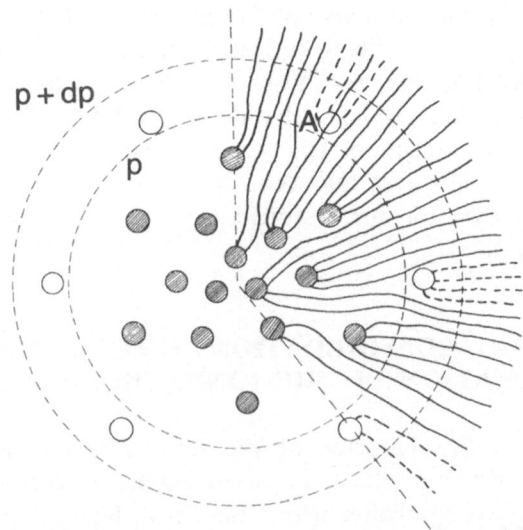

FIG. 1. Two-dimensional view of a cluster. Cross-hatched circles are multiplets inside a circle of radius p. Four chain segments are shown attached to each multiplet and represented by continuous lines. Open circles are the new multiplets in the shell between p and $p + dp$ from the centre. The 'new' segments are represented by broken lines. At point A a 'new' multiplet is located on a segment issuing from a multiplet further inside the cluster (from ref. 8).

From these considerations, Dreyfus proposed the model shown schematically in Fig. 1. The small circles correspond to the basic multiplets from which the chains emerge and tend to orientate outwards from the cluster. These multiplets are very probably formed by the association of two dipoles in the case of monovalent neutralised carboxylates. Their concentration decreases from the centre to the surface of the cluster. The radius of the cluster is obtained when the gain in energy arising from the attraction between a multiplet and the surface of a cluster equals the entropy loss resulting from the freezing of certain degrees of freedom, of the order of kT. The experimental scattering curves give both an ionomer peak and a zero-order scattering increase. Such results are explained by the model described above, by associating the scattering peak with a local order of the clusters, and the central peak with the presence of 'hypercrystallites' formed from the association of a finite number of clusters.

Such a model seems to explain, at least qualitatively, most of the available experimental results in ionomers. However, it is to be noted that, in opposition of what has been proposed by many authors, clusters already exist at very low ionic concentrations and there is no critical concentration for a multiplet–cluster transition.

4 STATISTICALLY CARBOXYLATED ELASTOMERS

4.1 General Comments on the Structure Determination

When considering the structure of 'multiphase' compounds and the relation between structure and properties, different levels of structure have to be considered. Macroscopic information, from which phase separation can be inferred, is obtained from experiments involving, for instance, dynamic mechanical and dielectric or thermodynamic measurements, but this is really indirect evidence which has to be confirmed by other results. More direct evidence of phase separation is obtained from small-angle scattering of X-rays or neutrons. Information about both the size and the geometry of the scattering particles can be gained from knowledge of different scattering domains. Such experiments are possible if the particles range from a few tens to a few hundred Ångströms in size and if the difference in electronic density permits direct observations by transmission electron microscopy; there are, however, severe limitations which will be discussed later. A third kind of information relates to the local structure of the complexes formed by the association of anions and cations. Such information on the local environments of cations can be obtained from experiments using the techniques of infrared, electron spin, nuclear magnetic resonance and Mössbauer spectroscopy. The dynamic properties of these ionomers have also to be considered, especially in these carboxylic rubbers in which temperature-dependent exchange reactions may define the mechanical properties.

4.2 Electron Microscopy Studies of Butadiene–Methacrylic Acid Copolymers

The first publication on the morphology of methacrylic acid copolymers appeared in 1971.[4] The authors used transmission electron microscopy to show the existence of two phases in these materials. Samples were prepared by putting a few drops of polymer solution onto a clean mercury surface at 23°C and allowing the solvent to evaporate. Methacryclic acid copolymers, containing different amounts by mass of methacrylic acid, were studied both in the acid and salt forms. Osmium tetroxide was used for staining the butadiene phase. High magnifications were used to get pictures from thin areas less than 50 Å thick. Evidence of unstained domains, corresponding to the ionic phase, was found in all samples including the acid forms. The size of these domains was not changed by neutralisation with sodium ions. A decrease in the size of these unstained regions was observed when going from the 6% methacryclic acid content (21 Å) to the 18% specimen (13 Å). These results have however to be examined in connection with more general work on transmission electron microscopy on ionomers.[9] In this paper, the authors have examined the different results obtained from solvent casting and ultramicrotomy of several ionomers such as ethylene–methacrylic acid copoloymers, sulfonated polypentenamers and polystyrene polymers. Several artifacts have been attributed to solvent casting. When the acid ionomer is cast into dilute solutions of base, 2–20 nm microcrystals of the base solution may be shown by bright and dark field imaging. A 'salt and pepper' structure similar to the picture shown in ref. 5 may appear in weak phase objects like amorphous carbon, depending on defocusing. Phase contrast has to be taken into account, in order to interpret the images obtained from thin films in which amplitude contrast is not important. Other factors have also to be considered when interpreting the results. Is the structure of the cast film representative of the structure of the bulk polymer? Since the image is a two-dimensional representation of the specimen volume, the presence of ionic clusters can only be demonstrated if their concentration is small or if their size is large compared with the sample thickness. Radiation damage has also to be taken in consideration. Further experiments seem necessary to confirm these transmission electron microscopy studies on butadiene–methacrylic copolymers.

4.3 Evidence and Characterisation of Bond Interchange in Ionic Crosslinks

Young's or torsion modulus measurements are very sensitive to crosslinking when temperatures are higher than the T_g. Crosslinks prevent the viscous flow that is observed at temperatures higher than T_g, and a plateau is obtained for highly crosslinked materials. With thermally reversible crosslinking, a second drop in the modulus is observed at higher temperatures, when the exchange reactions occur at such speeds that they do not contribute towards additional crosslinking. The temperature at which this second drop occurs will depend on the ionic associations and on the time scale of the measurements. A few experiments have been carried out on carboxylated polybutadiene. Young's modulus results obtained with an Instron apparatus have been reported by Otocka and Eirich[10,11] showing the influence of dipolar interactions on the modulus change (Fig. 2). With different percentages of lithium carboxylate groups, a plateau is observed over a large temperature range. The value of the modulus at the plateau increases with the molar concentration of salt groups. After the plateau, a drop in modulus occurs at the same temperature for all the samples. Similar behaviour is observed with carboxylated butadiene–methylpyridinium iodide copolymers, and also in mixtures of these two anionic and cationic copolymers. Quadrupoles or even larger dipolar associations give thermally reversible crosslinks. A crosslink density can be calculated from the value of the modulus at the plateau by using a simplified relationship obtained from the rubber elasticity theory: $E = 3RT (v/V)$. However, this equation is not valid for high degrees of crosslinking. The change of the crosslinking density with temperature may however give an idea of the energy of association of the dipoles involved in the crosslink. By using such a plot in the temperature range corresponding to the plateau where a small change in modulus is observed, the authors have obtained a network enthalpy.

The enthalpy found is around 3 kcal/mol for the lithium salts.

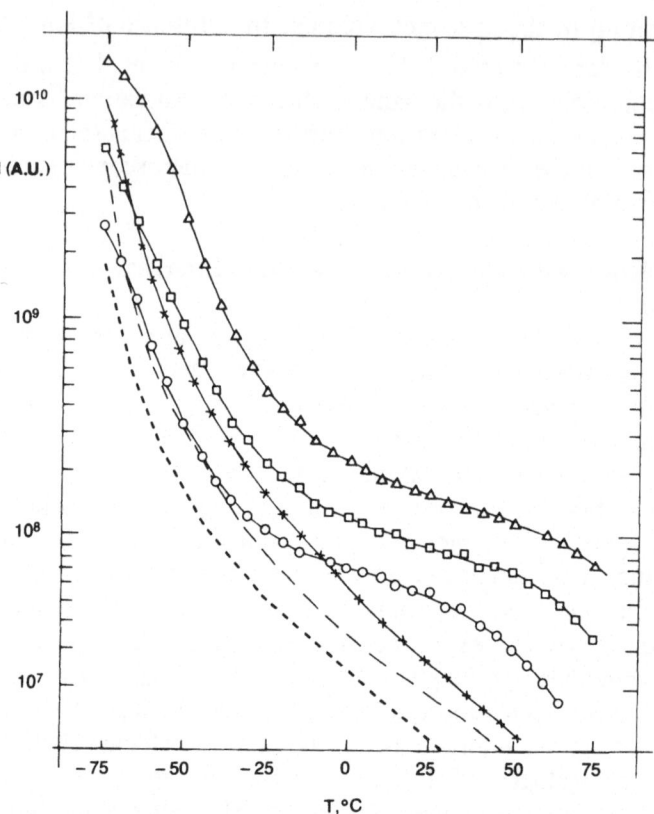

FIG. 2. Modulus temperature curves for butadiene–methacrylic acid copolymers and the corresponding lithium salts. The acid mole fractions are respectively 4·6 (– – –); 7·6 (——) and 11·5 (——) mol %. Titration was made with LiOCH₃ and the conversions were limited to 90%. The salt mole fractions of the corresponding samples are respectively 4·65 (○); 7·66 (□) and 11·52 (△).

Larger values would be obtained if considering the changes in modulus at higher temperatures where a large drop is observed, and may be associated with large changes in the 'effective' crosslink numbers due to rapid exchange between the dipolar associations. In a subsequent article by the same authors,[11] these bond interchanges have been shown to account for the relaxation behaviour of these copolymers. In this article, time–temperature superposition was found to be valid at temperatures higher than T_g in these polybutadienes containing different percentages of lithium carboxylate groups. The activation energy obtained from the dependence of the shift factor on temperature was found to decrease up to a temperature corresponding to $T_g + 50°C$, above which a constant value of 18 kcal/mol was obtained. A pertinent explanation was given to explain such behaviour; this $T_g + 50°C$ temperature should correspond to the point at which enough thermal energy is present to induce bond interchange migration. At lower temperatures, the ion-pair groups act as semi-permanent cross-linking points, but the diffusive relaxation is governed by the normal Williams–Landel–Ferry (WLF) dependence, with an apparent activation energy changing with the temperature. At high temperatures, the diffusion of polymeric chains is restricted only by the presence of the quadrupolar links. Relaxation therefore depends only on bond interchanges and the activation energy must reflect the association of the dipoles and therefore be constant above this transition. The glass transition temperature increases with the total number of carboxylate dipoles, as is normally found when increasing the density of crosslinks in polymers. Such behaviour in stress relaxation experiments has been observed in other ionomer rubbers containing butadiene, styrene and vinyl pyridine monomers.[12] Ionic crosslinks were obtained by complexation of the vinyl pyridine monomers with transition metal salts like nickel chloride. Analysis of stress relaxation versus temperature showed a double dependence of the shift factor: WLF dependence at low temperatures and Arrhenius at higher temperatures. A similar explanation was then proposed.

4.4 Thermally Stimulated Depolarisation (TSDC) and Polarisation (TSPC) Current Techniques

In recent publications,[13,14] results obtained with these techniques have been cited as evidence for clustering in N-butyl methacrylate ionomers. These dielectric experiments, indeed, permit us to study the dipolar and ionic relaxation properties in the solid state without any interference from the space-charge polarisation processes. Samples with ionic contents up to 12% methacrylic acid, neutralised to various extents to form potassium and lithium salts, have been studied. The most important conclusion is based on the absence of active species in the neutralised copolymers, which evidence implies that most of the carboxylate groups are incorporated into dielectrically inactive multiplets, because in the acid copolymers most of the carboxyl groups are dimerised through hydrogen bonding. Both kinds of association produce a marked increase of the glass transition temperature. Clustering in high ionic concentration samples is suggested because of the presence of a new relaxation peak, associated with either the glass transition of the chains into the clusters or with a space-charge polarisation on the boundaries of these clusters.

5 CARBOXYL TERMINATED ELASTOMERS

5.1 Origin and General Properties

Preparation of these telechelic polymers has been extensively described in the literature.[15]

These liquid carboxylic materials are available mainly as polybutadienes from different companies: Thiokol Chemical Corp.,[16,17] B. F. Goodrich Chemical Company,[18] General Tyre, Phillips[19] and Nippon Soda Company. Molar masses from 1000 to a few thousands have been obtained with functional groups. A wide range of applications has been developed for these materials. They are employed as binders in composite solid propellants. Because of the narrow weight distribution and a functionality close to two, the resulting network after curing is better defined than when randomly located reactive groups are linked together. Very short chains between crosslinks are avoided, thus increasing the elongation at break. Because of their liquid nature these carboxyl teminated polybutadienes can be readily processed into compounds for impregnation, casting, spraying and other different applications. A broad range of physical properties can be achieved upon curing; various agents can be used to obtain a cure including polyepoxides, polyimides and metal oxides or salts. In this chapter, we are mainly interested in curing by reaction of the acid end groups with metal compounds to form the corresponding carboxylate. These ionic interactions will expedite the formation of ionic crosslinks having thermally reversible behaviour. Such polymers constitute an approximate model for the understanding of the structure of ionomers, and explain why so much work has been done concerning both the local and long-range structures.

5.2 Evidence of Ionic Structures from Small-angle Scattering Experiments

Small-angle X-ray scattering (SAXS) experiments have been performed by different groups of workers on similar carboxylated polybutadienes with the object of establishing the extent of phase separation between the dipoles and the low dielectric constant matrix. The spectra commonly observed in ionomers contain a broad peak, also observed are significant increases in very small scattering angles in all neutralised samples. The position and amplitude of the peaks depend strongly on the cation, the molar mass of the telechelic polymer, the extent of neutralisation and the degree of swelling. These results are generally interpreted as evidence of the existence of scattering particles with a high electronic density compared with the matrix, because of the association of the ions to form multiplets or clusters. Analysis of several of these results has been undertaken to obtain the size of the scattering particles.

An initial series of experiments[20] was carried out, using a standard SAXS apparatus, on a Hycar CTB ($\bar{M}_w$ 4800) from B. F. Goodrich. Neutralisation was effected, using a Brabender plastograph as a mixer and the degree of neutralisation was varied. As shown in Fig. 3, as the degree of neutralisation is increased, the intensity of the peak increases, while its position does not change. This peak is associated with interference between the scattering particles with a mean separation of 70 Å at all concentrations. The size of these ionic domains has been analysed using a plot of $\log I(\varepsilon, x) - I(\varepsilon, 0)$ versus ε^2 ($I(\varepsilon, x)$ is the scattered intensity for the sample corresponding to $x\%$ of neutralisation and ε is the scattering angle). An average radius of 5·7 Å was found for degrees of neutralisation greater than 20%. A linear increase of the total scattering volume versus the degree of neutralisation was obtained. Whilst some criticism can be made of such an analysis made in an angular range corresponding to the high-angle side of the peak, the particle size so obtained is found to be in agreement with further analyses performed on similar samples. Because of the size of the scattering particles, it appears to be more accurate to use the term multiplets—corresponding in the Eisenberg terminology to the association of a few dipoles—rather than clusters. The interference distance of 70 Å, which does not change with the neutralisation ratio, would therefore correspond to distances between these multiplets. When varying the neutralisation ratio, the 70 Å distance remains constant but the relative volume changes. No drastic change in these parameters was found during annealing up to 160°C.

A more complete analysis was later performed on the samples neutralised with Na^+, K^+ and Cs^+.[21] Similar changes in the values of these parameters with the degree of neutralisation were found.

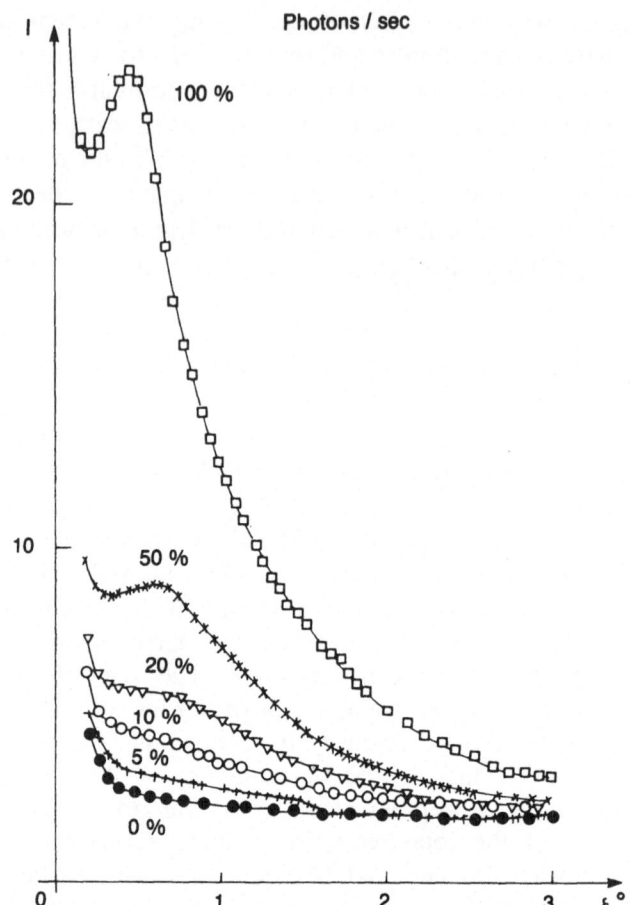

FIG. 3. Scattering intensity for a carboxylic telechelic polybutadiene neutralised to various extents expressed as %. I is the scattering intensity in photons/s and $\varepsilon°$ is the scattering angle.

Additional results have been obtained by swelling the butadiene segments with toluene. No change is observed for the radius of gyration of the scattering particles upon swelling. The interference peak is shifted to lower angles and the associated distance corresponds to the end-to-end distance of the chains. An analysis of the scattering curve at larger angles shows a $1/s^4$ relationship, corresponding to a well-defined border between the scattering particles and the matrix ($s = 2\pi \sin \theta/\lambda$, where θ is half the angle between the incident and scattered beams and λ is the wavelength). A more sophisticated analysis of the data is consistent with scattering particles formed by the association of two layers of COO^- groups as shown in Fig. 4. The average distance between these multiplets corresponds to an equilibrium between the tendency of the dipoles to form a separate phase and the elasticity of the polybutadiene chain.

A similar lamellar model has also been proposed, based on experiments on the same Hycar CTB carboxyl telechelic polymer and on a carboxyl telechelic polybutadiene with different molar mass, prepared by anionic polymerisation.[22] The X-ray diagrams obtained with these barium-, magnesium- and calcium-neutralised samples show two diffraction orders with Bragg's spacings in the ratio 1:2, characteristic of a lamellar structure. The thickness of the polybutadiene lamellae is obtained from the peak position and is shown not to depend on the cation but on the molar mass of the polymer (61 Å for a 4800 molar mass; up to 93 Å for a 14 000 molar mass).

In a very recent publication[23] a more systematic study of the small-angle X-ray scattering profiles has been described. Carboxyl telechelic polybutadiene of $\bar{M}_n$ 4600 (Hycar CTB 2000 × 156 from B. F. Goodrich) and carboxyl telechelic polyisoprenes of $\bar{M}_n$ 6000, 10 000, 24 000, 30 000 and 37 000 have been studied. The parameters taken

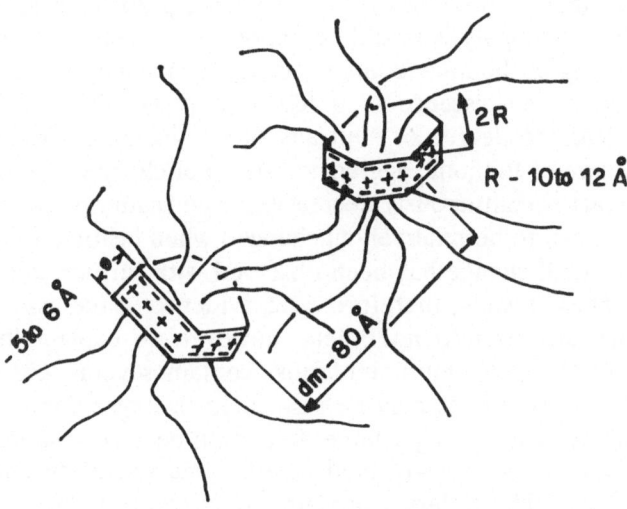

FIG. 4. Anisotropic association of ionic dipoles proposed by Moudden et al.[21]

into account in this work were the cation, molar mass, swelling of both the elastomeric and ionic phases, and temperature. High X-ray fluxes were obtained both from the Stanford synchotron Radiation Laboratory and from the French LURE-DCI Orsay facility. The radii of the scattering particles were deduced from the Porod and Debye Laws assuming a spherical shape. The distances between the scattering particles, obtained from Bragg's Law applied to the maximum position, decrease when plotted versus qp/m (q being the charge, p the density and m the atomic mass of the metallic cation) and tend to become constant for high qp/m values. There is no parallel systematic change in the size of the multiplets, therefore excluding the simple model of a uniform distribution of ionic multiplets dispersed in the polybutadiene matrix. When increasing the molar mass, the distance associated with the peak position tends to increase but no accurate correlation is possible because of the poor definition of the peak for high molar masses. Swelling by toluene—a good solvent of the ionic domains—has been used to obtain more information on the structure of these composite materials. The size of the multiplets has been found to remain the same when increasing the toluene volume fraction, while there is a continuous change of the distance between them. An increase of the multiplet size was observed after aqueous swelling of a polybutadiene, neutralised to different extents with Mg^{2+}. No systematic correlation between the amount of absorbed water and the observed change in the size of these diffusing particles has, however, been made. Such analysis would be important to define whether there is only swelling of the ionic domains, or reorganisation of the structure with a possible coalescence of a few multiplets. The authors have shown that as the degree of neutralisation is increased from 20% to 100%, the size of the ionic aggregates does not change, although their distance apart is continuously decreasing. The scattering profiles have also been shown to be relatively unchanged when heating up to 100°C, and only a small change has been observed with further annealing up to 200°C. These results, therefore, give evidence of the association of the primary dipoles into multiplets: however, to explain the radius values of 6–11 Å, the multiplets must contain several tens of these dipoles. These multiplets may be associated in larger domains with a paracrystalline order giving a large diffraction peak corresponding to a Bragg distance of 60 Å, associated with the end-to-end distance of the starting chain. These large domains may explain the zero-order scattering increase, observed in these ionomers as suggested by

Dreyfus in his model of clustering. This publication[8] provides good initial results, but more quantitative experiments have yet to be done.

5.3 Local Structures of Some Ionic Complexes

The local structures formed by different cations and the carboxylic anions have been explored using techniques such as electron paramagnetic resonance (EPR), Mössbauer spectroscopy and EXAFS. In an early publication[20] evidence for complexes corresponding to the association of two Cu^{2+} and four COO^- groups was obtained from EPR experiments (Fig. 5). Such structures were first established for monohydrated copper acetates and they form crosslinking points between four chains. Other structures involving —$COO^- Cu^{2+}$ ^-OOC— complexes have also been found, these later structures corresponding to a chain extension. A more complete study involving EPR, magnetic susceptibility and EXAFS measurements has recently been made,[24] providing a detailed description of these complexes. The exchange interaction in the copper dimers is antiferromagnetic. The ground state is a singlet state ($S = 0$) and the observed magnetic triplet state ($S = 1$) is an excited state only populated by thermally excited molecules. A dimerisation ratio, corresponding to the mass of the copper ions involved in copper pairs to the total copper mass, has been obtained from the temperature dependence of the EPR line and also from the magnetic susceptibility measurements. While in the low-temperature range, the dimer contribution to the total magnetic susceptibility is negligible; at higher temperatures a deviation from the Curie Law ($1/T$ proportionality of the magnetic susceptibility) is observed. A fitting procedure was performed, using a more realistic Brillouin Law to obtain dimerisation ratios which are 0·7 for the sample 75% neutralised compared with the stoichiometry, and 0·78 for the 50% neutralised sample. The decrease in the dimerisation ratio

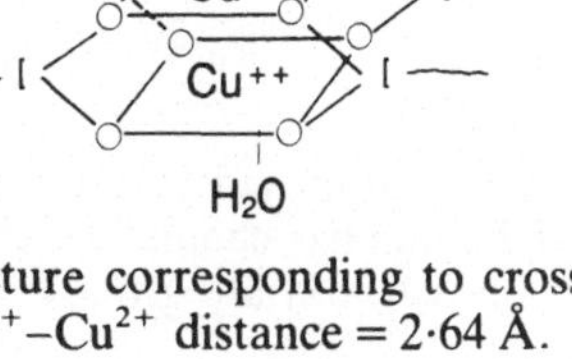

Fig. 5. Copper dimer structure corresponding to crosslinking of four chains. Cu^{2+}–Cu^{2+} distance = 2·64 Å.

when the neutralisation is increased may be due to the increasing difficulty of associating two Cu^{2+} and four COO^- end groups. No third species—besides the isolated and dimer complexes—has been demonstrated.

A similar analysis has been performed on samples neutralised to different extents with Fe^{3+}.[25] Analysis of the number of iron neighbours in the second shell and the iron–iron distances obtained from EXAFS measurements indicate a triangular trimeric structure of the iron complexes (Fig. 6). The magnetic data are also consistent with the presence of oxo-bridged trimeric species with a similar structure to the one proposed in ferric glycine.

The results so far obtained with Cu^{2+} and Fe^{3+} ions emphasise the importance of both the anion and cation with reference to the kind of

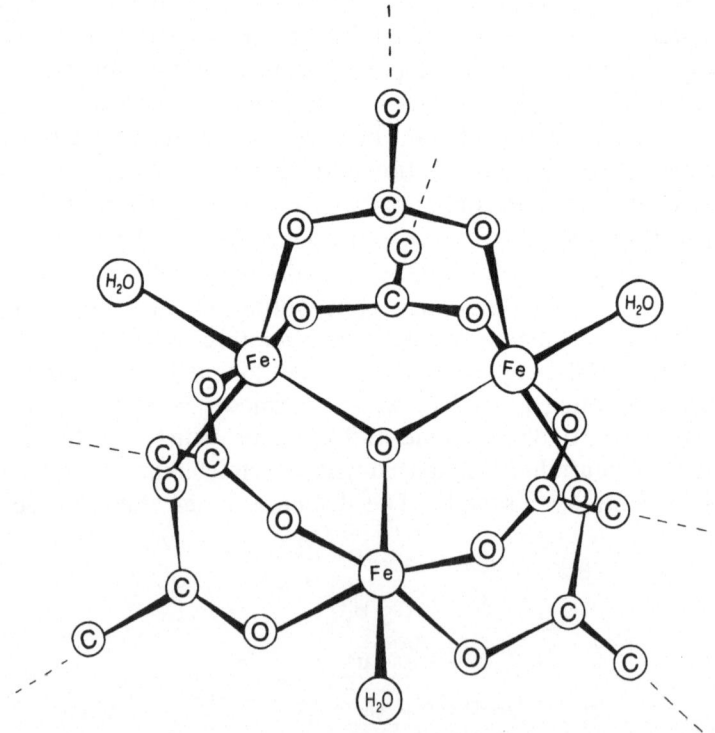

FIG. 6. Trinuclear iron centre proposed for Fe^{3+} telechelic polymers. Dashed lines represent the polybutadiene chains which are attached to the six carboxylate groups.

ionic associations obtained upon neutralisation of these telechelic polymers. Also of importance is the water content which seems to be present even in samples vacuum dried at room temperature.

EXAFS experiments have been extensively used to obtain information on the local structure of complexes in sulfonated polystyrene ionomers and in perfluorinated membranes.[26,27] Zinc carboxylate salts of carboxyl telechelic compounds have also been considered;[28] the main conclusion to be drawn from EXAFS spectra is the existence of very ordered structures over distances large enough for a third and fourth shell. However, such a result has to be analysed in respect of the possible arrangements of the carboxylate salts which are attached to the polybutadiene chain. Therefore it is difficult to interpret such ordered structures as corresponding to large isotropic ionic structures. An alternative interpretation would infer either the presence of the carboxylate complexes in lamellar or cylindrical structures, or the presence of zinc oxide. The Zn–O and Zn–Zn distances obtained from EXAFS are indeed the same as the corresponding distances in ZnO. A large change has only been observed in the number of neighbours in the first and second shell between the zinc oxide used as a reference compound and the telechelic salts. In spite of a number of points which have to be clarified this article demonstrates the potential of such measurements to describe the ionic associations formed in ionomers.

5.4 Glass Transition

The glass transition temperature in these halato telechelic polymers does not depend upon the extent of neutralisation. Such a result has been discussed in connection with the strong temperature depression observed in styrene or α-methylstyrene telechelic polymers upon neutralisation.[29] In the temperature range corresponding to the glass transition of these diene telechelic polymers (190 K) the ionic complexes interact to form permanent crosslinking points. The authors proposed that this insensitivity of T_g to both the extent of neutralisation and the type of cation may result from a cancelling-out of two opposing effects, i.e. restriction of chain end mobility arising from the crosslinking effect and an increase of free volume. A recent publication[30] has analysed the changes of the Mössbauer recoilless fraction (f) versus temperature in an α,ω-dicarboxylated polybutadiene, neutralised with ferric ions. The glass transition temperature obtained from quasi-static measurements has been associated with the temperature at which deviation occurs from Debye behaviour. The temperature at which f becomes unmeasurably small is identified as the glass transition temperature on a time scale of 10^{-7} s.

5.5 Viscoelastic Properties

The viscoelastic properties of these diene halato carboxyl polymers have been extensively studied by Jerome et al. In a Hycar (4600 $\bar{M}_n$) neutralised by magnesium methoxide, a secondary relaxation mechanism has been found and associated with the ionic crosslinks.[31] Time–temperature superposition has been verified in the temperature range over which this secondary relaxation is active and an Arrhenius dependence has been found when plotting the shift factor versus temperature. An activation energy of 30·5 kcal/mol has been calculated. The two relaxation processes, corresponding respectively to the glass transition of the polybutadiene chains and to the ionic associations, occur over well separated temperature ranges, thus allowing their characterisation. The secondary relaxation associated with the ionic phase may correspond to exchange reactions between the carboxylate end groups involved in the ionic clusters. The influence of the molar mass on these rheological properties has been examined in a further publication.[32] Carboxyl halato polyisoprene with molar masses ranging from 7000 to 70 000 have been neutralised by magnesium methoxide. Time–temperature superposition has been attempted for shear storage and loss moduli obtained at different temperatures (between 296 and 349 K) over a large frequency range. For molar masses larger than 2×10^4 and carboxylate group concentrations smaller than 0·7 mol %, time–temperature superpositions occur with a Williams–Landel–Ferry dependence for the shift factors. The influence of the ionic associations is therefore negligible, either because their life time is too small compared with the characteristic time of the dynamic measurements or because they constitute isolated multiplets, which only increase the molar mass of the polyisoprene chain. A thermal rheological complexity appears for carboxylate concentrations larger than 0·7 mol %. Time–temperature superposition is not possible, probably because of the existence of two relaxation mechanisms, both active in the observed temperature range. The relaxation of the ionic clusters overtakes that of the polymer backbone at high temperatures when the ionic concentration is large enough to form these associations. An Arrhenius dependence was also observed in the high-temperature region for the polymer ($\bar{M}_n = 7000$). An activation energy of 33·4 kcal/mol was obtained, a value which was in close agreement with the above-mentioned value of 30·5 kcal/mol for the polybutadiene with an $\bar{M}_n$ value of 4600.

The influence of the ionic radius of the cation has also been studied on barium, calcium, magnesium and zinc carboxylated telechelic polybutadienes of $\bar{M}_n$ 4600.[33] For all neutralised samples a secondary relaxation process has been demonstrated and associated with the ionic interactions. An Arrhenius dependence was found for the shift factor when plotted versus temperature. The activation energies depend on the cation as shown in Fig. 7, in which they are plotted versus the ionic radius; they increase with increasing electrostatic interactions. Addition of polar solvent has been found to decrease these activation energies as expected, since the ions are further apart. Such behaviour is therefore consistent with a picture involving the dissociation of the carboxylate salts and reformation of complexes after the carboxylates have moved from one ionic cluster or multiplet to another.

Polar additives such as water or alcohols can be absorbed on salts and therefore weaken the ionic interactions, resulting in a lower

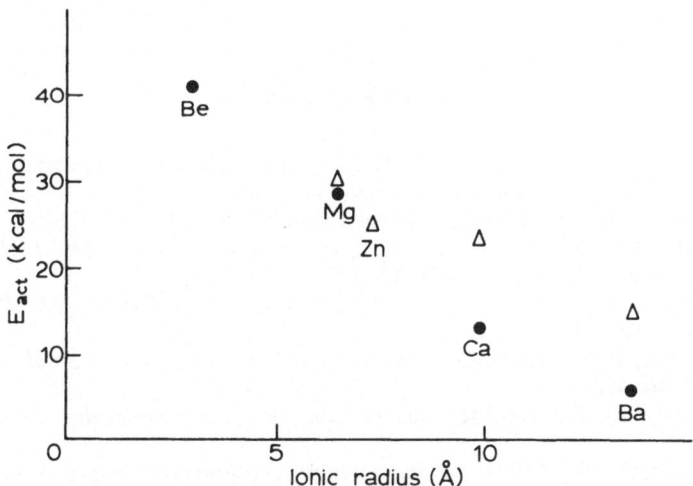

FIG. 7. Dependence of the activation energy of the ionic radius for a series of $\alpha-\omega$ alkaline earth dicarboxylato polybutadienes ($\bar{M}_n = 4600$) in bulk ($\triangle$) and in 10% xylene solution ($\bullet$). (From G. Broze et al., J. Polym. Sci. Polym. Phys., 1983, 21, 2205.)

modulus. This constitutes a considerable impediment to the possible industrial applications of these ionic crosslinks.

6 CONCLUSIONS

During recent years, a better knowledge of the structure of complexes formed by ionic crosslinks has been gained. More work is still necessary to correlate these results with those obtained using small-angle X-ray or neutron scattering experiments. Because of the importance of both anion and cation in the formation of the ionic associations, systematic studies have to be made. Theoretical approaches described in this chapter may permit a more comprehensive understanding of the relationship between the physical properties—including the thermal lability—and the structural properties. Water content of carboxylic elastomers is an important parameter which has still to be taken into account in the formation of the ionic complexes. Also, water sorption is of importance in industrial applications. It may be possible in the near future to prepare carboxylic elastomers, depending on the required industrial applications, by defining the mechanical properties and the temperature range over which lability is needed.

REFERENCES

1. JENKINS, D. K. and DUCK, E. W., in *Ionic Polymers* (L. Holliday, Ed.), Applied Science Publishers, London, 1975, Ch. 3.
2. DOLYOPLOSK, B. A. *et al.*, *Rubber Chem. Tech.*, 1959, **32**, 328.
3. PINERI, M., MEYER, C., LEVELUT, A. M. and LAMBERT, M., *J. Polym. Sci., Polym. Phys. Ed.*, 1974, **12**, 115.
4. MARX, C. L., KOURSKY, J. A. and COOPER, S. L., *J. Polym. Sci., Polym. Letters*, 1971, **9**, 167.
5. OTOCKA, E. P., HELLMAN, M. Y. and BLYLER, L. L., *J. Appl. Phys.*, 1969, **40**, 4221.
6. BROZE, G., JEROME, R. and TEYSSIE, P., *Macromolecules*, 1982, **15**, 920-7.
7. EISENBERG, A., 'Ion containing polymers', in *Polymer Physics*, Vol. 2 (R. S. Stein, Ed.), 1977, Academic Press, New York.
8. DREYFUS, B., *Macromolecules*, 1985, **18**, 280.
9. HANDLIN, D. L., MACKNIGHT, W. J. and THOMAS, E. L., *Macromolecules*, 1981, **14**, 795.
10. OTOCKA, E. P. and EIRICH, F. R., *J. Polym. Sci*, 1968, **6**, 921.

11. Otocka, E. P. and Eirich, F. R., *J. Polym. Sci. A2*, 1968, **6**, 933.
12. Meyer, C. and Pineri, M., *Polymer*, 1976, **17**, 382.
13. Aras, L., Vanderschueren, J., Niezette, J. and Vanderschueren, R., *J. Polym. Sci. B*, 1983, **21**, 799.
14. Vanderschueren, J., Aras, L., Boonen, C., Niezette, J. and Corapci, M., *J. Polym. Sci., Polym. Phys. Ed.*, 1984, **22**, 2261.
15. French, D. M., *Rubber Reviews*, 1969, **42**, 71.
16. Berenbaum, M. B. and Gobran, R. H., West German Patent 1 150 205, 12 Jun. 1963.
17. Thiokol Chemical Corp., British Patent 957 652, 6 May 1964.
18. B. F. Goodrich Co., Netherlands Appl. 6 504 684, 18 Oct. 1965.
19. Uraneck, C. A., Short, J. N. and Zelinski, R. P., US 3 135 716, 2 Jun. 1964.
20. Pineri, M., Meyer, C., Levelut, A. M. and Lambert, M., *J. Polym. Sci., Polymer Phys. Ed.*, 1974, **12**, 115, 130.
21. Moudden, A., Levelut, A. M. and Pineri, M., *J. Polym. Sci., Polymer Phys. Ed.*, 1977, **15**, 1707.
22. Broze, G., Jerome, R. and Teyssie, P., *J. Polym. Sci., Polymer Letters*, 1981, **19**, 415.
23. Williams, C. E., Russell, T. P., Jerome, R. and Horrion, J., Proc. American Chemical Society. Philadelphia Meeting, August 1984, Macromolecular Secretariat Symposium on Coulombic Interactions in Macromolecular Systems.
24. Galland, D., Belakhovsky, M., Merdrignac, F., Pineri, M. and Jerome, R., to be published in *Polymer*.
25. Meagher, A., Coey, J. M. D., Belakhovsky, M., Pineri, M., Jerome, R., Vlaic, G., Williams, C. and Van Dang, N., to be published in *Polymer*.
26. Yarusso, D. J., Cooper, S. L., Knapp, G. S. and Georgopoulos, P., *J. Polym. Sci.*, 1980, **18**, 557.
27. Pan, H. K., Meagher, A., Knapp, G. S. and Cooper, S. L., *J. Chem. Phys.*, 1985, **3**, 82.
28. Jerome, R., Vlaic, G. and Williams, C. E., *J. Phys. Lett.*, 1983, **44**, 717.
29. Jerome, R., Horrion, J., Fayt, R. and Teyssie, P., *Macromolecules*, 1984, **17**, 2447.
30. Meagher, A., Smyth, G., McBrierty, V. J., Coey, J. M. D. and Pineri, M., to be published in *J. Chem. Phys.*
31. Broze, G., Jerome, R., Teyssie, P. and Marco, C., *Polymer Bulletin*, 1981, **4**, 241.
32. Broze, G., Jerome, R., Teyssie, P. and Marco, C., *Macromolecules*, 1983, **16**, 1771.
33. Broze, G., Jerome, R., Teyssie, P. and Marco, C., *J. Polym. Sci., Polym. Phys. Ed.*, 1983, **21**, 2205.

Chapter 4

IONENE POLYMERS: PREPARATION, PROPERTIES AND APPLICATIONS

TETSUO TSUTSUI

Graduate School of Engineering Sciences, Kyushu University, Fukuoka, Japan

1 INTRODUCTION

Polyquaternary ammonium compounds with the ammonium ion integral in the backbone of the polymer chain have been called 'Ionenes' or 'Ionene polymers', and represent a unique class of cationic polymers among various ion-containing polymers;

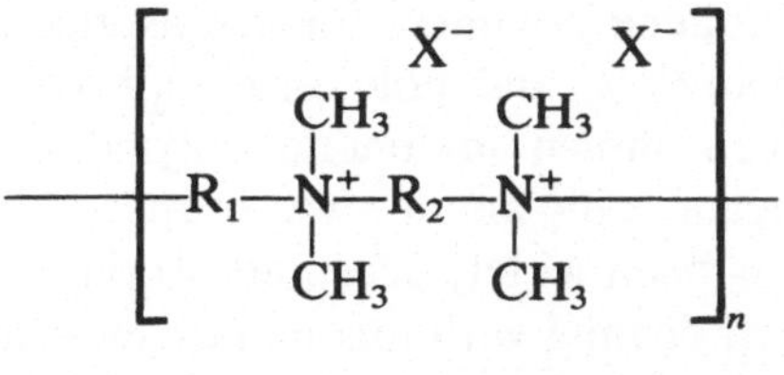

$$\left[-R_1-\overset{\overset{\displaystyle CH_3}{|}}{\underset{\underset{\displaystyle CH_3}{|}}{N^+}}-R_2-\overset{\overset{\displaystyle CH_3}{|}}{\underset{\underset{\displaystyle CH_3}{|}}{N^+}}-\right]_n \quad X = Br, Cl$$

They have been widely used as materials for fundamental research on ionic polymers as well as for industrial applications. The unique characteristics of ionene polymers may be summarized as follows.

(a) Ionic sites are located on the polymer skeletons, and the density of ionic sites is high compared with other ionic polymers.
(b) Ionic spacings on skeletal chains are regular and can be varied at will. In other words, charge densities on skeletal chains are controllable.
(c) The synthetic procedure for ionene polymers is relatively simple, and a variety of ionene polymers can be prepared easily.

163

(d) Functional groups other than the quaternary ammonium cation can be easily incorporated on the skeleton of ionene polymers.

(e) Usually, the counter anion, X^-, is a halogen, but in some cases it can be replaced by several other anionic species.

Simple chemical architecture and the ease of the preparation of ionene polymers with specified chemical structures have attracted many scientists, whose concern was the synthesis, solution properties, solid state properties and applications of ion-containing polymers.

In the early 1930s, Marvel and his coworkers[1-5] first reported the synthesis of aliphatic ionenes. In 1941, Kern and Brenneisen[6] reported the synthesis of aliphatic ionenes from tertiary diamines and dihalides, which, nowadays, serves as one of the most popular and important methods for the preparation of ionene polymers. Although several patents which claimed the use of ionene polymers as bactericides, dispersing agents and dye fixants had appeared,[7] polymer scientists had paid little attention to ionene polymers before the appearance of the work of Rembaum and his coworkers.

In 1968, Rembaum et al.[8] presented an epoch-making paper on the synthesis of a series of ionene polymers, including the elucidation of the polymerization mechanism and the molar mass determination of the resulting polymers. In the same paper, they pointed out the potential importance of ionene polymers for use as electroconducting materials, biomedical materials, and polyion complexes. Since then, growing interest has been shown in ionene polymers. The generic name 'ionene', which means ionic amine, was proposed by Rembaum and his associates and has been widely accepted. Today, more than 50 articles, including patents, dealing with ionene polymers are published every year.

In some articles, the cationic polymers which possess quaternary ammonium cations on their *side chains* are also classified as ionene polymers. However, the use of the name ionene should be restricted to the cationic polymers that possess quaternary ammonium cations on the *skeleton* of the polymer chains, because the cation integral in the backbone chain is one of the most important characteristics of ionene polymers.

2 PREPARATION OF IONENE POLYMERS

A variety of ionene polymers, such as aliphatic ionenes, aromatic ionenes, ionenes with cyclic structures or hetero atoms, ionenes with special functional groups, copolymers with ionene repeat units and other polymeric units, have been reported. First, synthesis of two classes of typical ionenes, aliphatic and aromatic, via Menschutkin polyaddition reactions will be described in detail. Then, various ionenes with heteroatoms will be reviewed. Miscellaneous preparation methods will also be mentioned. Finally, molecular characterization of those ionenes will be discussed.

2.1   Aliphatic Ionenes and Aromatic Ionenes

Aliphatic ionenes are prepared by successive Menschutkin reactions[9] between N,N,N',N'-tetramethyl α,ω-diaminoalkanes and α,ω-dihalides as shown in eqn (1).[6,8,10]

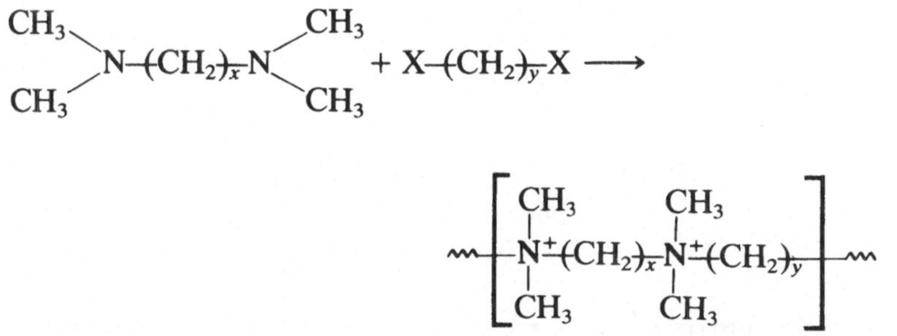

$$
\begin{matrix} CH_3 \\ \quad \\ CH_3 \end{matrix}\!\!N\!\!-\!\!(CH_2)_x\!\!-\!\!N\!\!\begin{matrix} CH_3 \\ \quad \\ CH_3 \end{matrix} + X\!\!-\!\!(CH_2)_y\!\!-\!\!X \longrightarrow
$$

$$
\sim\sim\!\!\left[\begin{matrix} CH_3 \\ | \\ N^+\!\!-\!\!(CH_2)_x\!\!-\!\!N^+\!\!-\!\!(CH_2)_y \\ | \\ CH_3 \end{matrix}\begin{matrix} CH_3 \\ | \\ \\ | \\ CH_3 \end{matrix}\right]\!\!\sim\sim \quad (1)
$$

I

One can obtain aliphatic ionenes with the desired cationic charge densities on the skeletons, if diamines and dihalides with suitable x- and y-values are selected. It should be noted that linear polymers cannot be obtained when numbers of methylene units in the diamine and the dihalides are too small. Thus, the minimum values for combinations of x and y that give linear polymers are $x = 2$, $y = 5$; $x = 2$, $y = 4$; $x = 3$, $y = 3$; and $x = 4$, $y = 3$.[10] Hereafter, the aliphatic ionenes will be frequently referred to with the notation of x,y-ionenes in which x and y denote the numbers of methylene sequences between cations.

Rembaum and his coworkers studied the detailed kinetics of polymerization in several polar solvent systems and reached several important conclusions, which are applicable, at least qualitatively, for almost all cases of the Menschutkin preparations of ionenes.[8]

(a) The reaction order changes from one in dimethylformamide (DMF) to two in DMF–methanol mixtures.

(b) The rates are, to a considerable extent, solvent-dependent; the higher the dielectric constant of the medium, the higher is the rate of reaction.

(c) The rates do not vary to any considerable extent as a function of the number of methylene groups in the amine or the halide.

(d) The rate of formation of ionene chloride is considerably lower than of ionene bromide.

The molar mass of these aliphatic ionenes lies between a few thousand and 40 000. Aliphatic ionenes with high molar mass (87 000) and high charge density ($x = 3$, $y = 4$) have been reported.[11] Also, aliphatic ionenes with low charge density (x or $y \geq 12$) can be prepared. Those ionenes were not completely soluble in water and form micelles or vesicles.[12,13]

Aliphatic ionenes can also be prepared from ω-haloalkyl dimethylamines, when the number of methylene groups, n, is larger than 6.[5] If

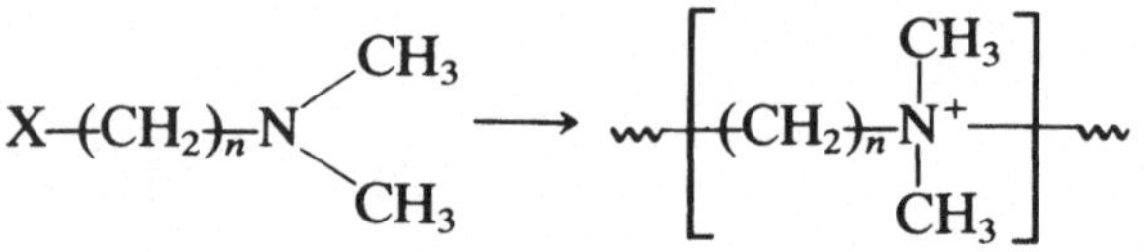

$$X\text{--}(CH_2)_n\text{--}N\begin{matrix}CH_3\\ \\CH_3\end{matrix} \longrightarrow \text{---}\left[(CH_2)_n\text{--}\overset{CH_3}{\underset{CH_3}{N^+}}\right]\text{---} \qquad (2)$$

$$\textbf{II}$$

$n < 6$, cyclic dimers or cyclic monomers were reported to be produced. However, Yen *et al.* subsequently succeeded in polymerizing dimethylaminopropyl chloride in a concentrated aqueous solution to get high molar mass 3-ionene.[14]

If one uses N,N-dimethylpiperazine[15] or 1,4-diaza[2.2.2]bicyclooctane[16] in place of N,N,N',N'-tetramethylalkylenediamines, ionenes with a cyclic or a bicyclic structure can be obtained.

The ionenes with xylylene units in the skeleton chains were synthesized from the reactions between *p*-xylylene dichloride and

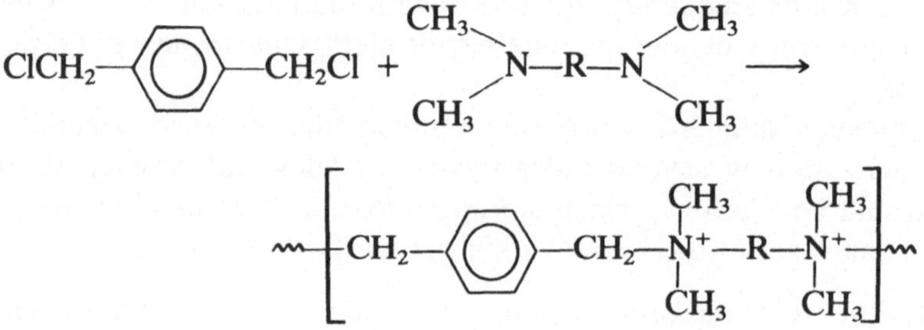

$$ClCH_2\text{--}\langle\bigcirc\rangle\text{--}CH_2Cl + \begin{matrix}CH_3\\ \\CH_3\end{matrix}N\text{--}R\text{--}N\begin{matrix}CH_3\\ \\CH_3\end{matrix} \longrightarrow$$

$$\text{---}\left[CH_2\text{--}\langle\bigcirc\rangle\text{--}CH_2\text{--}\overset{CH_3}{\underset{CH_3}{N^+}}\text{--}R\text{--}\overset{CH_3}{\underset{CH_3}{N^+}}\right]\text{---} \qquad (3)$$

$$\textbf{III}$$

several α,ω-di(tertiaryamine)s, such as N,N,N',N'-tetramethyl-p-xylylenediamine, N,N,N',N'-tetramethylethylenediamine, and N,N,N',N'-tetramethylphenylenediamine, by Tsuchida and his co-workers. They claimed that p-xylylenedichloride is one of the most effective monomers as a dihalogen compound for the formation of ionene polymers.[15,17]

Dipyridyl compounds can be used as another class of α,ω-di(tertiaryamine)s. Rembaum reported ionenes prepared from 1,4-dipyridylethane and several α,ω-dibromides.[8,18] Factor and Heinsohn prepared ionenes with viologen units called 'polyviologens', which were useful as redox polymers, from 4,4'-bipyridyl and xylylene dibromide.[19]

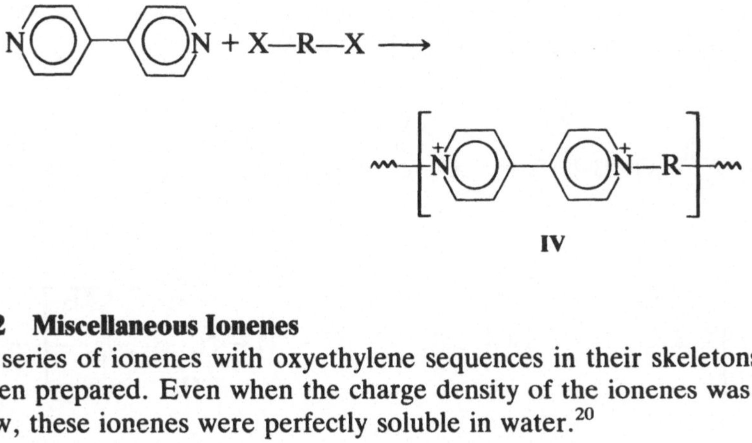

$$N\bigcirc\!\!\!-\!\!\!\bigcirc N + X\!-\!R\!-\!X \longrightarrow$$

$$\text{---}\!\left[\overset{+}{N}\bigcirc\!\!\!-\!\!\!\bigcirc\overset{+}{N}\!-\!R\right]\!\!\text{---} \quad (4)$$

IV

2.2 Miscellaneous Ionenes

A series of ionenes with oxyethylene sequences in their skeletons has been prepared. Even when the charge density of the ionenes was very low, these ionenes were perfectly soluble in water.[20]

$$Br\!-\!CH_2CH_2\!\!-\!\!(OCH_2CH_2)_x\!\!-\!\!Br + \underset{CH_3}{\overset{CH_3}{\diagdown}}N\!-\!CH_2CH_2\!\!-\!\!(OCH_2CH_2)_y\!\!-\!\!N\underset{CH_3}{\overset{CH_3}{\diagup}}$$

$$\longrightarrow \text{---}\!\left[CH_2CH_2\!\!-\!\!(OCH_2CH_2)_x\!\!-\!\!\overset{\overset{CH_3}{|}}{\underset{\underset{CH_3}{|}}{N^+}}\!\!-\!\!CH_2CH_2\!\!-\!\!(OCH_2CH_2)_y\!\!-\!\!\overset{\overset{CH_3}{|}}{\underset{\underset{CH_3}{|}}{N^+}}\right]\!\!\text{---} \quad (5)$$

V $\quad x, y = 1 \sim 4$

From 1,4-bischloroacetylpiperidine and N,N,N',N'-tetramethylhexamethylenediamine, an ionene polymer with a very high molar mass was prepared.[21]

$$\text{ClCH}_2\overset{\overset{\text{O}}{\|}}{\text{C}}-\text{N}\underset{\smile}{\frown}\text{N}-\overset{\overset{\text{O}}{\|}}{\text{C}}\text{CH}_2\text{Cl} + \begin{matrix}\text{CH}_3\\ \end{matrix}\diagdown\text{N}-(\text{CH}_2)_6\text{N}\diagup\begin{matrix}\text{CH}_3\\ \\ \text{CH}_3\end{matrix}$$

$$\longrightarrow \text{www}\left[\text{CH}_2\overset{\overset{\text{O}}{\|}}{\text{C}}-\text{N}\underset{\smile}{\frown}\text{N}-\overset{\overset{\text{O}}{\|}}{\text{C}}\text{CH}_2-\overset{\overset{\text{CH}_3}{|}}{\underset{\underset{\text{CH}_3}{|}}{\text{N}^+}}(\text{CH}_2)_6\overset{\overset{\text{CH}_3}{|}}{\underset{\underset{\text{CH}_3}{|}}{\text{N}^+}}\right]\text{www} \quad (6)$$

VI

Tazuke and Suzuki[22-24] have described the preparation of 'functionalized ionene polymers'. They used bischloroethanoate compounds, in which specific functional groups, such as a pendant (anthryl)methyl group, can be introduced.

$$\text{ClCH}_2\overset{\overset{\text{O}}{\|}}{\text{C}}-\text{OCH}_2\overset{\overset{\text{R}_1}{|}}{\text{C}}\text{HCH}_2-\text{O}-\overset{\overset{\text{O}}{\|}}{\text{C}}-\text{CH}_2\text{Cl} + \begin{matrix}\text{CH}_3\\ \end{matrix}\diagdown\text{N}-\text{R}_2-\text{N}\diagup\begin{matrix}\text{CH}_3\\ \\ \text{CH}_3\end{matrix}$$

$$\longrightarrow\text{www}\left[\text{CH}_2\overset{\overset{\text{O}}{\|}}{\text{C}}-\text{O}-\text{CH}_2\overset{\overset{\text{R}_1}{|}}{\text{C}}\text{HCH}_2-\text{O}-\overset{\overset{\text{O}}{\|}}{\text{C}}-\text{CH}_2-\overset{\overset{\text{CH}_3}{|}}{\underset{\underset{\text{CH}_3}{|}}{\text{N}^+}}\text{R}_2-\overset{\overset{\text{CH}_3}{|}}{\underset{\underset{\text{CH}_3}{|}}{\text{N}^+}}\right]\text{www} \quad (7)$$

VII

Ionene polymers with counter anions other than bromide or chloride ions can be easily prepared by exchange of counter anions. Bromide ions in aliphatic ionenes have been exchanged with I^-, BF_4^-, ClO_4^-, or SCN^- by adding concentrated aqueous solutions of the corresponding potassium salts to aqueous ionene polymer solutions.[25] Even the ionene polymers with large tetraphenylboron anions, which exhibited no hygroscopic nature, could be prepared.

Several ionic polymers with skeletal cations which possess very similar skeletal structures to the dimethylammonium cation have been reported. The ionene polymers with diethylammonium cations were synthesized, although the molar masses were rather low.[26]

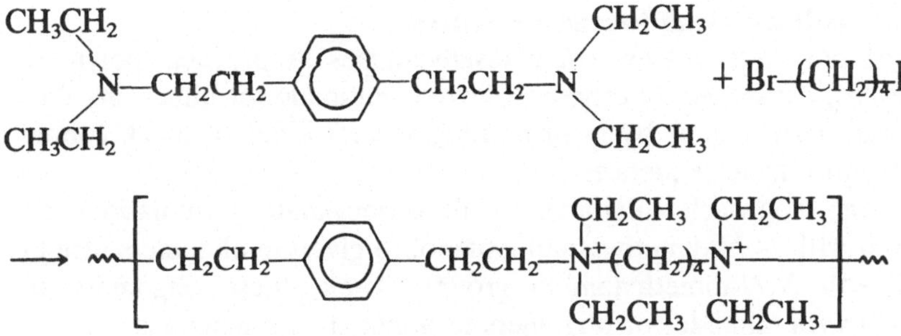

$$\begin{array}{l} CH_3CH_2 \\ \qquad\qquad N-CH_2CH_2-\underset{}{\bigcirc}-CH_2CH_2-N \\ CH_3CH_2 \end{array} \begin{array}{l} CH_2CH_3 \\ \\ CH_2CH_3 \end{array} + Br-(CH_2)_4Br$$

$$\longrightarrow \ \ \left[CH_2CH_2-\underset{}{\bigcirc}-CH_2CH_2-\overset{CH_2CH_3}{\underset{CH_2CH_3}{N^+}}(CH_2)_4\overset{CH_2CH_3}{\underset{CH_2CH_3}{N^+}} \right] \ \ (8)$$

VIII

Cationic polymers with diphenylphosphonium cations in their skeletons were prepared from alkylenediphenylphosphine and α,ω-dibromides.[27]

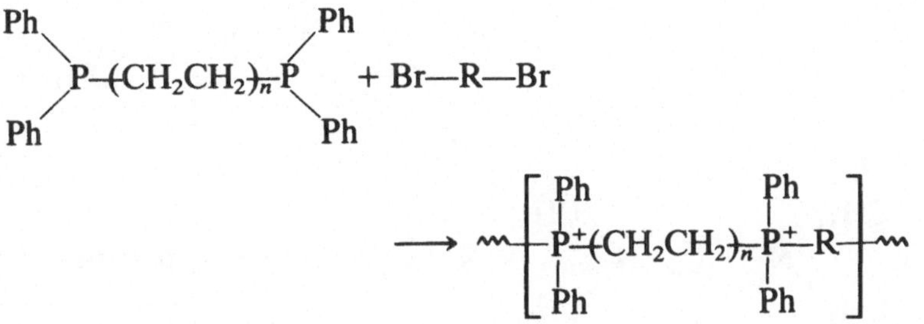

$$\begin{array}{l} Ph \\ \\ \underset{}{\diagdown}P-(CH_2CH_2)_n-P \\ Ph \end{array} \begin{array}{l} Ph \\ \\ Ph \end{array} + Br-R-Br$$

$$\longrightarrow \ \ \left[\overset{Ph}{\underset{Ph}{P^+}}(CH_2CH_2)_n\overset{Ph}{\underset{Ph}{P^+}}-R \right] \ \ (9)$$

IX

Cationic polymers with methylsulfonium cations in their skeletons have been prepared by methylation of polysulfides with methyl iodide.[28,29]

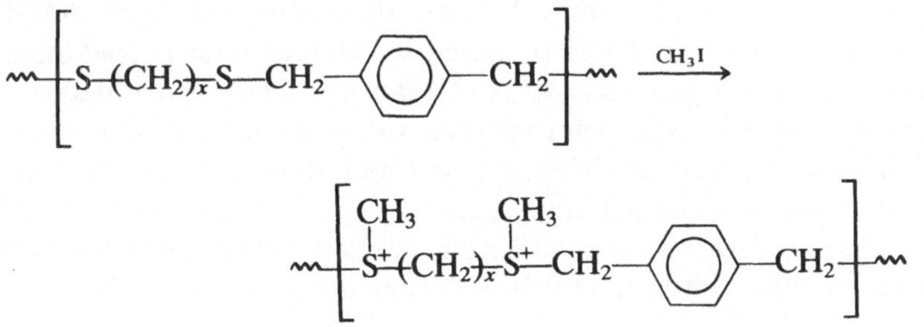

$$\left[S-(CH_2)_x S-CH_2-\underset{}{\bigcirc}-CH_2 \right] \ \xrightarrow{CH_3I}$$

$$\left[\overset{CH_3}{\underset{}{S^+}}(CH_2)_x\overset{CH_3}{\underset{}{S^+}}-CH_2-\underset{}{\bigcirc}-CH_2 \right] \ \ (10)$$

X $x = 3-10$

2.3 Copolymers with Ionene Sequences

Ionene polymers possess some disadvantages as practical polymeric materials; consequently proposals have been made for improving their material performance by incorporating several kinds of block copolymers with ionene sequences.

In one approach, oligomers with diisocyanate terminations are reacted with N,N-dimethylaminoethanol to give the oligomers terminated with N,N-dimethylamino groups. Then, these oligomers are reacted with dihalides or oligomers terminated with halogens.

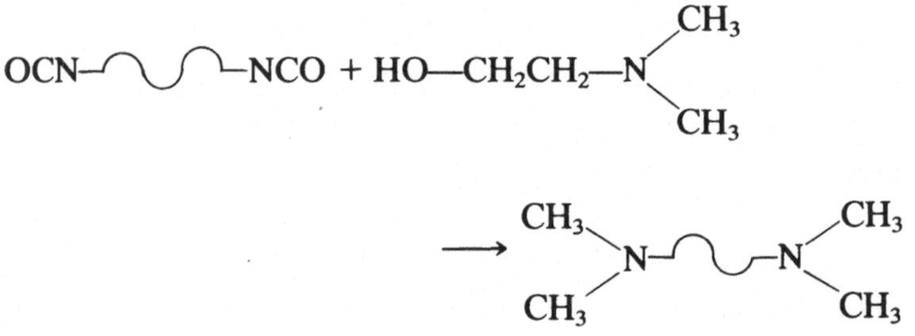

$$OCN-\!\!\!\bigcirc\!\!\!\bigcirc\!\!\!-NCO + HO\!-\!CH_2CH_2\!-\!N\!\!\begin{array}{c}CH_3\\ \\CH_3\end{array}$$

$$\longrightarrow \quad \begin{array}{c}CH_3\\ \\CH_3\end{array}\!\!N\!\!\!\bigcirc\!\!\!\bigcirc\!\!\!-N\!\!\begin{array}{c}CH_3\\ \\CH_3\end{array} \qquad (11)$$

$$\begin{array}{c}CH_3\\ \\CH_3\end{array}\!\!N\!\!\!\bigcirc\!\!\!\bigcirc\!\!\!-N\!\!\begin{array}{c}CH_3\\ \\CH_3\end{array} + X\!\!\!\bigcirc\!\!\!\bigcirc\!\!\!-X \longrightarrow \text{copolymer} \qquad (12)$$

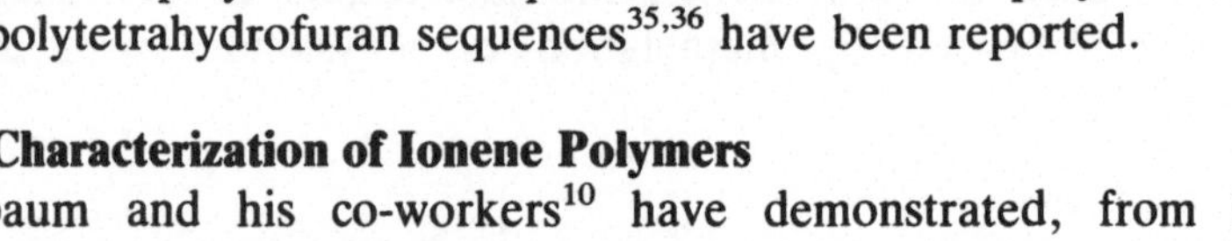

Elastomeric ionenes of the polyurethane type,[30–33] block copolymer ionenes with polybutadiene sequences,[34] and block copolymer ionenes with polytetrahydrofuran sequences[35,36] have been reported.

2.4 Characterization of Ionene Polymers

Rembaum and his co-workers[10] have demonstrated, from NMR spectra, that ionene polymers prepared by Menschutkin polyaddition reactions are linear polymers without branches. Only a few detailed reports dealing with the determination of molar masses and molar mass distributions are available. Casson and Rembaum[37] performed detailed studies on the solution properties of aliphatic ionenes, and found the intrinsic viscosity–molecular weight relationship for two aliphatic ionenes measured in 0·4M KBr aqueous solutions, to be:

Aliphatic 3,4-ionene

$$[\eta] = 2\cdot94 \times 10^{-4}M^{0\cdot61} \qquad (13)$$

Aliphatic 6,6-ionene

$$[\eta] = 6.22 \times 10^{-4} M^{0.58} \tag{14}$$

These two equations have been widely used, and some investigators have consulted them for a rough evaluation of molecular weights of other kinds of ionene polymers.

Gel permeation chromatography has been applied for the estimation of molar mass distributions of ionene polymers.[38,39] Dubin and Levy used a special hydrophilic gel column (Toyo Soda PW) and succeeded in evaluating molar mass distributions of some cationic polymers including ionene polymers.

3 PROPERTIES OF IONENE POLYMERS

3.1 Water Sorption, Solubility and Crystallinity

Ionene polymers are very hygroscopic, as are many other ion-containing polymers. Therefore, the solid state properties of ionene polymers must be determined on dried materials which should be carefully examined under dry atmospheres, to obtain the properties inherent to ionene polymers. Figure 1 shows the water sorption isotherms of the aliphatic 6,10-ionenes with several counter anions.[25] The 6,10-ionene (bromide counter anion), for example, has an equilibrium water content of about 10 wt% (1.4 H_2O molecules per ion) at 40% relative humidity at room temperature. Water content, of course, is largely dependent on the counterion species as well as on the chemical structure of the ionene backbones. Some reports on water sorption isotherms of ionene polymers are available.[25,40,41]

Generally, ionene polymers are soluble in water and in polar solvents such as methanol, DMF, and dimethyl sulfoxide (DMSO), but insoluble in almost all other common solvents. The solubilities, however, are very much dependent upon charge densities, skeletal chemical structures, and counterion species. Several ionenes with low charge densities and hydrophobic skeletal structures are insoluble in water and partly soluble in chloroform or tetrahydrofuran. One should consult the literature on the solubility of each ionene polymer individually.[10,15,25,42,43]

From X-ray diffraction experiments, aliphatic ionenes have been shown to be crystalline, although their crystallinities have not been quantitatively estimated.[10,25,44,45] The crystallinities of aliphatic ionenes depend on the numbers of methylene units between cations

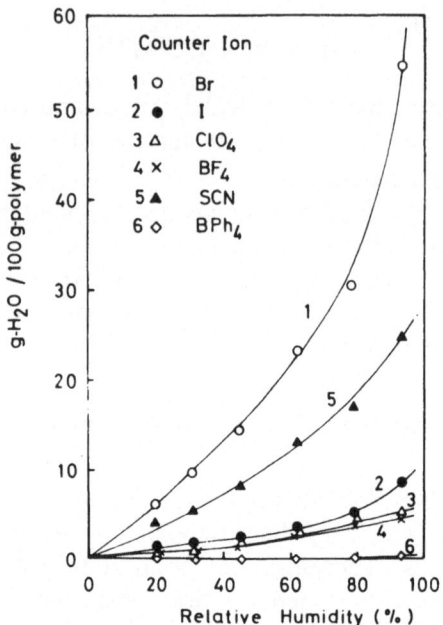

FIG. 1. Water sorption isotherms of aliphatic 6,10-ionenes with several counter anions at 25°C.[25] (Reproduced by permission of John Wiley & Sons, Inc.)

(charge density); the ionenes with low charge densities are nearly amorphous. From bulk densities and mechanical relaxation peaks (primary absorption), Tsutsui et al.[45] discussed qualitatively the crystallinities of aliphatic ionenes and showed that crystallinities of some aliphatic ionenes could be varied by changing the preparation conditions of the bulk samples. The ionene polymers with oxyethylene sequences (V) are also crystalline, but their crystallinites are lower than those of aliphatic ionenes.[20] Only scattered data on the crystallinity of other ionene polymers can be found in the literature,[46] although solid state properties of polymers are primarily determined by whether they are crystalline or amorphous. When one tries to estimate the crystallinity of ionene polymers, attention should be paid to the possible presence of absorbed water. Some amorphous ionene polymers became crystalline when they absorbed water.[25]

3.2 Thermal and Mechanical Properties

The glass transition temperatures (T_g) of aliphatic ionenes which were plasticized with several polar solvents were determined by the use of differential scanning calorimetry (DSC) by Eisenberg *et al.*[47] They estimated the T_g values of several aliphatic ionenes by extrapolation from the T_g values of the plasticized systems and found those of the unplasticized ionenes were largely dependent on the dielectric constants of the plasticizers. The T_g value of unplasticized 6,8-ionene, obtained by extrapolation to zero plasticizer content from the T_g data on the polymer plasticized with glycerine, was $-80°C$, while that from the polymer plasticized with dimethylformamide was $0°C$. They explained that this difference probably originated from the difference in conformation of chains in plasticized ionenes, which persisted in the polymer without plasticizer.

The use of a torsional braid analysis technique (TBA) made possible the direct determination of the T_g values of aliphatic ionenes and the ionene polymers (**V**) containing oxymethylene sequences.[20,23,45,48] The T_g values of aliphatic ionenes determined by TBA were much higher than the extrapolated values reported by Eisenberg *et al.*,[47] although the T_g values for the ionenes plasticized with water were consistent with each other. Although no satisfactory explanation for this discrepancy has been offered, one can assume that the plasticizing effect in ionene–plasticizer systems is very much different from that in common plasticized polymer systems, owing to the specific solvation of quaternary ammonium cations.

The T_g values of ionene polymers increase with the increase of charge densities on their skeletons. This relation can be rationalized by the use of the quantity q/a (q = charge of counterions; a = sum of the ionic radii of ions on polymer chains and of counterions) proposed by Eisenberg,[49] who claimed that the T_g values of ionic polymers are primarily determined by the strength of anion–cation interactions in the systems, q/a.[49] Shapes and ionic radii of the counter anions are other important factors that determine the T_g values of ionene polymers. The T_g values of aliphatic ionenes with counter anions, Br^-, I^-, ClO_4^-, BF_4^-, SCN^-, and BPh_4^-, were investigated. For spherical counter anions, the q/a theory gave a good explanation of the relationship between T_g values and ionic radii. In the case of large and non-spherical anions, however, the ionic radius is not the only factor that determines T_g values.[25] Reports have appeared which discuss the quantitative relationship between the T_g values of ion-containing polymers, including ionene polymers, and the strength of ionic interactions.[50-52]

The glass transition temperatures of ionene polymers with aromatic rings or cyclic structures in their skeletons, such as ionenes (**III**) and (**IV**), have not been investigated. One can assume that the T_g values of ionene polymers with rigid backbones are much higher than those of aliphatic ionenes, which possess more flexible skeletons.

Mechanical relaxation behaviour in ionene polymers is of much interest, because ammonium cations in skeletal chains are expected to restrict the segmental movements of skeletal chains. Figure 2 shows the dynamic mechanical loss spectra of aliphatic 12,n-ionenes obtained by the use of a TBA technique.[45] Three mechanical relaxation peaks, denoted α, β, and γ, were found. The α peaks were the primary

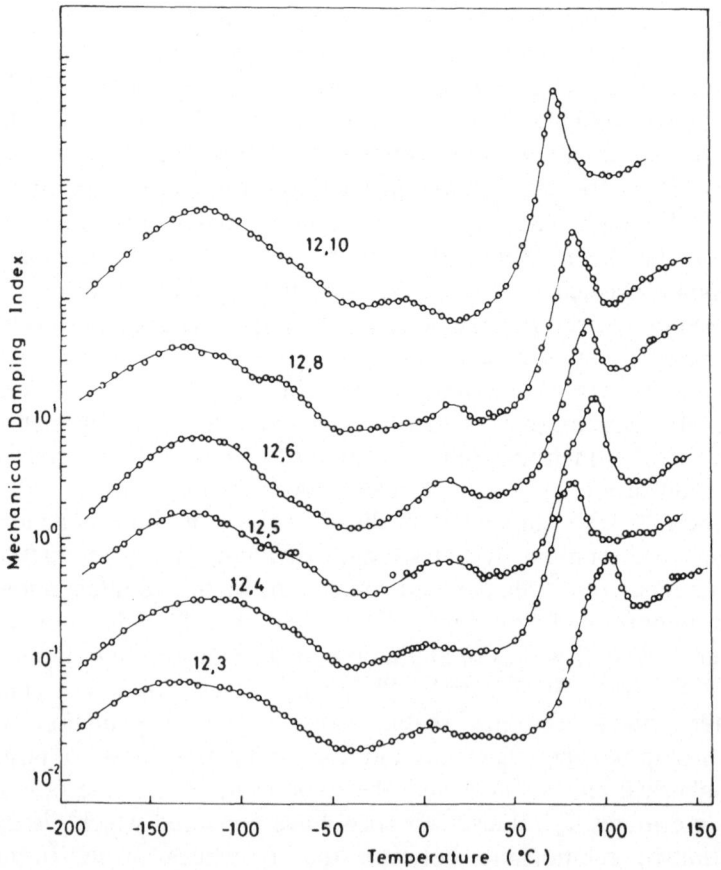

FIG. 2. Dynamic mechanical loss spectra of aliphatic 12,y-ionenes.[45] Numerals in the figure indicate numbers of methylene sequences. (Reproduced by permission of John Wiley & Sons, Inc.)

absorption due to the micro-Brownian motions of the polymer segments in amorphous regions, which we discussed above in terms of the glass transition temperatures. The β peaks were assumed to relate to the molecular motions of ionic parts, but the detailed mechanism was not clear. The γ peaks were attributed to the local segmental motions of methylene sequences, which are well known in polymers with skeletal methylene sequences such as polyethylene, aliphatic polyamides, and aliphatic polyesters. The relationships between the peak temperatures of those relaxations and the average number of methylene units between cations are shown in Fig. 3. Comparison of these results with the abundant data on mechanical relaxation of polymers that possess both methylene sequences and polar groups on skeletons reveal rather surprising parallelism. Hence, one may conclude that ionene polymers do not exhibit very specific mechanical properties, even though their backbones contain cations. Long-range interactions, due to ion–ion or dipole–dipole interactions, do not play a major role in segmental movement of polymeric chains. Figure 4 shows the dynamic mechanical storage and loss spectra of 6,10-ionenes with counter anions, ClO_4^-, BF_4^-, and SCN^-.[25] The small influence of

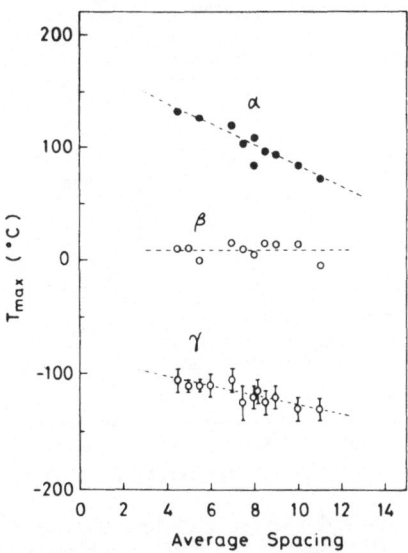

FIG. 3. The relation between the peak temperatures of α, β, and γ relaxations and the average number of methylene sequences in aliphatic x,y-ionenes.[45] (Reproduced by permission of John Wiley & Sons, Inc.)

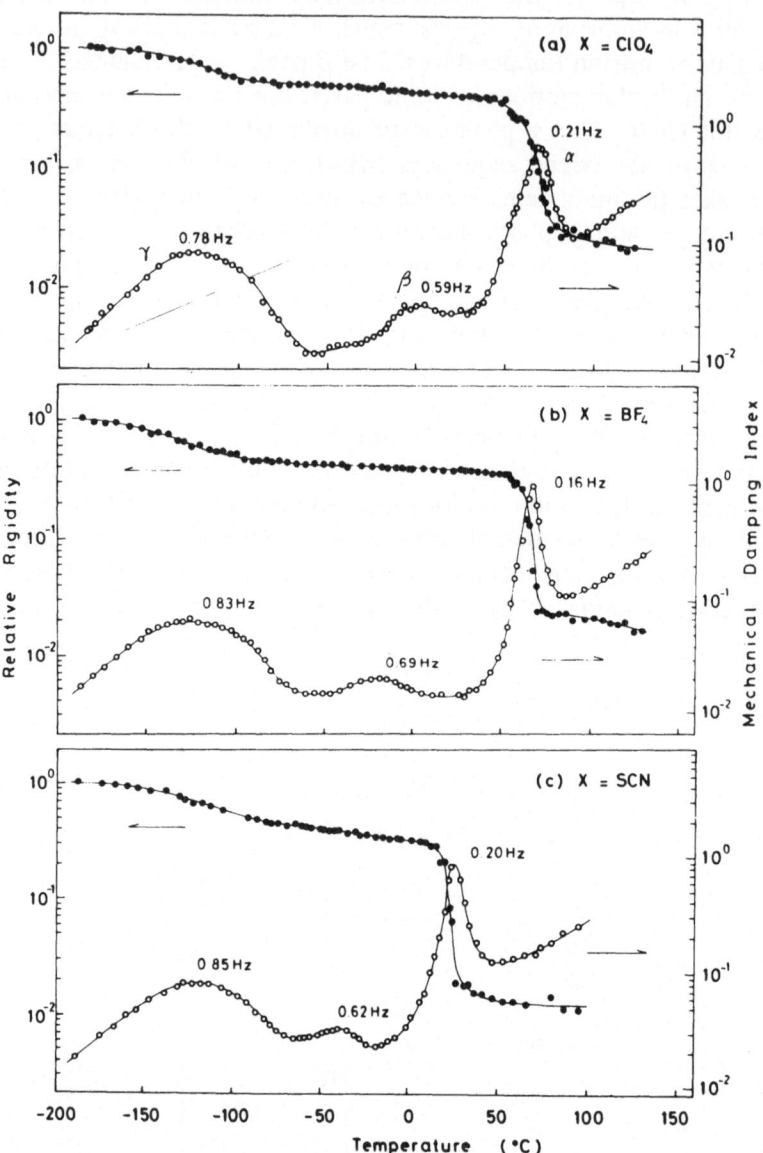

FIG. 4. Dynamic mechanical storage and loss spectra of 6,10-ionenes with counter anions, ClO_4^-, BF_4^-, and SCN^-.[25] (Reproduced by permission of John Wiley & Sons, Inc.)

counterion species on the α, β, and γ relaxations was found.

Dynamic viscoelasticity data on elastomeric ionenes with polyurethane sequences and their 7,7',8,8'-tetracyanoquinodimethane (TCNQ) salts have been reported.[53,54] Primary dispersions were observed at around −50°C, which corresponded to the T_g value of polyurethane segments. Tensile and dynamic mechanical properties of polybutadiene ionenes have also been reported.[34,55] These elastomeric ionenes possess microheterogeneous structures, and therefore their mechanical properties are dominated by the properties of the polyurethane or polybutadiene sequences.

4 APPLICATIONS

4.1 Conductive Polymers

Ionene polymers exhibit high electric conductivity when they are combined with 7,7',8,8'-tetracyanoquinodimethane (TCNQ), although ionene polymers themselves are considered to be insulators. A typical preparation of ionene polymer–TCNQ complexes with high conductivities is as follows. Methanol solutions of ionene polymers and a LiTCNQ salt are mixed in an inert atmosphere. The resulting precipitates which contain equimolar amounts of TCNQ⁻ anions and quaternary ammonium cations on polymer skeletons are called 'simple salts'. The simple salts are soluble in polar solvents such as DMF and N-methylpyrolidone. In many cases, neutral TCNQ is added to the simple salts to get more conductive 'complex salts', in which electronic carriers are believed to move along TCNQ and TCNQ⁻ molecules which are stacked to form conducting paths. On the other hand, carriers must hop via TCNQ⁻ molecules which distribute rather randomly in polymer matrices in the simple salts.

In 1967, Lupinski et al.[56] reported that polycation polymer–TCNQ complexes, which were prepared by mixing a poly(vinylpyridinium salt) and LiTCNQ followed by the addition of neutral TCNQ, exhibited high conductivity, up to $10^{-3}\,\mathrm{S\,cm^{-1}}$. Prompted by this discovery, Rembaum and his coworkers[8,18,57,58] prepared ionene polymer–TCNQ complexes and examined their electric conductivities.

There are several advantages in utilizing ionene polymers in place of common polycation polymers. Firstly, regular cation spacings and high charge densities on the polymer skeletons can be realized. Secondly, spacings between charges can be adjusted through easy synthetic procedures. Thirdly, the flexibility of skeletal chains, which are expected to serve to accommodate the stacking of TCNQ molecules, can be changed.

In the early work by Rembaum and his coworkers, aliphatic ionenes, some aromatic ionenes with a pyridinium cation on the skeletons, and elastomeric ionenes were utilized. Figure 5 shows the relationship between specific resistivity and the numbers of methylene units in aliphatic ionene–TCNQ salts, reported by Hadek et al.[57] The resistivity of the complex salts (with neutral TCNQ) was much higher than that of the simple salts (without neutral TCNQ). The relationship between cationic spacings and resistivities was not straightforward, and these workers could not deduce a universal trend between ionic spacings on the ionene skeletons and conductivities of the TCNQ complexes. The conductivities of the TCNQ complexes of the ionene polymers which carried pyridinium cations on their skeletons were a little higher than those of corresponding complexes of aliphatic ionenes, although it was expected that pyridinium rings on the polymer backbone would work to enhance the conductivity of the

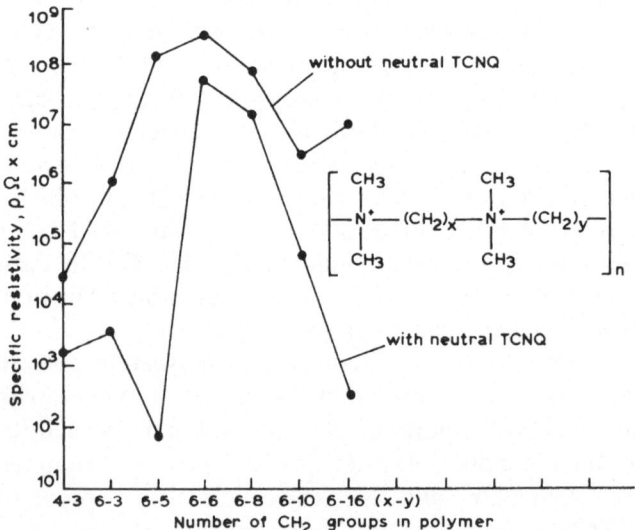

FIG. 5. Relation between specific resistivity and number of methylene units in aliphatic ionene–TCNQ salts.[57] (Reproduced by permission of the American Chemical Society.)

TCNQ complexes, due to the increase of electronic interactions between TCNQ and polymer skeletons.[18]

Later, Shinohara[59–67] and his coworkers studied extensively the conductivity of various kinds of ionene polymer–TCNQ complexes. Firstly, they re-examined aliphatic ionene complexes, in order to clarify the relationship between ionic spacings and the degree of TCNQ stacking.[59,61] They pointed out the importance of reporting both the conductivities and the activation energies of conduction as a function of the molar ratio of neutral $TCNQ^0$ and $TCNQ^-$ anions in the complexes. Figure 6 shows the relationship between specific

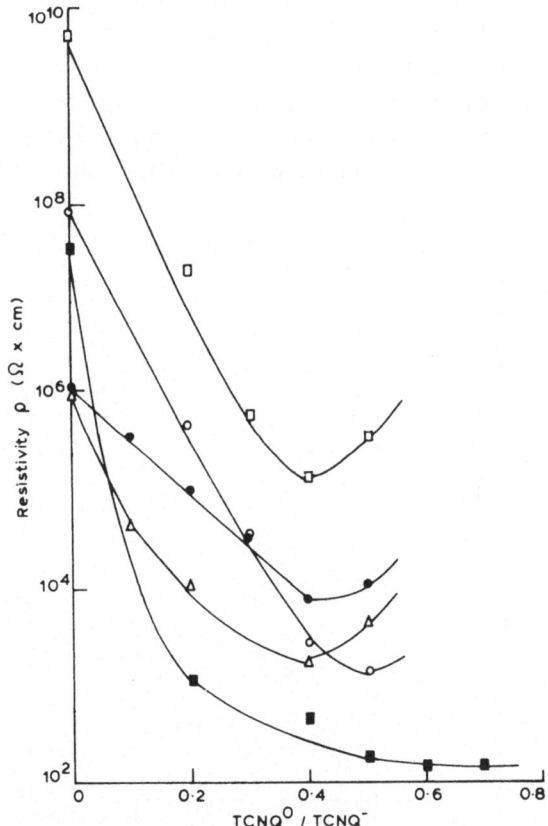

FIG. 6. Relation between specific resistivity and the molar ratio ($TCNQ^0/$ $TCNQ^-$) in several aliphatic ionene–TCNQ complexes at 25°C:[59] ○, 3,3-ionene; △, 4,4-ionene; ●, 5,5-ionene; ■, 6,5-ionene; □, 6,6-ionene. (Reproduced by permission of Waseda University, Japan.)

resistivity and the molar ratio (TCNQ^0/TCNQ$^-$) in several x,y-ionene–TCNQ complexes. One should note that specific resistivities did not always continue to decrease to give the lowest values at the molar ratio of 1·0. This fact suggests that alternating stacking of TCNQ^0 and TCNQ$^-$ should not always be postulated. In the case of the 6,5-ionene, both resistivity and activation energy of conduction decrease until the molar ratio approaches 1·0 (Fig. 7). Therefore, it can be reasonably assumed that alternating stacking of TCNQ^0 and TCNQ$^-$ is realized in the case of the 6,5-ionene. However, the x,y-ionenes with x and y larger than 6 were found to be unsuitable for regular stacking of TCNQ$^-$ and TCNQ^0 molecules. The 6,6-ionene, which was expected to possess the most favourable ionic spacing, did not show the highest conductivity among aliphatic ionene complexes, when its TCNQ complex was prepared. This abnormal behavior was examined in detail and was ascribed to the high crystallinity of that polymer.[68] Recently, profound effects of heterogeneous structures on the conductivities of the complex salts have been shown.[69]

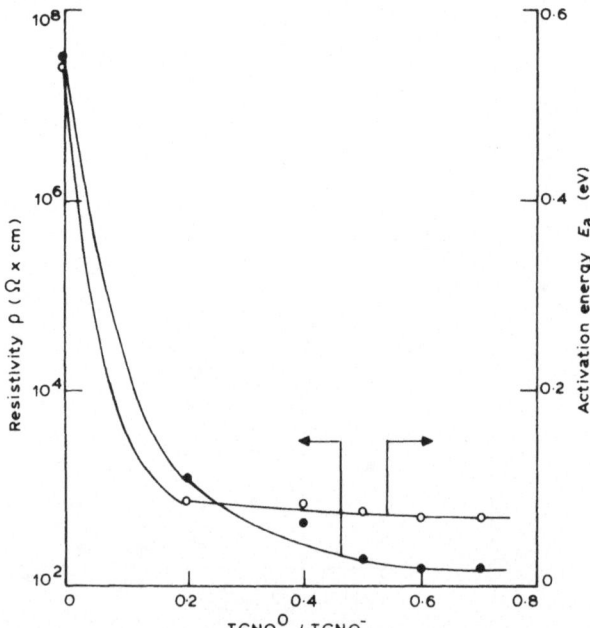

FIG. 7. Specific resistivity and activation energy of conduction versus molar ratio (TCNQ^0/TCNQ$^-$) in 6,5-ionene.[59] (Reproduced by permission of Waseda University, Japan.)

The second factor studied by Shinohara's group was the steric effect of the direct substituents on quaternary ammonium cations. They prepared ionene polymers carrying piperidine, morpholine, and tetrahydro-1,4-thiazine rings on nitrogen atoms.[60,62]

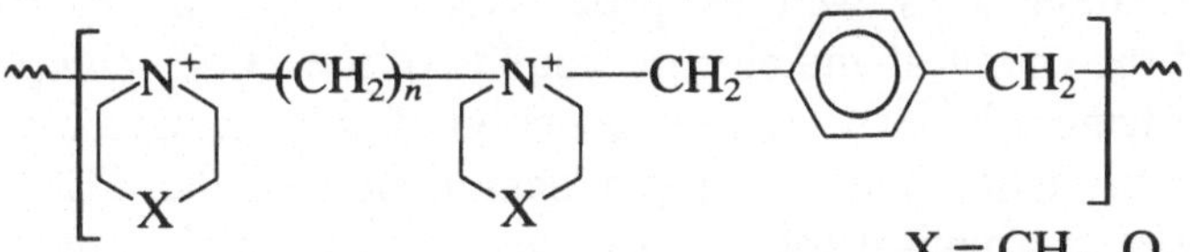

$$X = CH_2, O, S$$

XI

These workers expected that the ring structures on ammonium cations could contribute to the improvement of the regularity of $TCNQ^0$ and $TCNQ^-$ stacking. They found, however, that the complex salts of those polymers were less conductive than those of aliphatic ionenes, although the simple salts exhibited markedly high conductivity compared with other simple salts of ionene polymers.

Thirdly Shinohara's group studied the effect of rigidity of skeletal structures on the conductivity of the TCNQ complexes. They synthesized a number of ionene polymers with piperidinium rings, cyclohexane rings, and 4,4'-bipyridinium ions in their skeletons.[63,65,70,71] In all the examples of the ionene polymers that they prepared, the conductivities of the complex salts exhibited maxima at the molar ratio $(TCNQ^0/TCNQ^-)$ of 1·0. This result implied that regular stacking of $TCNQ^-$ and $TCNQ^0$ molecules was realized in those systems. In fact, the conductivities of those complex salts were very high, ranging from $6·7 \times 10^{-4}$ to $0·08 \, S \, cm^{-1}$. The highest value, $0·08 \, S \, cm^{-1}$, was observed in the complex salt with molar ratio of 1·0 prepared from the ionene polymer **XII**.

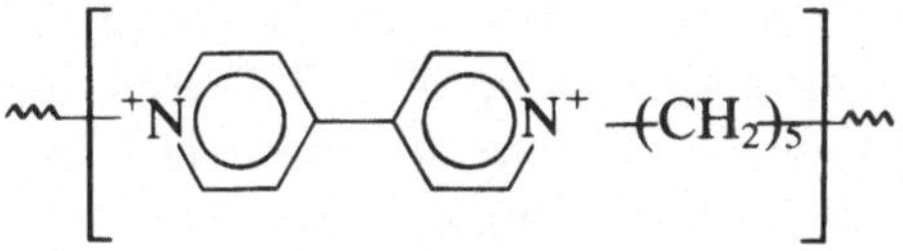

XII

The chemical structure of the ionene polymer **XII** is very similar to the polyviologens prepared by Factor and Heinsohn[19] in 1971. They reported that the complex salt (molar ratio, 0·16) from the ionene polymer **IV** with R = o-xylylene exhibited a conductivity of

$0.02 \, \text{S cm}^{-1}$. Therefore, electronic interactions between viologen rings and TCNQ molecules might be important for further enhancement of the conductivity of the complex salts.

Conducting polymers for practical usage must possess chemical and electrical stability as well as processability. Several reports have discussed the chemical and electric stability of the TCNQ complexes of ionene polymers.[62,63,72-74] The introduction of polar groups into ionene polymers brought about the rapid degradation of the TCNQ complexes. The mechanism for the decrease of conductivity with time in the TCNQ complexes was studied and the decomposition of TCNQ molecules was found to be responsible.

The TCNQ complexes prepared from ionene polymers discussed above possess poor processability except in a few cases. Solvent-cast films could not be prepared, and consequently, compacted disks prepared from powder samples have been used for conductivity measurements. Processable and tough TCNQ complex films prepared from elastomeric ionenes, which could be cast from solutions, were first proposed by Rembaum, and recently several investigators have succeeded in improving their mechanical and electrical properties.[30-32,53,54,75] The simple salts of elastomeric ionenes exhibited low conductivity of $\sim 10^{-9} \, \text{S cm}^{-1}$, owing to very low TCNQ concentrations. Unexpectedly, however, the addition of neutral TCNQ brought about large increases of conductivity without losing film-forming characteristics. The occurrence of micro-phase separation caused the creation of continuous conducting paths in the films. By varying the structures and lengths of both polyether (urethane) sequences and ionene sequences, tough and strong films with conductivities as high as $0.04 \, \text{S cm}^{-1}$ were produced.[33] Anisotropy of conductivity was observed in stretched films of elastomeric ionene–TCNQ complexes.[54]

From the cationic polymers with S^+ ions on skeletons (**X**), conductive TCNQ complexes have also been prepared. The interrelation between ionic spacing and stacking of TCNQ molecules was found to be very similar to that of ionene polymer–TCNQ complexes. The conductivities of this type of complex were reported to be somewhat higher than those of the complexes with corresponding chemical constitutions without S^+ ions in the skeleton. This difference was attributed to the large polarizability of the polycation polymer with S^+ ions.[28,29]

Dielectric constants of the ionene polymer–TCNQ complexes are very high. For example, the dielectric constants of the complex salts prepared from the polymer (**III**) with $R = -(CH_2)_6$ was 70–7000 at 30 kHz–1 MHz at room temperature, and it increased with increasing temperature and decreased with increasing frequency. These abnormally high dielectric constants were interpreted to originate from interfacial polarization among micro-domains induced by mobile electronic carriers in the isolated conductive phases. The presence of both continuous and isolated conductive paths in the complex salts was thus demonstrated.[75] Dielectric properties of elastomeric ionene–TCNQ complexes were also investigated and discussed in terms of the presence of micro-heterogeneous structures.[53,76]

Elastomeric ionenes have been used for a component of polymeric solid electrolytes.[77] A thermoswitching phenomenon in elastomeric ionene–TCNQ complexes has also been reported.[78]

4.2 Polyelectrolyte Complexes

Ionene polymers are important as components of polyelectrolyte complexes, which are prepared simply by mixing cationic polymers and anionic polymers in aqueous solutions. Much attention has been paid to polyelectrolyte complexes as potential practical materials, such as dialysis membranes, ultrafiltration membranes, antistatic materials and biomedical materials, since the early reports on the mechanism of their formation and their properties by Michaels.[79] Also, the formation of polyelectrolyte complexes has been of interest in providing simple models for biological systems.

Rembaum[80] pointed out that ionene polymers were promising as the cationic polymer component of polyelectrolyte complexes, although he presented no experimental results on preparation and properties of the polyelectrolyte complexes using ionene polymers. Ionene polymers have been expected to contribute novel features to the properties of polyelectrolyte complexes which are composed of common pendant-type anionic and cationic polymers, because they carry cations, integral to their backbone chains.

Tsuchida *et al.*[81] investigated the formation of polyelectrolyte complexes from poly(styrene sulphonate) and the ionene polymer (**III**) with $R = -(CH_2)_3$, and found that water-soluble complexes with non-equimolar composition were formed. When cationic polymers (pendant-type) were mixed with anionic polymers (pendant-type), polyelectrolyte complexes with equimolar compositions precipitated, independent of the contents of cationic and anionic polymers in solutions. On the other hand, in the cases of the combination of cationic polymers of the integral type (ionene polymer) and anionic polymers (pendant-type), polyelectrolyte complexes with the equimolar composition became soluble when excess cationic polymer molecules were added to the solutions. They assumed that 2 moles of extra ionene polymer chains were able to adsorb onto the equimolar complex chains and make them soluble, because free 'exposed positions' still remained on ionene polymer chains, even after they were complexed with equimolar anionic polymer chains. These workers also studied the complexation phenomenon between the same ionene polymer and some anionic polymers with pendant carboxylic acid groups. The compositions of the complexes were dependent on the degree of dissociation of the poly(carboxylic acid)s. They found that the complexation was cooperative and therefore assumed that complexes with 'loop' and 'ladder' structures were formed.[82,83]

In some cases, when anionic polymers and cationic polymers are mixed phase separation called 'coacervation', a well-known phenomenon in biopolymer systems, occurs instead of precipitation of solid complexes; this behavior depends on both the concentrations of component polymers and the ionic strength. Phase diagrams of some ionene polymer–anionic polymer systems have been studied.[84,85] Non-stoichiometric water-soluble polyelectrolyte complexes were found, and solution properties of those complexes were studied.[86–88]

Tsuchida and his coworkers[89,90] reported the formation of the polyelectrolyte complexes with higher order structures from the ionene polymers (III) with $R = -(CH_2)_x$ (where $x = 2$, 3 and 6) and poly(methacrylic acid). Well-defined fibrous structures, which exhibited optical anisotropy, were observed. The complexes with higher order structures could be grown by standing precipitates in solutions at room temperature for several days. They indicated the importance of hydrophobic interactions for the formation and growth of the complexes with higher order structures. Figure 8 shows the mechanism of formation and growth of interpolymer polyelectrolyte complexes proposed by Tsuchida. Immediately after the mixing of polycations and polyanions in solutions, random primary complexes are formed. Then they transform to secondary, more ordered, complexes through intra-chain complex re-formation. Finally, fibrous aggregates start growing.

Matrix polymerization on an ionene template has been examined by Blumstein and his coworkers. Using ionene polymers with a rigid backbone, they polymerized p-styrenesulfonic acid fixed on the

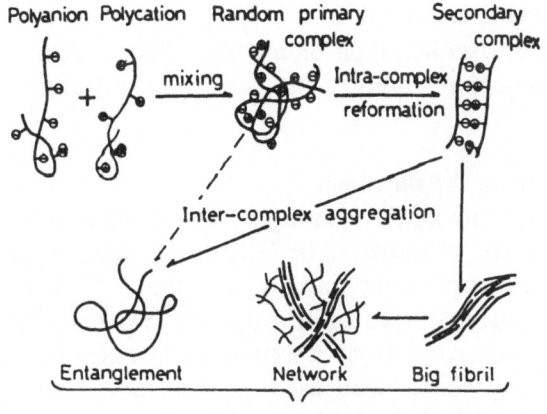

Fig. 8. Schematic diagram of the formation of interpolymer complex.[90] (Reproduced by permission of the American Chemical Society.)

template. Thus, they succeeded in obtaining highly ordered poly-electrolyte complexes through matrix polymerization.[46] The kinetics of matrix polymerization in ionene polymer–vinyl monomer systems have been also studied in detail.[91–93] (See also Chapter 5.)

4.3 Biomedical Applications

Both low molar mass and cationic polymers containing quaternary nitrogen atoms in their structures have been known to possess important biological activity. In the 1940s, patents describing the use of linear poly(quaternary ammonium salt)s for fungicides and bacterio-cides appeared.[7]

Rembaum and his associates,[8,11] in their first paper on ionene polymers, indicated that ionene polymers exhibited bacteriostatic and bactericidal activity. They also proposed that elastomeric ionenes could be utilized as nonthrombogenic structural materials when they were combined with heparin to yield insoluble polyelectrolyte complexes. They studied the viscoelastic properties of heparinized elas-tomeric ionenes in detail and proposed that elastomeric ionenes were promising for tailor-made biomedical materials.[94,95] Ohno et al.[96] discussed the interactions between human serum proteins and synthe-tic polymers including ionene polymers in solutions and indicated that high selectivity against proteins might be attained by controlling the types and amounts of reaction sites in synthetic polymers.

Many patents relating to ionene polymers and their electrolyte complexes have described their biomedical effects, such as antiheparin activity, haemocompatibility, bile acid sequestration, and antimicrobial activity.

4.4 Miscellaneous Applications

Among various kinds of flocculating agents, cationic polymers can be one of the most effective. The effects of ionene polymers for flocculating clay suspensions have been tested.[97,98] Ionene polymers with various structures have been proposed in patents for flocculants which would be effective for a variety of waste water treatments.

Ionene polymers have been used for antistatic agents for papers, textiles, and plastics. They are also important in various fields of application, such as hair conditioning agents, dye-fixing agents, templates for zeolite or aluminosilicate crystallization.[99]

REFERENCES

1. LITTMANN, E. R. and MARVEL, C. S., *J. Amer. Chem. Soc.*, 1930, **52**, 287.
2. GIBBS, C. F., LITTMANN, E. R. and MARVEL, C. S., *J. Amer. Chem. Soc.*, 1933, **55**, 753.
3. LEHMAN, M. R., THOMPSON, C. D. and MARVEL, C. S., *J. Amer. Chem. Soc.*, 1933, **55**, 1977.
4. GIBBS, C. F. and MARVEL, C. S., *J. Amer. Chem. Soc.*, 1934, **56**, 752.
5. GIBBS, C. F. and MARVEL, C. S., *J. Amer. Chem. Soc.*, 1935, **57**, 1137.
6. KERN, W. and BRENNEISEN, E., *J. Prakt. Chem.*, 1941, **159**, 193.
7. HOOVER, M. F., *J. Macromol. Sci.—Chem.*, 1970, **A4**, 1327.
8. REMBAUM, A., BAUMGARTNER, W. and EISENBERG, A., *Polymer Letters*, 1968, **6**, 159.
9. MENSCHUTKIN, N., *Z. Physik. Chemie*, 1890, **5**, 589.
10. REMBAUM, A. and NOGUCHI, H., *Macromolecules*, 1972, **5**, 261.
11. REMBAUM, A., RILE, H. and SOMOANO, R., *Polymer Letters*, 1970, **8**, 457.
12. SONNESSA, A. J. and CULLEN, W., *Macromolecules*, 1978, **13**, 195.
13. KUNITAKE, T., NAKASHIMA, N., TAKARABE, K. and NAGAI, M., *J. Amer. Chem. Soc.*, 1981, **103**, 5943.
14. YEN, S. P. S., CASSON, D. and REMBAUM, A., *Water Soluble Polymers*, (N. M. Bikales, Ed.), 1973, Plenum Press, New York, p. 291.
15. TSUCHIDA, E., SANADA, K. and MORIBE, K., *Makromol. Chem.*, 1972, **151**, 207.
16. SALAMONE, J. C. and SNIDER, B., *J. Polym. Sci.*, A-1, 1970, **8**, 3495.
17. TSUCHIDA, E., SANADA, K. and MORIBE, K., *Makromol. Chem.*, 1972, **155**, 35.

18. REMBAUM, A., HERMANN, A. M., STEWART, F. E. and GUTTMANN, F., *J. Phys. Chem.*, 1969, **73**, 514.
19. FACTOR, A. and HEINSOHN, G. E., *Polymer Letters*, 1971, **9**, 289.
20. TSUTSUI, T., TANAKA, R. and TANAKA, T., *J. Polym. Sci., Polym. Phys. Ed.*, 1975, **13**, 2091.
21. WEISS, J. and SCHINDLBAUER, H., *Makromol. Chem.*, 1977, **178**, 2573.
22. TAZUKE, S. and SUZUKI, Y., *J. Polym. Sci., Polym. Lett. Ed.*, 1978, **16**, 223.
23. SUZUKI, Y. and TAZUKE, S., *Macromolecules*, 1980, **13**, 25.
24. SUZUKI, Y. and TAZUKE, S., *J. Polym. Sci., Polym. Lett. Ed.*, 1982, **20**, 259.
25. TSUTSUI, T., TANAKA, R. and TANAKA, T., *J. Polym. Sci., Polym. Phys. Ed.*, 1976, **14**, 2273.
26. DONKAI, N. and INAGAKI, H., *Makromol. Chem.*, 1979, **180**, 65.
27. NOHIRA, H., TANIGUCHI, M. and SAITO, Y., *Kobunshi Ronbunshu, Japan*, 1975, **31**, 391.
28. PECHERZ, J., CIESIELSKI, W. and KRYSZEWSKI, M., *Macromolecules*, 1981, **14**, 1139.
29. PECHERZ, M. and KRYSZEWSKI, M., *Polym. J.*, 1983, **15**, 401.
30. SOMOANO, R., YEN, S. P. S. and REMBAUM, A., *Polym. Lett.*, 1970, **8**, 467.
31. IKENO, S., YOKOYAMA, M. and MIKAWA, M., *J. Polym. Sci., Polym. Phys. Ed.*, 1978, **16**, 717.
32. WATANABE, M., KAMIYA, T., GOTO, K. and MATSUBARA, N., *Makromol. Chem.*, 1981, **182**, 2659.
33. WATANABE, M., TONEAKI, N. and SHINOHARA, I., *Polym. J.*, 1982, **14**, 189.
34. YAMASHITA, S., ITOI, M. and KOHJIYA, S., *Kobunshi Ronbunshu, Japan*, 1981, **38**, 189.
35. TAKAHASHI, A., KAWAGUCHI, M., KATO, T. and MATSUMOTO, S., *J. Macromol. Sci.—Phys.*, 1980, **B17**, 747.
36. KAWAGUCHI, M., OOHIRA, M., TAJIMA, M. and TAKAHASHI, A., *Polym. J.*, 1980, **12**, 849.
37. CASSON, D. and REMBAUM, A., *Macromolecules*, 1972, **5**, 75.
38. DUBIN, P. L. and LEVY, I. J., *J. Chromatography*, 1982, **235**, 377.
39. LEVY, I. J. and DUBIN, P. L., *Ind. Eng. Chem., Prod. Res. Div.*, 1982, **21**, 59.
40. TSUCHIDA, E., SANADA, K., and MORIBE, K., *Nippon Kagaku Kaishi, Japan*, 1972, 2161.
41. KUROKAWA, Y., SHIRAKAWA, N., TERADA, M. and YUI, N., *J. Appl. Polym. Sci.*, 1980, **25**, 1645.
42. KOHJIYA, S., OHSUKI, T. and YAMASHITA, S., *Makromol. Chem., Rapid Commun.*, 1981, **2**, 417.
43. HIRAOKA, K. and YOKOYAMA, T., *J. Polym. Sci., Polym. Lett. Ed.*, 1978, **16**, 401.
44. RAZVODOVSKII, Y. F., NEKRASOV, A. V. and YEMIKOLOPYAN, N. S., *Vysokomol. Soed.*, 1971, **A13**, 1980.
45. TSUTSUI, T., TANAKA, R. and TANAKA, T., *J. Polym. Sci., Polym. Phys. Ed.*, 1976, **14**, 2259.

46. BLUMSTEIN, A., KAKIVAYA, S. R. and SALAMONE, J. C., *J. Polym. Sci.*, *Polym. Lett. Ed.*, 1974, **12**, 651.
47. EISENBERG, A., MATSUURA, H. and YOKOYAMA, T., *Polym. J.*, 1971, 2, 117.
48. TSUTSUI, T., SATO, T. and TANAKA, T., *Polym. J.*, 1973, **5**, 332.
49. EISENBERG, A., *Macromolecules*, 1971, **4**, 125.
50. BECKER, R., Z. *Physik Chem.* (*Leipzig*), 1976, **257**, 667.
51. TSUTSUI, T. and TANAKA, T., *Polymer*, 1977, **18**, 817.
52. PFEIFFER, D. G., *Polymer*, 1980, **21**, 1135.
53. WATANABE, M., TONEAKI, N., TAKIZAWA, Y. and SHINOHARA, I., *J. Polym. Sci.*, *Polym. Chem. Ed.*, 1982, **20**, 2669.
54. WATANABE, M., TAKIZAWA, M. and SHINOHARA, I., *J. Polym. Sci.*, *Polym. Chem. Ed.*, 1983, **21**, 2397.
55. YAMASHITA, S. and KOHJIYA, S., *Colloid and Polymer Sci.*, 1981, **259**, 574.
56. LUPINSKI, J. M., KOPPLE, K. D. and HERTZ, J. J., *J. Polym. Sci.*, 1967, **C16**, 1561.
57. HADEK, V., NOGUCHI, H. and REMBAUM, A., *Macromolecules*, 1971, **4**, 494.
58. REMBAUM, A., *J. Polym. Sci.*, 1970, **C29**, 157.
59. MIZOGUCHI, K., SUZUKI, T., TSUCHIDA, E. and SHINOHARA, I., *Nippon Kagaku Kaishi, Japan*, 1973, 1751, 1756, 1760.
60. MIZOGUCHI, K., SUZUKI, T., TSUCHIDA, E. and SHINOHARA, I., *Nippon Kagaku Kaishi, Japan*, 1973, 1765.
61. SANADA, K., EDA, N., TSUCHIDA, E. and SHINOHARA, I., *Nippon Kagaku Kaishi, Japan*, 1974, 584.
62. SANADA, K., IWASAWA, A., TSUCHIDA, E. and SHINOHARA, I., *Nippon Kagaku Kaishi, Japan*, 1974, 955.
63. SANADA, K., IWASAWA, A., TSUCHIDA, E. and SHINOHARA, I., *Nippon Kagaku Kaishi, Japan*, 1974, 961.
64. MIZOGUCHI, K., KITAJIMA, Y., KAJIURA, S., TSUCHIDA, E. and SHINOHARA, I., *Nippon Kagaku Kaishi, Japan*, 1974, 1751.
65. MIZOGUCHI, K., TABATA, A., NAKANO, A., TSUCHIDA, E. and SHINOHARA, I., *Nippon Kagaku Kaishi, Japan*, 1974, 1974.
66. MIZOGUCHI, K., KAJIURA, S., NAKANO, A. T., TSUCHIDA, E. and SHINOHARA, I., *Nippon Kagaku Kaishi, Japan*, 1975, 1403.
67. MIZOGUCHI, K., TOGOU, H., TSUCHIDA, E. and SHINOHARA, I., *Nippon Kagaku Kaishi, Japan*, 1975, 2211.
68. MOSTOVOI, R. M., GLAZKOVA, I. V., KOTOV, B. V., VASHKEVICH, V. A., VARAKINA, Ye. N., TVERSKOI, V. A., ZUBOV, Yu. A., SAFRONOV, S. N., GASYUK, O. V. and PRAVEDNIKOV, A. N., *Vysokomol. Soed.*, 1978, **A20**, 1042.
69. MOSTOVOI, R. M., BERENDYAYEV, V. I., VASHKEVICH, V. A., KOTOV, B. V., ZUBOV, Yu. A., OVCHINNIKOV, S. Yu., TVERSKOI, V. A. and PRAVEDNIKOV, A. N., *Vysokomol. Soed.*, 1982, **A24**, 801.
70. KAMIYA, T., GOTO, K. and SHINOHARA, I., *J. Polym. Sci.*, *Polym. Chem. Ed.*, 1979, **17**, 561.
71. KAMIYA, T. and SHINOHARA, I., *J. Polym. Sci.*, *Polym. Lett. Ed.*, 1979, **17**, 641.

72. MIZOGUCHI, K., TSUJI, S., TSUCHIDA, E. and SHINOHARA, I., J. Polym. Sci., Polym. Chem. Ed., 1978, **16**, 3259.
73. KAMIYA, T., TSUJI, S., OGATSU, K. and SHINOHARA, I., Polym. J., 1979, **11**, 219.
74. PECHEZ, J. and KRYSZEWSKI, M., Polym. Bull., 1982, **8**, 87.
75. IKENO, S., MATSUMOTO, K., YOKOYAMA, M. and MIKAWA, H., Polym. J., 1977, **9**, 261.
76. IKENO, S., YOKOYAMA, M. and MIKAWA, H., Polym. J., 1978, **10**, 123.
77. WATANABE, M., NAGAOKA, K., KANBA, M. and SHINOHARA, I., Polym. J., 1982, **11**, 877.
78. TAKIZAWA, Y., AIGA, H., WATANABE, M. and SHINOHARA, I., J. Polym. Sci., Polym. Chem. Ed., 1983, **21**, 3145.
79. MICHAELS, A. S., Ind. Eng. Chem., 1965, **57**, 32.
80. REMBAUM, A., J. Macromol. Sci.—Chem., 1969, **A3**, 87.
81. TSUCHIDA, E., OSADA, Y. and SANADA, K., J. Polym. Sci., Polym. Chem. Ed., 1972, **10**, 3397.
82. TSUCHIDA, E., OSADA, Y., and ABE, K., Makromol. Chem., 1974, **175**, 583.
83. TSUCHIDA, E. and OSADA, Y., Makromol. Chem., 1974, **175**, 593.
84. ABE, K., OHNO, H., and TSUCHIDA, E., Makromol. Chem., 1977, **178**, 2285.
85. TSUCHIDA, E., OSADA, Y., and OHNO, H., J. Makromol. Sci.—Phys., 1980, **B17**, 683.
86. GULYAYEVA, Z. G., POLETAYEVA, O. A., KALACHEV, A. A., KASAIKIN, V. A., and ZEZIN, A. B., Vysokomol. Soed., 1976, **A18**, 2800.
87. KHARENKO, O. A., KHARENKO, A. V., KASAIKIN, V. A., ZEZIN, A. B., and KABANOV, V. A., Vysokomol. Soed., 1979, **A21**, 2726.
88. KHARENKO, O. A., KHARENKO, A. V., KALYUZHNAYA, R. I., IZUMRUDOV, V. A., KASAIKIN, V. A., ZEZIN, A. B. and KABANOV, V. A., Vysokomol. Soed., 1979, **A21**, 2719.
89. TSUCHIDA, E., Makromol. Chem., 1974, **175**, 603.
90. TSUCHIDA, E., ABE, K. and HONMA, M., Macromolecules, 1976, **9**, 112.
91. BLUMSTEIN, A. and WEILL, G., Macromolecules, 1977, **10**, 75.
92. BLUMSTEIN, A., PONRATHNAM, S. and BELLANTONI, E., J. Polym. Sci., Polym. Lett. Ed., 1980, **18**, 299.
93. PONRATHNAM, S., MILAS, M. and BLUMSTEIN, A., Macromolecules, 1982, **15**, 1251.
94. YEN, T. F., DAVEY, M. and REMBAUM, A., J. Macromol. Sci.—Chem., 1970, **4**, 693.
95. REMBAUM, A., YEN, S. P. S., LANDEL, R. F. and SHEN, M., J. Macromol. Sci.—Chem., 1970, **4**, 715.
96. OHNO, H., ABE, K. and TSUCHIDA, E., Makromol. Chem., 1981, **182**, 1253.
97. CASSON, D. and REMBAUM, A., Polymer Letters, 1970, **8**, 773.
98. SERITA, H., OHTANI, N. and KIMURA, C., Kobunshi Ronbunshu, Japan, 1981, **38**, 415.
99. DANIELS, R. H., KERR, G. T. and ROLLMANN, L. D., J. Amer. Chem. Soc., 1978, **100**, 3097.

Chapter 5

POLYELECTROLYTE COMPLEXES

Eishun Tsuchida

Department of Polymer Chemistry, Waseda University, Tokyo, Japan

and

Koji Abe

Department of Functional Polymer Sciences, Shinshu University, Ueda, Japan

1 INTRODUCTION

In this review the authors lay emphasis on recent trends and new topics as well as on fundamental findings on polyelectrolytes and their complexes called polyelectrolyte complexes (PEC), polyion complexes or polysalts.

Oppositely charged polyelectrolytes associate to form polyelectrolyte complexes which, in general, are phase-separated in aqueous media by precipitation and complex coacervation, accompanied by desolvation.

A knowledge of molecular structure is essential for the characterization of these materials. It is well known that various morphological differences observed in micelles, liposomes, non-isotropic liquid crystals and crystallites are caused by the mode of molecular aggregation, leading to higher ordered structure. The connection between changes in physicochemical properties and structural changes caused by the effect of temperature, pressure, electromagnetic field and microenvironment is of considerable interest and this knowledge is widely used in practical applications.

In addition, the considerable progress made in studies on biological systems has established that macromolecules and hybrid intermacromolecular aggregates are fundamental constituents of the tissues of living bodies. Moreover, now that a detailed account of the important role of macromolecular assemblies in the *in vivo* bioreactions has been obtained, it can be used to establish a deeper understanding of the mutual relationship between intermacromolecular interactions and physical properties within such systems.

Complex formation between macromolecules is unusual in involving many fundamental factors. These include the nature of the interaction forces and aggregation processes, particularly cooperative and concerted phenomena, the dynamics of polymer chain formation and the behaviour of two-state structures. New applications based on hybrid formation with controlled compositions are expected in mechanical devices, membrane technology, and biomedical materials, such as immunomicrospheres, separation and purification processes for serum proteins, and immobilized enzymes and artificial cells.

Polyelectrolytes are considered to adopt rod-like configurations in water by intrachain electrostatic repulsion, but in reality synthetic polyelectrolytes with random configurations contain statistically folded conformations of relatively expanded chains in certain domains. Thus, whereas it would be expected that reaction between polyelectrolytes should take place only on the surface of the domains by the excluded volume effect, in fact it takes place in almost the whole domain, with the destruction of the domain structure. Such a reaction is usually very rapid and leads to the formation of the scrambled structure proposed originally by Michaels.[1]

If component polyelectrolytes have rigid chains or the reaction is carried out slowly in dilute solution, there are some occasions when polyelectrolyte complexes with ladder-like structures are formed. The structures of the primary complexes obtained by the initial interaction between the polyelectrolytes are controlled by the kinetics of the reaction. Under certain conditions these complexes transform themselves into thermodynamically more stable states, sometimes with supermolecular structure. These transformations are time-dependent.[2]

In the past, studies have centred on complexes that, when initially formed, were approximately stoichiometric, i.e. the ratio of cationic to anionic sites in the [polyanion]/[polycation] polyelectrolyte complex was 1:1. However, in recent years more attention has been paid to non-stoichiometric polyelectrolyte complexes where the chain lengths of the component polyelectrolytes are very different.[3] In the reaction between polyelectrolytes, other secondary binding forces, namely hydrogen bonding, van der Waals forces and, especially in an aqueous medium, hydrophobic interactions contribute to stabilize the complexes.[4] Thus, the reactions between polyelectrolytes are dependent upon the molecular parameters of the component polyelectrolytes, preparation conditions and thermodynamic parameters. By controlling these factors, molecular structures can be identified and selective complexation can be carried out even in the case of artificial macromolecules.[5]

Physical and chemical properties of the polyelectrolyte complexes are characterized by the coexistence of hydrophilic microdomains, consisting of ionic sites of high density with bound water, and hydrophobic microdomains. The physicochemical properties of these complexes reflect, in part, those of the component polyelectrolytes and, in part, specific new properties resulting from the delicate and complicated balance between the hydrophilic and hydrophobic microdomains.

For the background to the history and fundamental science of polyelectrolytes and their complexes, readers are referred to several excellent books[6-8] and reviews.[9-12]

2 CHARACTERIZATION OF POLYELECTROLYTE COMPLEXES

2.1 Polyelectrolytes

Polyelectrolytes are polymers which contain many ionic groups. They dissolve in water and dissociate into macro-ions and micro-ions. For this reason, the solution properties of polyelectrolytes are quite different from those of the non-electrolyte polymers and microelectrolytes. These properties include high viscosity, decreased surface tension, formation of acidic and basic gels and Donnan equilibria. These properties enable polyelectrolytes to be used in a number of applications, for example as ion-exchange resins, flocculants, ionomers and colloid stabilizers. Moreover, most natural polymers are considered to be polyelectrolytes, and polyelectrolyte interactions are often exhibited in biological systems. For this reason, knowledge of polyelectrolytes and their complexes can assist basic investigations aimed at elucidating the interactions between biopolymers in biosystems.

TABLE 1

POLYELECTROLYTES

Polymers	Polyanions	Polycations	Polyampholytes
Natural polymers	Heparin, chondroitin sulfate, alginic acid, pectinic acid, gum arabic, DNA, RNA, agar	Chitin, gelatin	Proteins
Artificially modified natural polymers	Carboxymethylcellulose, sulfated cellulose, phosphorylated cellulose	Diethylaminoethyl-dextran, glycol chitosan	
Artificial polymers	Poly(methacrylic acid), poly(acrylic acid), poly(styrenesulfonate), poly(L-glutamic acid), poly(phosphoric acid), sulphonated poly(vinyl alcohol)	Poly(L-lysine), poly(ethyleneimine), poly(vinylpyridine), poly(vinylbenzyltrimethyl-ammonium chloride), poly(diallyldimethyl-ammonium chloride), Ionenes	Copolymers of anionic and cationic comonomers

Polyelectrolytes are generally divided into three classes: (1) poly-anions, i.e. polymers with acid groups that undergo dissociation such as the carboxylic acids, the sulfonic acids and the phosphonic acids; (2) polycations, i.e. polymers with basic groups such as primary, secondary, tertiary and quaternary amines, and phosphonium moieties; (3) polyampholytes with both acid and basic groups. Table 1 shows typical examples of these polyelectrolytes. They can be further classified into sub-groups depending on whether ionic groups are strong or weak.

The specific properties of polyelectrolytes in aqueous medium result from three main causes; the shapes of macro-ions in water, the interactions between macro-ions and other macro-ions and micro-ions, and the hydration of macro-ions. Counterions that have dissociated are restrained in the region of the macro-ions by a high electrostatic potential. This phenomenon can be measured experimentally by determining the changes of activity or activity coefficient (γ) of micro-ions. For example, in the presence of partially sulfated poly(vinyl alcohol), KPVS, (ca 10^{-2} mol of ionic groups of KPVS/litre), γ_K is reduced to about 0·1, i.e. the substantial concentration of free K^+ ions is decreased.[13] On adding micro-salts to polyelectrolytes, micro-ions carrying an electrical charge with the same sign as that of the macro-ions are excluded from the domains of macro-ions, while those carrying a charge of the opposite sign tend to be concentrated into the macro-ion domains. This latter effect leads to a reduction of the electrostatic potential of the macro-ion. Finally, the effect of electric fields on macro-ions is contrasted with the effect on non-electrolyte polymers. Electrostatic potentials of polyelectrolytes have been analyzed by assuming hard sphere and rod models.[14] The surface potentials (ψ) of these models are respectively:

$$\psi = \frac{q}{\varepsilon r_1}\left(1 - \frac{\kappa r_1}{1 + \kappa r_2}\right) \tag{1}$$

$$\psi = \frac{2q}{\varepsilon L}\left(\frac{K_0(\kappa r_1)}{\kappa r_1 K_1(\kappa r_1)} + \ln\frac{r_1}{r_2}\right) \tag{2}$$

where r_1 is the radius of the domain where micro-ions are excluded, r_2 is the sum of the radius of this domain and that of a micro-ion, q is the surface charge, ε is the dielectric constant of the solvent, $1/\kappa$ is the radius of the ionic atmosphere where the probability of micro-ions existing is the highest, L is the unit length of rod, and K_0 and K_1 are zero-order and first-order modified Bessel functions, respectively. In these equations, one must note that q and ε, that is the charge density of macro-ions and ionic strength, are important factors. Experimental findings obtained by using globular proteins (sphere model) and DNA (rod model), support these theories.[15]

Generally organic acids are dissociated in water, thus:

$$AH \overset{K}{\rightleftharpoons} A^- T—T + H^+$$

$$K = \frac{[A^-][H^+]}{[AH]} \qquad (3)$$

where K is the dissociation constant.

However, in polyacid systems, individual dissociation is not observed. The dissociation of ionizable groups of a polyelectrolyte is affected by the states of adjacent ionizable groups, the type of counterion, ionic strength and conformations of the polyelectrolyte. Therefore, a polyacid is not to be regarded as an assembly of monomeric acid units where the relationship between pH and the degree of neutralization $(\bar{\alpha})$ is given by the equation:

$$pH = pK_a + \log\{\bar{\alpha}/(1 - \bar{\alpha})\} \qquad (4)$$

For polyacids this equation cannot be applied directly, for the reasons given above, and has to be replaced by a modified equation which holds over a wide range of values:

$$pH = pK_a' + n \log\{\bar{\alpha}/(1 - \bar{\alpha})\} \qquad (5)$$

where pK_a' is an apparent pK_a and n is a parameter of the interaction between adjacent ionizable groups in eqn (5), the Henderson–Hasselbach equation.[16] In the poly(methacrylic acid) (PMAA) and poly(acrylic acid) (PAA) systems, the pK_a' and n values are respectively 7·3 and 2·3 for PMAA and 6·0 and 2·2 for PAA (see Fig. 1), which contrast with 3·5 and 1·0 for the methacrylic acid (MAA) monomer.[17] This discrepancy is explained by reduction of the dissociation of polyelectrolytes by the interactions between adjacent ionic groups. By contrast, in the PMAA system, a plateau is observed at pH 6. This phenomenon is caused by a structural transition:

Compact form $\rightleftharpoons$ extended form

Similar phenomena are observed with poly(α-amino acid)s such as poly(L-glutamic acid) (PGA) and poly(L-lysine) (PLL), i.e. α-helix→ random coil. These structural changes are thought to be caused by a

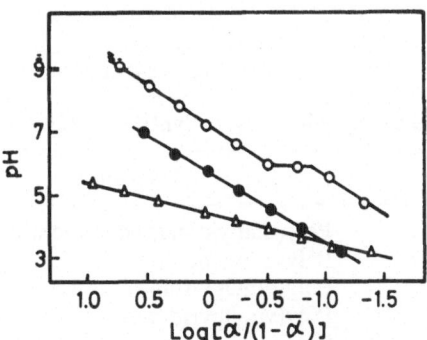

FIG. 1. Henderson–Hasselbach plots of weak polyacids: ○, poly(methacrylic acid); ●, poly(acrylic acid); △, methacrylic acid.

change in the balance of electrostatic repulsion, intramolecular hydrophobic interaction and hydrogen bonding.

So far, the effects of mainly monovalent and inorganic counterions have been discussed. However, multivalent or organic counterions are more strongly bound into the domains of polyelectrolytes. The reaction between oppositely charged polyelectrolytes can be regarded as an extension of counterion–polyelectrolyte reactions. Thus, counterions and complementary polyelectrolytes react competitively, with the latter reacting preferentially, to form polyelectrolyte complexes, probably because of cooperative interaction.

2.2 Complexation between Polyelectrolytes

Table 2 summarizes reports on PECs since 1981. These can be conveniently classified according to the nature of their component polymers, i.e. whether they are synthetic or natural. There are three such classes of PECs if those composed only of biopolymers are excluded. The essential and typical difference between synthetic and biological polymers is said to lie in their heterogeneities, e.g. molar mass distribution, monomer sequence and higher ordered structure. Thus, in PEC systems composed only of synthetic polymers, it is necessary to discuss the reactions in statistical terms, and this makes theoretical considerations difficult. Very simple theories based on colloidal chemistry were proposed by Oosawa,[126] Nakajima,[127] and Polderman.[128] These workers considered very simple and specific models, such as parallel rods and hard spheres, and assumed a homogeneous distribution of molar mass with no contribution from interaction forces other than electrostatic interactions. The actual

TABLE 2
POLYELECTROLYTE COMPLEXES

A. Synthetic polyelectrolyte–Synthetic polyelectrolyte

Polyanion	Polycation	Reference
	Poly(diallyldimethylammonium)	18–20
	Poly(ethyleneimine)	21–23
	Poly(vinylpyridine)	24–27
	Poly(vinylpyridinium)	28–32
	Ionene	33–36
Poly(carboxylic acid)s	Polycations containing piperidine rings	37,38
	Polyamine	39,40
	Polyampholytes	41
	Polyviologens	42
	Other polycations	43,44
Poly(styrenesulfonate)	Polyviologens	45–48
	Poly(acrylamide)	49
Poly(phosphate)	Poly(diallyldimethylammonium)	50,51
	Poly(1,4-pyridiniumdiylethylene)	52
Sulfated poly(vinyl alcohol)	Poly(diallyldimethylammonium)	53
	Weak polycations	54
Polysilicilic acid	Poly(diallyldimethylammonium)	55
Anionic copolymers	Cationic copolymers	56–61

B. Natural polyelectrolyte–Synthetic polyelectrolyte

Natural	Synthetic	Reference
Heparin	Polycations	62–64
Carboxymethylcellulose	Poly(diallyldimethylammonium)	65–67
Lignosulfonate	Polycations	68–70
Acidic polysaccharides	Weak polybases	71–73
Chitosan	Polyanions	74,75
Diethylaminoethyldextran	Polyanions	76,77
Cellulose	Polyampholytes	78
Proteins	Strong polyelectrolytes	79,80
Haemoglobin	Sulfated poly(vinyl alcohol)	81
Albumin	Poly(carboxylic acid)s	82,83
Albumin	Poly(vinylpyridinium)	84
Globulin	Polyampholyte	85
Lipoproteins	Polyanions	86
C-reactive proteins	Poly(L-lysine)	87,88
Enzymes	Weak polyacids	89,90
Enzymes	Strong polyacids	91,92

TABLE 2 (*contd*)

B. Natural polyelectrolyte–Synthetic polyelectrolyte

Natural	Synthetic	Reference
Enzymes	Weak polybases	93
Enzymes	Strong polybases	94,95
Gelatin	Polycations	96–98
DNA	Ionenes	99
DNA	Polyamines	100
DNA	Poly(L-lysine)	101–103
Polynucleotides	Poly(L-lysine)	104,105
Cells	Poly(acrylic acid)	106
Cells	poly(L-lysine)	107,108
Bacteria	Poly(vinylpyridinium)	109
Microtubules	Poly(L-lysine)	110
Tobacco mosaic virus	Poly(L-lysine)	111
Vessel wall	Poly(L-lysine)	112

C. Natural polyelectrolyte–Natural polyelectrolyte

Polysaccharides	Polysaccharides	113–115
Proteins	Polysaccharides	116–123
Proteins	Proteins	124
Proteins	DNA	125

complexation reactions cannot be accurately forecast using such theories and, consequently, the preparation and physicochemical properties of PECs have, unfortunately, to be discussed qualitatively.

A general equation for the formation of PECs can be represented as follows:

$$
\begin{array}{l}
\left|\!\!\begin{array}{l}
-A^-H^+(M^+) \quad (OH^-)X^- {}^+B- \\
-A^-H^+(M^+) + (OH^-)X^- {}^+B- \\
-A^-H^+(M^+) \quad (OH^-)X^- {}^+B-
\end{array}\right. \overset{K}{\rightleftharpoons} \\
\quad \text{Polyanion} \quad\quad \text{Polycation}
\end{array}
$$

$$
\left|\!\!\begin{array}{l}
-A^- \cdots {}^+B- \\
-A^- \cdots {}^+B- \\
-A^- \cdots {}^+B-
\end{array}\right. + H^+(M^+)X^-(OH^-) \quad (6)
$$

$$
\underset{\text{PEC}}{} \quad\quad\quad \underset{\text{Micro-electrolytes}}{}
$$

where A^- is an acid residue of a polyanion, B^+ is a basic residue of a polycation, and M^+ and X^- are an alkaline metal and halogen ion, respectively. The formation of PECs is considered to be a neutralization reaction accompanied by the release of micro-electrolytes and

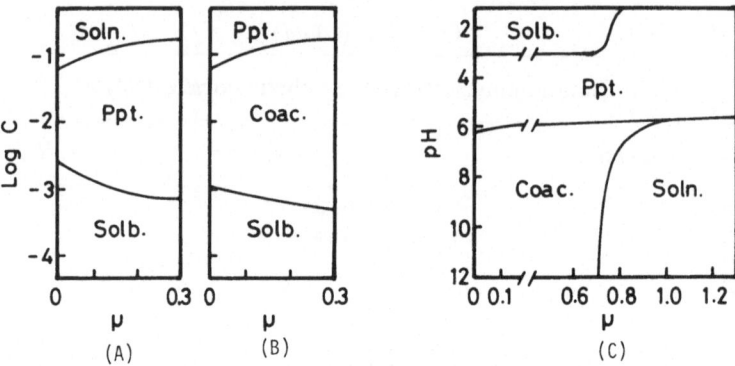

FIG. 2. Phase diagrams of poly(methylacrylic acid)(PMAA)–2X complex. (A) Degree of neutralization of PMAA ($\bar{\alpha}$) = 0; (B) $\bar{\alpha}$ = 1; (C) concentration of PEC is constant (5×10^{-3} unit mol/litre). $T = 25°C$. Ppt., precipitate; Coac., coacervate; Solb., soluble complex; soln., PEC is not formed.

water. This reaction can be regarded as a specific case of the exchange reaction of counterions. The equilibrium constant (K) may be dependent on the concentration of micro-electrolytes, but when the molar masses of component polyelectrolytes are high enough and exceed a certain critical chain length, eqn (6) is usually shifted to the right, apparently irreversibly, because of cooperative and concerted interactions;[129] these are discussed in detail in Section 3. Stoichiometric PECs, with an equimolar composition of dissociated ionic sites (i.e. in which neutralization between a polycation and a polyanion takes place completely) are obtained as precipitates, complex coacervates, gels or water-soluble complexes; the exact nature of the product depends on molecular, thermodynamic and preparation parameters. The important molecular parameters of the component polyelectrolytes are their chemical structure—especially whether ionic sites are on the main or side chains; molar mass, molar mass distribution, crystallinity and hydrophobicity. The thermodynamic parameters include composition and temperature, and preparation parameters include ionic strength and solvent. Figure 2 shows the phase diagrams of a PEC composed of PMAA and ionene-type* (2X) polycations.[130] Since PMAA is a weak polyacid, the degree of dissociation of PMAA

* Ionenes

$$
\left[\begin{array}{c} CH_3 \quad X^- \quad CH_3 \, X^- \\ | \qquad\qquad | \\ -N^+ -R_1 -N^+ -R_2 - \\ | \qquad\qquad | \\ CH_3 \qquad\quad CH_3 \end{array}\right]_n
$$

$$
\left(\begin{array}{l} mm\text{-ionene; } R_1 = R_2 = -(CH_2)_{\overline{m}} \\ mm'\text{-ionene; } R_1 = -(CH_2)_{\overline{m}}-, \ R_2 = -(CH_2)_{\overline{m'}} \\ mX\text{-ionene; } R_1 = -(CH_2)_{\overline{m}}-, \ R_2 = -CH_2C_6H_4CH_2- \\ XX\text{-ionene; } R_1 = R_2 = -CH_2C_6H_4CH_2- \end{array}\right)
$$

$X^- = Cl^-, \ Br^-$

changes with the changes of pH, the concentration of the repeating unit (mol/litre) in the solution and ionic strength (μ), hence causing a change in the composition of PEC.

The increase of μ also reduces the electrostatic interaction between complementary polyelectrolytes and increases the hydrophobicity of the polyelectrolytes as a result of the contraction of their chains. In addition the degree of hydration of the polar regions of ionic linkages, rigidity of the polymer chains, interpenetration of polymer chains and temperature determine the state of PECs. Additionally, for non-stoichiometric PECs with excess anionic or cationic sites, precipitates or water-soluble complexes are formed. The nature of the complex formed depends on the steric conformities between complementary chains, and the combination of the difference of the chain length of each component polyelectrolyte and the composition of the poly-electrolyte solutions[131] (see Section 4.2).

The composition of the stoichiometric complexes (r) is calculated from the following equation.

$$r = \frac{[\text{Ionizable groups of PA}]}{[\text{Ionizable groups of PC}]} = \frac{\alpha_{PA}}{\alpha_{PC}} \tag{7}$$

where PA and PC are a polyanion and a polycation, and α_{PA} and α_{PC} are the degrees of dissociation of PA and PC, respectively. For example, in the systems containing weak polyacid–strong polybase (3X) combinations, r varies from about 0·2 to 1·0 with changes of $\bar{\alpha}$, as shown in Fig. 3.[132] It is well known that the α_{PA} value in the absence

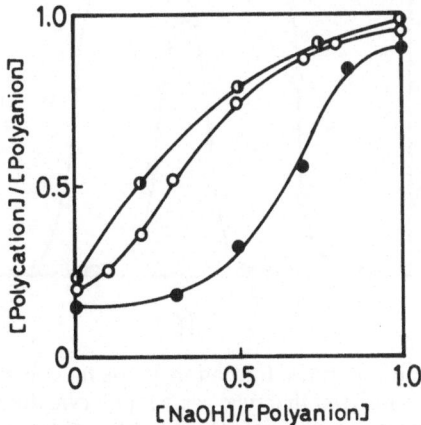

FIG. 3. The compositions of complexes of weak polyacids with 3X: O, poly(methacrylic acid); ◑ Poly(acrylic acid); ●, poly(L-glutamic acid).

of 3X at $\bar{\alpha} = 0$ is only a few per cent. A polycation induces an α_{PA} of about 20% while the pK_a' of PMAA changes from 7·3 to 4·3 with the addition of 3X.[133] In contrast to PAA, both PMAA and PGA, whose conformations are transformed dramatically with pH change of the solution, show S-shape curves corresponding to those conformational changes. Values of $1/r$ for PECs of reverse pairs of polyelectrolytes, i.e. strong polyacid–weak polybase, also vary from about 0·1 to 1·0 for the same reason. On the other hand, the r value of a stoichiometric PEC of a strong polyacid–strong polybase pair is always unity irrespective of pH.[134] In a weak polyacid–weak polybase system, r is variable and corresponds with the changes in the dissociation of both weak polyelectrolytes.[135] Figure 4 shows the pH dependence of turbidometric changes of a solution prepared by mixing solutions of the complementary polyelectrolytes, poly(4-vinylpyridine) (P4VP), a weak polybase, and PMAA in a ratio of 1:1.[25] In pronounced acidic (pH < 1) and alkaline (pH > 10) conditions the individual weak polyelectrolytes have an insufficient number of dissociated ionizable groups to form stable complexes. In the pH regions of 1–2 and 9–10 PMAA-rich and P4VP-rich non-stoichiometric complexes are formed, respectively. In the pH range 2–9 stoichiometric PECs are obtained, but their composition varies according to α_{PA} and α_{PC}. For reference, the pH dependence of the turbidity of a random copolymer of 4-vinylpyridine and methacrylic acid (composition = 49:51) is also

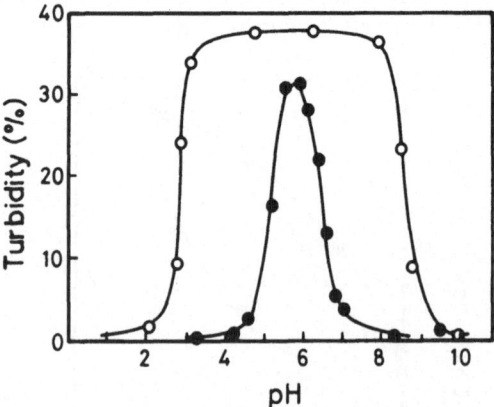

FIG. 4. Dependence of complex formation in weak polyacid–weak polybase system on pH: ○, poly(methacrylic acid)–poly(vinylpyridine) (1:1); ●, poly(methacrylic acid-co-vinylpyridine) (51:49). Solvent, Water/methanol (1:1).

shown in Fig. 4. Compared with the mixed system of P4VP and PMAA, the precipitation of a stoichiometric PEC occurs over a narrower pH range. This result is considered to arise from the presence of an excess of dissociated groups in one chain of a copolymer, making it soluble in water.

So far, the effects of pH, ionic strength, molar mass and composition of the component polyelectrolytes on the complexation have been discussed. There are other factors which affect the reactions of the complementary polyelectrolytes and which will now be discussed briefly. Molecular parameters, especially (1) the conformations and configurations of the component polyelectrolytes, e.g. relative positions of complementary ionic sites, (2) the distribution of ionizable groups along the chains and (3) rigidity and crystallinity, e.g. restriction of internal rotations and bonds, affect complexation. It was reported by Nakajima et al.[136] that polysaccharides with rigid polymer chains formed non-stoichiometric PECs with a ladder-like structure. The most typical systems reflecting steric effects are composed of poly(α-amino acid)s or polynucleotides. Figure 5 represents the circular dichroism spectra of PGA and its complexes with several polycations (A), and the relation between reduced molar ellipticities at 222 nm ($[\theta']_{222}$), which belongs to the α-helical structure, and

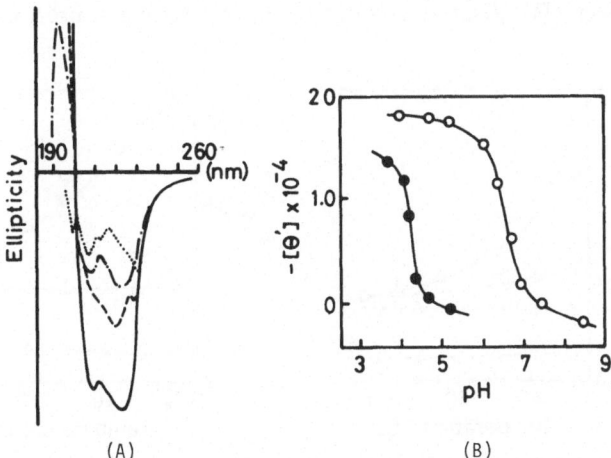

FIG. 5. Conformational changes observed by circular dichroism spectra in the formation of polyelectrolyte complexes. (A) Poly(L-glutamic acid) (—) and its complexes with 2X (– – – –), XX (·····) and poly(oxymethyl-1-methyl-ene-N,N,N-trimethylammonium chloride) (—·—). (B) Helix–coil transition of poly(L-glutamic acid) (○) and its complex with 6,6-ionene (●).

pH (B).[137] It is clear that the α-helical structure of PGA is destabilized and partly destroyed by complexation with polycations and that the helix–coil transition pH is shifted from 7 to 4. The degree of destabilization of the α-helical structure varies with the type of polycation. In the PLL–PMAA system, this destabilizing effect was found to be proportional to the atactic content of PMAA, i.e. isotactic $\ll$ syndiotactic $<$ atactic.[138] There have been many studies on the conformation of PECs composed of PLL or poly-(L-arginine) and various polyanions,[139–142] where such conformations as an α-helix, a β-form, a super-helix and a random coil were investigated in relation to the structure of polyanions. These results indicate that steric conformity plays an important role in the reactions of polyelectrolytes and in the higher ordered structures of the PECs.

Organic solvents, with the lower dielectric constants, reduce the dissociation of the component polyelectrolytes and weaken hydrophobic interactions, but strengthen the electrostatic interaction. By contrast, increasing temperature increases the hydrophobic interaction in aqueous media and the kinetic energy of the polymer chains, but has little effect on the electrostatic interaction. Figure 6 shows the effects of temperature and of organic solvent on the complexation of PMAA with poly(4-vinylbenzyltrimethylammonium chloride) (PVBMA).[4] The viscosities of PEC aqueous solution (A) and methanol solution (B) decrease with increasing temperature. Changes of

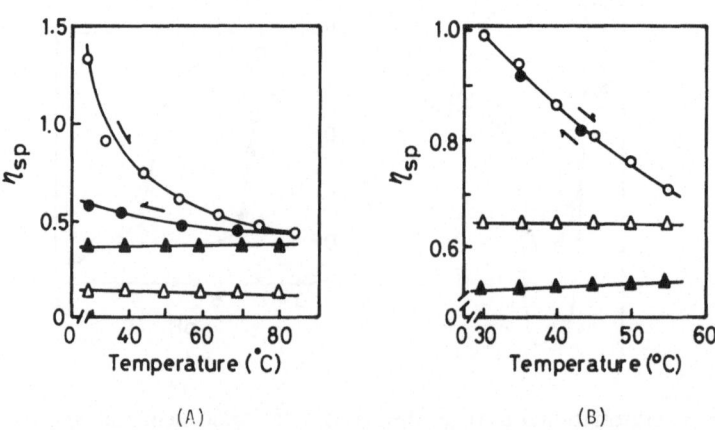

(A)   (B)

FIG. 6. Hydrodynamic properties of polyelectrolyte complex in water (A) and in methanol (B): $\triangle$, poly(methacrylic acid); $\blacktriangle$, poly(4-vinylbenzyltrimethylammonium chloride); ($\bigcirc$, $\bullet$), polyelectrolyte complex.

viscosity with temperature are irreversible for aqueous solutions but reversible for methanol solutions. These phenomena suggest that the hydrophobic interaction stabilizes a PEC and that once formed, it is difficult to dissociate a PEC into its component polyelectrolytes, probably because of an entropy term. Moreover, the optimum [PVBMA]/[PMAA] ratio for gel formation at room temperature in these systems is 0·5 in an aqueous medium and 0·17 in a methanol solution; this is because an organic solvent reduces the dissociation of polyelectrolytes, especially that of PMAA which is a weak polyelectrolyte.

2.3 Physicochemical Properties

Physicochemical properties of PECs are determined mainly by two factors; one reflects the properties of the component polyelectrolytes and the other relates to the mixing at a molecular level of the polymer chains. A mixing rule was proposed by Nielsen[143] in order to estimate the physicochemical characteristics of the system composed of two different polymers, described by the following equation:

$$P^n = (P_A)^n C_A + (P_B)^n C_B + k C_A C_B \qquad (8)$$

where P is a given property, C_A and C_B are the concentration fractions of the component polymers, k is an interaction parameter (or an interfacial parameter in the case of a phase separation) and n is a morphological parameter (for homogeneous mixing $n = 1$, as in the case of a PEC, and for phase separation $-1 < n < 1$). The first two terms on the righthand side of eqn (8) together represent the additive characteristics of the individual component polymers and the last term represents the mixing effects, which can induce specific new characteristics, different from those of each component polymer. Compared with other interpolymer complexes which are hydrogen bonded, charge-transfer, stereo-complexed or polymer hybrids, PECs have unique properties, because the main interaction force is very strong, and hydrophilic and hydrophobic domains coexist. Some of these unique characteristics are as follows.

(1) Stoichiometric PECs are insoluble in common organic and inorganic solvents; the exceptions are coacervates of stoichiometric PECs which are dissociated to the component polyelectrolytes above $\mu = 0·7$ (see Fig. 2) and non-stoichiometric

PECs which are usually soluble in water containing certain concentrations of microsalts.

(2) Changes in the water and micro-ion contents, molecular para-meters and composition of the PECs yield materials ranging from rubber-like or leather-like latices to hard plastics.

(3) PECs show specific electrical properties, probably due to the existence of microdomain structures.

(4) PECs have good transparency with definite refractive indices, good permeabilities for various solutes, solvents, ions and gases and anticoagulant properties.

(5) However, mechanical properties are not good.

Figure 7 shows a phase diagram of a stoichiometric PEC of poly(sodium styrenesulfonate), NaSS, with an ionene-type polycation (3X). This PEC is soluble in only a narrow region of ternary solvent mixtures, i.e. water/water-compatible organic solvent/micro-salts, such as water/methanol (acetone, dioxane)/KBr (NaBr, KCl). This fact suggests that an organic solvent weakens the hydrophobic interactions and micro-salts weaken the electrostatic interactions and water acts as a medium dissolving the dissociated polyelectrolytes.[144]

Table 3 lists the mechanical properties of various PECs, in com-parison with a cellophane and with a partially crosslinked poly(2-hydroxyethyl methacrylate) (PHEMA). PECs 1–3 and 8–9 are com-plexes of a pair of strong polyelectrolytes, and PECs 4–7 are of weak ones. As shown in this Table, it is found that there are few basic

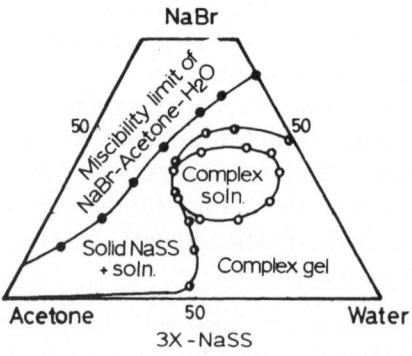

FIG. 7. Solubility in ternary solvent system of polyelectrolyte complex of 3X–poly(sodium styrenesulfonate) at 30°C.

TABLE 3

PHYSICAL PROPERTIES OF POLYELECTROLYTE COMPLEXES

No.	Sample	Strength (dyn/cm^2)	Modulus (dyn/cm^2)	Elongation $(\%)$	Ref.
1	NaSS–PVBMA (40%)[a]	8.5×10^7	4.0×10^9	10	
2	NaSS–PVBMA (60%)	5.2×10^7	3.2×10^8	25	145
3	NaSS–PVBMA (80%)	0.8×10^7	5×10^7	220	
4	PAA–PEPP (35%)	1.6×10^7	1.3×10^9	250	146
5	PAA–PEPP (50%)	5.0×10^7	3.0×10^7	400	
6	PC–PA (65% RH)[b]	2.1×10^7	8.3×10^8	201	147
7	PC–PA (in H_2O)	1.4×10^6	2.8×10^5	179	
8	PTC–PSA (65% RH)	3.2×10^8	5.6×10^9	139	147
9	PTC–PSA (in H_2O)	2.8×10^6	0.7×10^5	600	
10	Cellophane (45%)	2.8×10^8	3.0×10^8	92	148
11	PHEMA (45%)	4.1×10^6	$<6.9 \times 10^7$	140	

Abbreviations: NaSS, poly(sodium styrenesulfonate); PVBMA, poly(vinyl-benzyltrimethylammonium chloride); PAA, poly(acrylic acid); PEPP, poly(ethylenepiperazine); PC, aminoacetalized poly(vinyl alcohol); PA, carboxymethylated poly(vinyl alcohol); PTC, poly(vinyl alcohol) acetalized with 2,2-diethoxyethyl trimethyl ammonium; PSA, sulfated poly(vinyl alcohol); PHEMA, crosslinked poly(hydroxyethylmethacrylate).
[a] Water content,
[b] Relative humidity.

differences between weak and strong polyelectrolytes, but the mechanical properties are dramatically affected by the water content and composition rather than by the molecular characteristics of the component polyelectrolytes. In a pair of weak polyelectrolytes, however, hydrogen bonds as well as electrostatic interactions occasionally increase strength. Such variations in mechanical properties have both advantages and disadvantages; the advantages include application as humidity sensors and ease in tailor-making hydrogels with specific properties; the disadvantages are dimensional instability and difficulty in reproducing mechanical properties. In order to avoid these problems, resort has to be made to chemical crosslinking or interpenetration of the polymer chains effected by heat treatment, addition of crosslinking agents or development of new preparation methods. From the dynamic mechanical measurements, the master strain–relaxation curves of these PECs are found to be similar to those of conventional glassy amorphous polymers.[148]

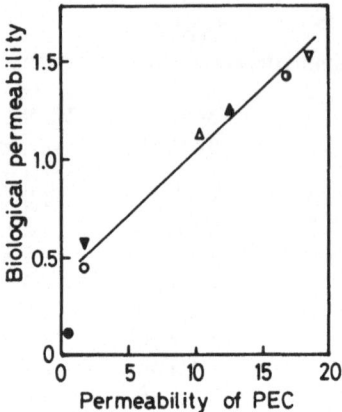

FIG. 8. Relationship between permeability of biological membrane and membrane composed of poly(sodium styrenesulfonate) and poly(4-vinylbenzyltrimethylammonium chloride): △, Aminopyrine; ▲, aniline; ◐, antipyrine; ▼, propylene glycol; ▽, barbital; ○, salicylic acid; ●, thiourea.

PECs have extremely high and controllable permeabilities towards gases, water, micro-electrolytes and micro-organic solutes and high resistivity to high molar mass compounds. Utilizing these characteristics, PECs composed of PVBMA–NaSS have been developed as membranes for ultrafiltration, dialysis, fuel-cell and battery separators; also environmental sensors and chemical detectors are available on a commercial basis. Observations show that the PEC membranes are permeated by biological micro-solutes in a similar manner to biomembranes, as shown in Fig. 8.[149] The ratio of permeabilities of CO_2 to O_2 is about 20 (values for general polymer membranes are about 6–8).[150] These observations show that PEC membranes behave like 'immobilized water' membranes similar to lung alveoli and biological lipid bimolecular membranes; this is because the presence of many anionic and cationic sites in PEC membranes can effectively destroy the structure of water in the membranes. Therefore, selective permeabilities with respect to solutes are controlled by the water and micro-electrolyte contents and the topology of ionic crosslinking. The biocompatibility of PECs having a slight excess of polyanion[151] is discussed in detail in Section 5.. PECs find applications as biomedical polymers, for example in contact lenses,[152] haemodialysis membranes and anticoagulant materials,[153] because they are both transparent and permeable.

3 EFFECTS OF MACROMOLECULES ON COMPLEXATION OF POLYELECTROLYTES

3.1 Cooperative and Concerted Interactions

Deviations of properties of polymers from those expected if they were the simple sum of the repeating units are generally called 'polymer effects'. In discussing these so-called polymer effects, one must take into account inter- and intra-chain interactions. The complexation effects are considered to be characterized by the following four factors:

(1) cooperative interaction—the effects caused mainly by the entropy term since the active sites are set closely together;
(2) concerted interaction—the effects caused by the co-existence of different kinds of interaction forces;
(3) component polymer chains having complementary higher ordered structures, for example matching distributions of active sites; and
(4) the appearance of specific microdomains caused by the excluded volume, concentration of polymer chains and active sites, e.g. hydrophobic and electrostatic effects.

A qualitative representation of the cooperative interaction between two complementary polyelectrolytes and their relationship with thermodynamic parameters is illustrated schematically in Fig. 9. If one of the ionic sites reacts with a complementary one, then by fixing the

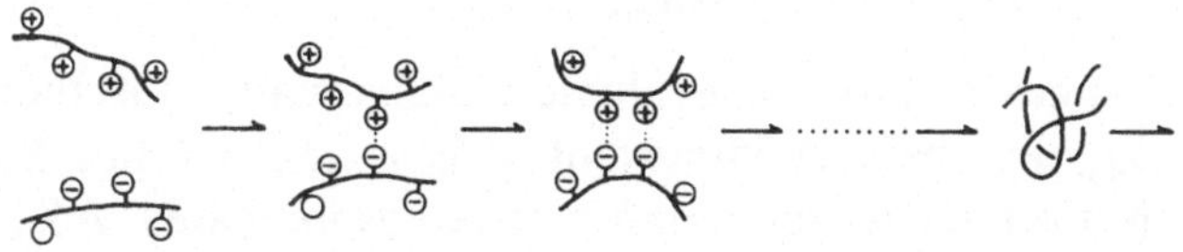

$$\Delta S_1^{\ddagger} \gg \Delta S_2^{\ddagger} > \Delta S_3^{\ddagger} > \Delta S_i^{\ddagger} < \Delta S_{i+1}^{\ddagger} \quad (\approx \Delta S_n^{\ddagger})$$

$$\Delta S_1 < \Delta S_2 < \Delta S_3 < \Delta S_i > \Delta S_{i+1} \quad (\approx \Delta S_n)$$

$$\Delta H_1 \geq \Delta H_2 \geq \Delta H_3 \geq \Delta H_i \simeq \Delta H_{i+1} \quad (\approx \Delta H_n)$$

$$\Delta F_1 > \Delta F_2 > \Delta F_3 > \Delta F_i < \Delta F_{i+1} \quad (\approx \Delta F_n \approx 0)$$

$$\Delta K_1 < \Delta K_2 < \Delta K_3 < \Delta K_i \simeq \Delta K_{i+1} \quad (\approx \Delta K_n)$$

FIG. 9. Schematic diagram of the difference of thermodynamic parameters of each bonding in the formation of a polyelectrolyte complex.

ionic sites at sterically appropriate positions, the neighbouring active sites are more favourably placed to form the additional bonds. Thus, the activation entropy $(\Delta S^{\ddagger})$ is the highest for the formation of the first bond, since the active sites have to approach within a limited distance of each other—in contrast to low molar mass ions; additionally, the diffusion of polyelectrolytes is slow and the matching of steric features more difficult. The activation entropy is dramatically reduced when the second bond is formed and thereafter progressively decreases as the bonds increase, but if there are distortions of component polyelectrolytes as fresh bonds are formed by complexation, $\Delta S_i^{\ddagger}$ increases again. The absolute value of enthalpy change (ΔH_i) is not very large, because the PEC formation may be regarded as a substitution reaction of counterions as shown by eqn (6). When the number of bonds increases, the electrostatic potential of an ionic site is lowered by the decrease of electrostatic repulsions from neighbouring ionic sites. This phenomenon is enhanced as weak polyelectrolytes are induced to dissociate. Therefore, since the energy for dissociation of the undissociated ionizable groups is lowered, ΔH_i decreases gradually with the advance of complexation but the difference of the absolute values of ΔH_i may be small. On the other hand, the entropy change (ΔS_i) is the sum of entropy changes resulting from the fixation of polymer chains (ΔS_{ip}), an increasing freedom of released micro-ions (ΔS_{im}), hydrophobic interactions (ΔS_{ihp}), configurational changes of polyelectrolytes (ΔS_{ic}), and the changes of solvation and solvent structures (ΔS_{is}). Thus:

$$\Delta S_i = \Delta S_{ip} + \Delta S_{im} + \Delta S_{ihp} + \Delta S_{ic} + \Delta S_{is} \qquad (9)$$

These entropy changes cannot be evaluated individually. Of them, ΔS_{ip}, ΔS_{im} and ΔS_{ihp} may be very important. ΔS_{ip} is negative but ΔS_{im} and ΔS_{ihp} are both positive. As the number of bonds increases, ΔS_{im} is scarcely changed whereas ΔS_{ihp} increases and ΔS_{ip} decreases, resulting in the gradual increase of ΔS_i. Consequently, the enthalpy for breaking one bond of the inner repeating unit of PEC is not compensated for by an increase in entropy, because this process is not accompanied by additional translational degrees of freedom. The free energy change $(\Delta F_i$, where $\Delta F_i = \Delta H_i - T\Delta S_i)$ decreases initially with the increase in the number of bonds and then increases to about zero because of the effect of entropy–enthalpy compensation. As a result, the total free energy change (ΔF^0) for the formation of PECs may decrease with the increase in the number of ionic sites and chain lengths of the component polyelectrolytes until it reaches a certain critical chain length.

The theoretical treatment of the cooperative interaction in polymer–oligomer systems resulting from hydrogen bonding was reported by Kabanov and Papisov.[10] The following equilibrium constant relationship was proposed:

$$K_n = K_1^n = \exp\left(-n\Delta F_1^0/RT\right) \tag{10}$$

where K_n is the overall equilibrium constant for the formation of interpolymer complexes composed of a high polymer and an oligomer with n repeating units, and K_1 and ΔF_1^0 are the equilibrium constant and the free energy change respectively for the reaction of one repeating unit.

Another important factor in the formation of an interpolymer complex is concerted interaction. It has already been stated that many kinds of interaction forces contribute concertedly to the formation of PECs. Thus the total free energy change may be represented by the following equation:

$$\Delta F^0 = \Delta F_e^0 + \Delta F_{hb}^0 + \Delta F_{hp}^0 + \Delta F_v^0 + \ldots \tag{11}$$

where ΔF_e^0, ΔF_{hb}^0, ΔF_{hp}^0 and ΔF_v^0 represent the free energy changes due to electrostatic interaction, hydrogen bonding, hydrophobic interaction and van der Waals forces, respectively. Although van der Waals forces are weak, they take place between all molecules. In PEC systems, ΔF_e^0 is, of course, the predominating term. In an aqueous medium, however, the contribution of ΔF_{hp}^0 becomes important, and in PEC systems composed of weak polyelectrolytes, ΔF_{hb}^0 is of significance in the stabilization of PECs.

The adequacy of such theoretical considerations of cooperative and concerted interactions was demonstrated experimentally in the system of PMAA and quaternized oligo(ethyleneimine)s (QOEI) with different degrees of polymerization (n). The degree of polymerization corresponds to the number of cationic sites.[154] As shown in Fig. 10, the relationships between K_n or $-\Delta F^0$ and n below $n = 4$ are described by the following equations:

$$K_n = A \exp\left(Bn\right) \tag{12}$$

$$-\Delta F^0 = \alpha n + \beta \tag{13}$$

where α is comparable with a cooperative coefficient, β is the basic binding constant and A and B are constants corresponding to α and β,

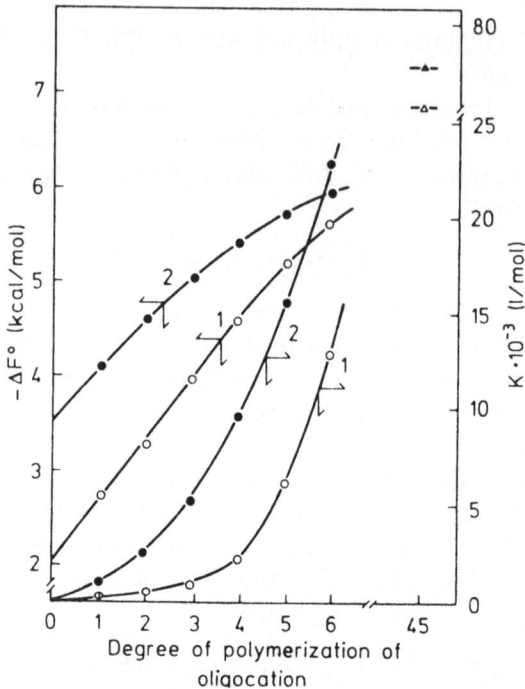

FIG. 10. Effect of chain length on the polyelectrolyte complex formation in poly(methacrylic acid) with quaternized oligo(ethyleneimine) (QOEI) system: 1, QOEI quaternized by methyl iodide; 2, QOEI quaternized by benzyl chloride.

respectively. These values are compiled in Table 4. Equation (12) is in good agreement with eqn (10) and $-\Delta F_1^0$ is calculated to be about 0·5–0·7 kcal/mol. This value is smaller than that of the Coulombic force. This result may be explained by the entropic disadvantages. The difference of β values between the two different types of QOEI might be attributed to dissimilar hydrophobicities of the benzyl and methyl groups. When $n > 5$, $-\Delta F^0$ exceeds the kinetic energies of component polyelectrolytes before the formation of a stable PEC. Under such conditions, the reactions of complementary polyelectrolytes become apparently irreversible and, for the most part, the formation of PECs leads to phase separation. The chain length required to form a stable interpolymer complex is called 'the critical chain length'. When this chain length is reached, discontinuities in various physicochemical properties, such as hydrodynamic properties and phase rule, are

TABLE 4

COOPERATIVE PARAMETERS FOR THE
FORMATION OF PECs IN PMAA–QOEI
SYSTEM^a AT 25°C

QOEI	A	B	α	β
R = CH$_3$	31·9	1·0	0·63	2·05
R = CH$_2$C$_6$H$_5$	49·0	0·8	0·47	3·66

^{a} Abbreviations: PMAA, poly(methacrylic acid), $M_w = 5·3 \times 10^4$, $\bar{\alpha} = 1·0$; QOEI, quaternized oligo(ethyleneimine)s.

observed. Critical chain lengths $(\overline{Pn})$ in some systems are listed in Table 5. In comparison with the PMAA–QOEI system, where $\overline{Pn} = 4$, $\overline{Pn}$ of PAA–QOEI is larger (5), because PAA is less hydrophobic than PMAA. Using oligo(ethyleneimine) (OEI) as the component polyelectrolyte instead of QOEI, $\overline{Pn}$ increases to 5 since OEI is a weak polyelectrolyte.[135] Moreover, when PGA–OEI is in a water/methanol medium (1:1 by volume) $\overline{Pn}$ equals 6 since the organic solvent, methanol, weakens hydrophobic interactions and reduces the dissociation of both weak polyelectrolytes.[137] By contrast, it is found that the differences between the conformations of component polyelectrolytes, e.g. the α-helix of PGA, packed coils of PMAA and random coils of PAA, have no effect on $\overline{Pn}$.[154] Similar results have been reported for

TABLE 5

CRITICAL CHAIN LENGTH $(\overline{Pn})$ FOR THE FORMATION
OF PECs AT 25°C

System^a	Solvent	$\overline{Pn}$
PMAA–QOEI	Water	4
PAA–QOEI	Water	5
PGA–QOEI	Water	4
PGA–OEI	Water	5
PGA–OEI	Water–methanol(50:50)	6

^{a} PMAA, poly(methacrylic acid), $M_w = 5·3 \times 10^4$; PAA, poly(acrylic acid), $M_w = 4·6 \times 10^4$; PGA, poly(L-glutamic acid), $M_w = 5·0 \times 10^4$; OEI, oligo(ethyleneimine); QOEI, quaternized OEI (R = CH$_3$), $\bar{\alpha} = 1·0$.

hydrogen-bonded interpolymer complexes, although $\overline{Pn}$ values in these systems are larger than those in PEC systems, since hydrogen bonds are much weaker than electrostatic interactions.[155] Figures 2 and 4 show that the formation or dissociation of a PEC takes place in very narrow regions of pH and ionic strength. These facts suggest a cooperative interaction. These values of pH and ionic strength are considered to be critical values corresponding to the critical chain length.

3.2 Molecular Selection in Complexation

It is well known that selective complex formation, which is a function of molecular stereochemistry, plays a very important role in controlling various *in vivo* phenomena, e.g. antigen–antibody reactions, enzymatic reactions and self-assemblies of proteins. In these systems, the so-called complexation effect, mentioned in the preceding section, is the main cause of molecular selection. Synthetic polymers can also form interpolymer complexes, but the accurate molecular selection in the *in vivo* situation has not yet been realized. Reasons for this are that their molecular parameters are intrinsically non-uniform and there are inevitably statistical factors in the reactions. However, even in synthetic polymer systems, there may be a possibility that, under suitable conditions, selective interpolymer complexation takes place. These conditions may be obtained by utilizing the factors controlling the formation of PECs, including interaction forces, ionic strength, pH, temperature, solvent, steric conformity and chain length. In the simplest case, the criteria for complexation among three different polymers are as follows:

$$P_1 + P_2 \xrightarrow{K_1^{n_1}} (P_1-P_2)\text{comp.} \tag{14}$$

$$P_3 + P_2 \xrightarrow{K_3^{n_3}} (P_3-P_2)\text{comp.} \tag{15}$$

$$P_1 + P_2 + P_3 \begin{cases} \xrightarrow{K_1^{n_1} > K_3^{n_3}} (P_1-P_2)\text{comp.} + P_3 \\ \xrightarrow{K_1^{n_1} < K_3^{n_3}} (P_3-P_2)\text{comp.} + P_1 \end{cases} \tag{16}$$

where $K_1^{n_1}$ and $K_3^{n_3}$ are the overall equilibrium constants in eqn (10), n_1

and n_3 are degrees of polymerization of P_1 and P_3 (where the chain length of P_2 is much greater than that of P_1 and P_3), and K_1 and K_3 are equilibrium constants of the reactions between one repeating unit in P_1–P_2 and P_3–P_2, respectively. If three component polyelectrolytes coexist under certain conditions, then one or other of P_1 and P_3 preferentially forms a PEC. The ratio $K_1^{n_1}/K_3^{n_3}$ is a coefficient, w, which characterizes the degree of selectivity; Kabanov and Papisov[10] have theoretically considered the evaluation of w. When $K_1^{n_1}$ is larger than $K_3^{n_3}$, the following cooperative interchain substitution reaction takes place:

$$(P_3-P_2)\text{comp.} + P_1 \rightarrow (P_1-P_2)\text{comp.} + P_3 \qquad (17)$$

Typical examples of selective complexation in PEC systems are compiled in Table 6.[5] In a system consisting of a weak polybase (P_1), poly(ethyleneimine) (PEI), a weak polyacid (P_2), PMAA, and a proton-accepting non-ionic polymer (P_3), poly(ethylene glycol) (PEG), the complexation is governed by the pH of the solution. At acidic pH (<2), whereas almost all of the amino groups of PEI are protonated, carboxylic acid groups of PMAA are scarcely dissociated. K_1, controlled by an electrostatic interaction, is larger than K_3,

TABLE 6
MOLECULAR RECOGNITION AND SELECTION IN PEC SYSTEMS

Polymers[a]			Solvent (pH)	Complex	Main controlling factor
P_1	P_2	P_3			
P2VP	PMAA	PEG	H_2O (3) H_2O (6)	PEG–PMAA P2VP–PMAA	pH
PEI	PMAA	PEG	H_2O (2) H_2O (5) H_2O (9)	PEG–PMAA PEI–PMAA None	pH
P4VP	PMAA	P2VP	H_2O/MeOH	P4VP–PMAA	Steric effect
PMAA	2X	NaSS	H_2O (9) $(\mu > 0.7)$	NaSS–2X	Ionic strength

[a] PMAA, poly(methacrylic acid); P2VP, poly(2-vinylpyridine); PEG, poly(ethylene glycol); PEI, Poly(ethyleneimine); P4VP, Poly(4-vinylpyridine); 2X, ionene-type polycation; NaSS, poly(sodium styrenesulfonate).

determined by hydrogen bonding, but n_3, corresponding to the number of undissociated carboxylic acid groups of PMAA, is much larger than n_1, corresponding to the number of dissociated ones. Therefore, the PMAA–PEG complex is preferentially formed by hydrogen bonding ($K_1^{n_1} < K_3^{n_3}$). In the middle pH region ($3 < pH < 8$), where both PMAA and PEI are partly ionized, n_1 and n_3 are changed. Then, if $K_1^{n_1}$ is greater than $K_3^{n_3}$ the PMAA–PEI complex will be selectively formed. Under alkaline conditions ($pH > 9$), where PEI is scarcely protonated and PMAA is completely dissociated, no complex is formed.

The cooperative interchain substitution reaction was observed in the 2X–PMAA–PEG system. Since 2X is a stronger polybase than PEI, 2X can form a PEC with PMAA even in alkaline conditions.[5] This substitution reaction seems to be explained by the following scheme:

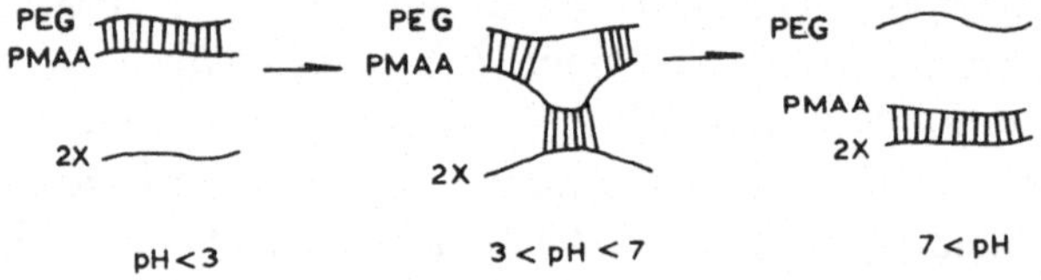

$$ (18) $$

Before the initial binary complex is completely dissociated, another component polymer interacts with the complex to form a ternary complex. Then, with the changes of conditions, one of the component polymers in the initial complex dissociates to complete the substitution reaction. Free non-bonding chains in the ternary complex may make it soluble in the aqueous solution. Since the substitution reaction generally takes place by way of the ternary complex formation, it seems to be accelerated by strengthening the interaction between the parent polymer matrix and a third component polymer.

Steric conformity also plays an important role in the selective complexation in the poly(2-vinylpyridine) (P2VP) (P_1)–PMAA (P_2)–P4VP (P_3) system in Table 6. All of these component polyelectrolytes are weak, so that both electrostatic interaction and hydrogen bonding coexist. The former is a comparatively long-range interaction force and non-directional, but the latter is rather short-range with an optimum direction for the bond. Therefore, the active sites which are located at nitrogen atoms of P2VP may be less active in forming the complex than those in P4VP because the active sites on P2VP are so close to the main chain. In the PMAA ($\bar{\alpha} = 1$) (P_1)–2X (P_2)–NaSS

(P_3) system, the selective complexation is realized by controlling the ionic strength (μ) of the solution as shown in Table 6. Below $\mu = 0.7$, the complexes obtained are ternary ones or mixtures of PMAA–2X and NaSS–2X complexes. At $\mu > 0.7$, however, PMAA–2X complex is not formed as shown in Fig. 2, selective complexation of NaSS–2X occurs. Hydrogen bonding is the chief mechanism which controls the molecular selection and formation of interpolymer complexes.[5]

4 SPECIFIC CHARACTERISTICS OF POLYELECTROLYTE COMPLEXES

4.1 Higher Ordered Structure

Higher ordered structures of PECs are affected strongly by the molecular parameters of the component polyelectrolytes. When two synthetic polyelectrolytes with random structures are mixed together, they commonly form scrambled network structures.[1] This result suggests that the formation of PECs proceeds very rapidly, so that it is completed instantaneously on first contact between the complementary polyelectrolytes. However, PECs with ladder-like structures have been formed either by matrix polymerization, where ionic monomers in very dilute solution are polymerized in the presence of an oppositely charged polyelectrolyte, or by using polyelectrolytes with rigid chains such as polysaccharides and nucleic acids as the components of the PECs. PECs with scrambled network structures may exist in a metastable state, since an instantaneous complexation will give rise to distortions in conformations and configurations of the component polymer chains. Then, the PEC that is formed initially—the primary complex—may be gradually transformed into more stable structures—secondary complexes—by the reorientation of the polymer chains as existing bonds are rearranged or new bonds are formed by electrostatic interactions and other secondary binding forces. Moreover, under certain conditions, primary and secondary complexes may aggregate spontaneously, probably through a hydrophobic interaction, to form stable assemblies with some ordered supermolecular structures (as shown in the schematic diagrams of Fig. 11).[2]

When a clear aqueous solution of 2X–PMAA ($\bar{\alpha} = 1$) PEC is prepared, by removing an amorphous precipitate of the primary complex from the solution by centrifuging, or by forming the PEC in

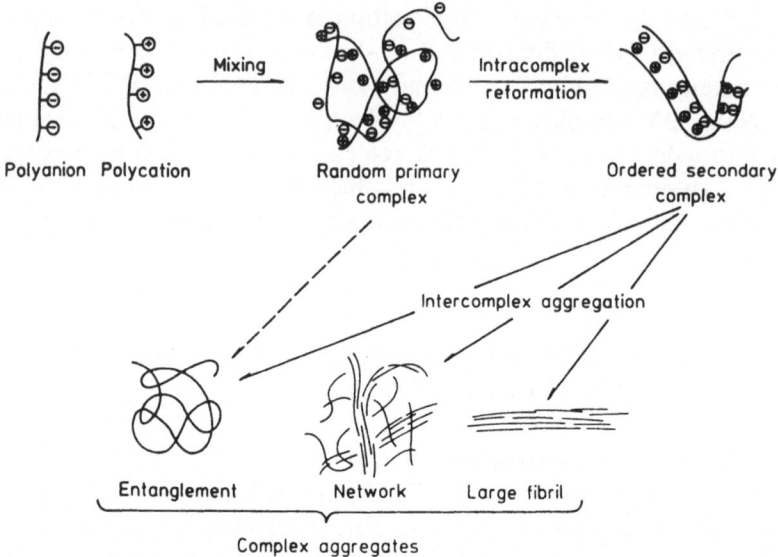

Fig. 11. Schematic diagrams showing the assembling process of a poly-electrolyte complex in solution.

very dilute solution, and is subsequently stored undisturbed under a nitrogen atmosphere at room temperature, then, within four days, fibrous aggregate with a network structure are formed and separated from solution.[2] This process is accelerated by elevating the temperature[156] and by adding small amounts of micro-salts. These results suggest that the hydrophobic interaction may play an important role in the formation of these aggregates. The compositions of primary and fibrous aggregates have a 2X PMAA mole ratio (r) of 1:1.

Results from X-ray diffraction and polarized light microscopy have shown that PECs obtained immediately after mixing in an aqueous medium are usually completely amorphous, even though PMAA is amorphous and 2X has some crystallinity. By contrast, fibrous aggregates of PEC have good crystallinity. Furthermore, Tsuchida and Matsuda[157] have found that fibrous aggregates with supermolecular structures were obtained immediately after mixing methanol solutions of PVBMA and poly(lithium styrenesulfonate) at 65°C.[157] Crystalline PECs have also been obtained in the poly(vinylsulfonic acid)–P4VP system by Blumstein and coworkers[158] and in the isotactic PAA–

poly(1,4-pyridiniumdiylethylene) (PPE) system by Kabanov and coworkers.[159]

Salamone and coworkers[160] reported a PEC with a higher ordered structure, composed of the amorphous polyelectrolytes PVBMA and PMAA with a low molar mass. Depending on the preparation and purification conditions, a needle-like structure, a radially extended 'fuzzy sphere' or an amorphous powder can be obtained. The amorphous form is obtained by mixing freshly prepared aqueous solutions of each polyelectrolyte. The needle-like and fuzzy-sphere PECs are formed on ageing. The needle-like form is obtained when micro-salts are added to polyelectrolyte solutions aged for at least three to four days. The fuzzy-sphere PECs are obtained when PEC gels formed by mixing freshly prepared polyelectrolyte solutions are left undisturbed for a period of at least two weeks. The needle-like PECs are well-defined crystallites 2 mm in length and $0.1-0.2$ mm in diameter and are soluble only in basic media. The crystallites contain PVBMA/PMAA in the mole ratio (r) $1:1.4$ and the syndiotactic sequence of PMAA lies in the range 10–20% of the total polymer content.

Kabanov et al.[52] have reported the preparation of PECs by mixing aqueous solutions of completely amorphous poly(oxyphosphinato-oxytrimethylene) or partly crystalline poly(sodium phosphate) and PPE, and also by polymerization of 4-vinylpyridine in the presence of these polyanions. The schematic representations of X-ray diagrams of the component polyelectrolytes and their PECs are shown in Fig. 12. It is evident that the X-ray patterns of the PECs differ from those of the component polyelectrolytes, and that PECs obtained by matrix polymerization are considered to have a lamella-type structure and to be more crystalline than PECs obtained by mixing two polyelectrolyte solutions.

In order to relate the conformation of the component poly-electrolytes to the formation and structure of PECs, it is important to study the PECs of polyelectrolytes with specific conformations such as the nucleic acids, the poly(α-amino acid)s, the polysaccharides and the proteins. For example, the double helical structure of DNA is, in general, stabilized by complexation with ionene-type polycations.[99] The melting point (T_m) of DNA shifts from 65–70°C to 90–95°C on complexation. These stabilization phenomena are affected by the chemical structure of the ionene, the ratio in which the two com-ponents are mixed (Z), ionic strength and amount of added urea. When $Z = 0.23$ two T_m values, attributed to free DNA and complexed

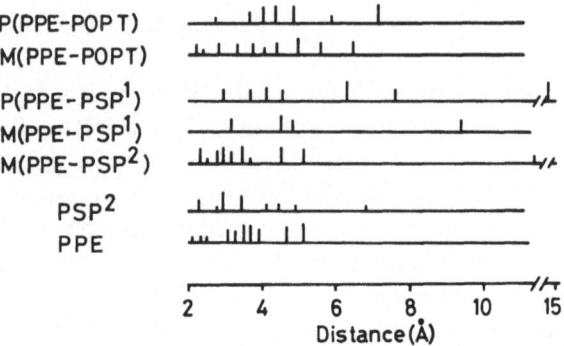

F IG. 12. Schematic representations of X-ray diagrams of crystalline poly-electrolyte complexes (PEC). P( ) is PEC from the assembly in solution. M( ) is PEC from matrix polymerization of 4-vinylpyridine in the presence of polyanions. PPE, poly(1,4-pyridiniumdiylethylene); POPT, poly(oxyphosphinato-oxytrimethylene); PSP^1, poly(sodium phosphate) $(M_v = 5 \times 10^5)$; PSP^2, poly(sodium phosphate) $(M_n = 3200)$.

DNA, are observed, but when $Z = 0.46$ only one T_m, that of complexed DNA, is observed. On increasing the ionic strength, the T_m value of complexed DNA is raised and it exhibits a sharp melting profile reflecting a higher degree of cooperativity; this is because micro-salts facilitate the formation of a more uniform type of interaction. On the other hand, urea also causes a rearrangement of the initially formed random PEC to a thermodynamically favoured ordered form as shown in Fig. 13. These results demonstrate that hydrophobic and electrostatic interactions between ionenes and DNA, and intramolecular hydrogen bonding of DNA, are important in determining the structure of the complex. Ionenes in this system exhibit a large extrinsic Cotton effect at 232·5 nm, which is attributed to exciton interactions of certain ordered ionenes. A left-handed structure of Z-DNA with alternating dG–dC sequence is found to be stabilized by complexation with poly(L-lysine) (PLL) and poly(L-arginine) (PLA), even in aqueous solution near the physiological ionic strength.[103] Klevan and Schumaker[103] found that the PEC of PLL–poly(I–C), where poly(I–C) is an equimolar complex of poly(inosinic acid) and poly(cytidilic acid), was more effective as an interferon inducer than poly(I–C), since the double helical structure of poly(I–C) might be stabilized by the complexation; these conclusions were based on an investigation of the triple helix structure. Additionally, the permeability through biomembranes may be increased due to the

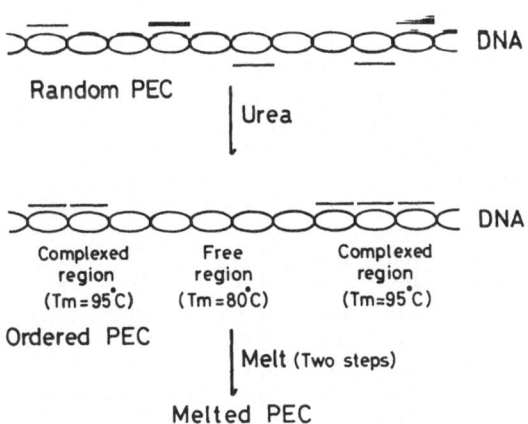

FIG. 13. Reorientation of polyelectrolyte complex composed of DNA and 2X.

reduction of anionic charges of poly(I–C).[104] PECs of bovine serum albumin (BSA) with PAA were studied by Sakamoto and co-workers,[82] who suggested that cooperative interactions played an important role in PEC formation and that the structure of BSA (mainly α-helix) was not changed by the complexation. These results suggest that carboxylate anions on PAA are bound non-specifically to the positive charged sites located on the exterior of BSA.

So far, PEC formation between oppositely charged polyelectrolytes has been discussed. However, in recent years, Ise and Okubo[161] suggested that electrically charged solutes having the same sign as the polyelectrolyte, polymer lattice, ionic micelle or micro-electrolyte, may also form ordered structures in dilute solutions. According to accepted theories in the fields of colloid and solution chemistry of polyelectrolytes, such as the Debye–Hückel theory, macro-ions with the same charges always repel each other. However, results obtained by means of light scattering,[163] quasi-elastic light scattering,[164] small-angle neutron scattering[165] and small-angle X-ray scattering[166] are not explained by existing theories. Table 7 shows the inter-macro-ion distances calculated from the small-angle X-ray scattering data for the PLL hydrobromide system, by assuming the Bragg equation $(2D_{exp})$ or a simple cubic distribution of macro-ions in the solution $(2D_0)$.[167] The following phenomena were found: $2D_{exp}$ is always smaller than $2D_0$ but increases with the concentration of added micro-salts and the degree of polymerization of PLL, and decreases when the concentration of PLL and the degree of dissociation increase. Similar results have been

TABLE 7

INTER-MACRO-ION DISTANCE OF POLY(L-LYSINE) HYDROBROMIDE IN DILUTE AQUEOUS SOLUTIONS AT 25°C

Concn (g/ml)	$\overline{Pn}^a$	pH	[NaBr] (M)	$2D_{exp}$ (Å)^c	$2D_0$ (Å)^d
0·02	406	6·4	0	100	192
0·04	406	6·4	0	77	152
0·06	406	6·4	0	58	133
0·04	406	6·4	0·03	83	152
0·04	406	6·4	0·1	140	152
0·04	406	6·4	0·3	No peak	152
0·04	406	3·0	0	74	152
0·04	406	9·8	0	114	152
0·04	13	6·4	0	$(57)^b$	48

^{a} Weight-average degree of polymerization.
^{b} Calculated from the vague shoulder.
^{c} Inter-macro-ion distance calculated from small-angle X-ray scattering data by assuming the Bragg equation.
^{d} Inter-macro-ion distance calculated theoretically from the concentration by assuming simple cubic distribution.

obtained in PAA,[168] NaSS and BSA[161] systems. The effect of ionic strength on $2D_{exp}$ demonstrates that the shielding effect of the micro-ions promotes an attraction between the macro-ions. This fact may be explained by the presence of an intermediate layer of counter micro-ions surrounding the macro-ion domains. When the charge density of a macro-ion is small and its concentration is very low, the conventional Debye–Hückel theory is obeyed. Therefore, Ise and coworkers[168] proposed a two-state structure in dilute solution. This structure consisted of ordered regions, with densely concentrated macro-ions at distances of $2D_{exp}$, and disordered regions, where macro-ions are sparsely and randomly distributed. A structure consisting of two states is not static but dynamic. In fact, the two-state structure of a system containing anionic lattices of copolymers of styrene and styrenesulfonate was confirmed by microscopy. The factors that determine whether macro-ions adopt a two-state or one-state structure are the charge density of the macro-ion, the concentrations of macro-ions and micro-ions, the temperature and the dielectric constant.

4.2   Quantitative Bindings between Polyelectrolytes

Almost all colloids hold surface charges, regardless of whether they are organic or inorganic materials. Flocculation phenomena are considered to be related to the formation of PECs and the neutralization of surface charges, and crosslinking by polyelectrolytes may be the cause of the aggregation of colloids. When the neutralization of charges on colloids by polyelectrolytes takes place stoichiometrically, then the number of ionic sites can be determined by so-called colloid titrations, using oppositely charged polyelectrolytes.[169] The end point of the colloid titration is usually indicated by measuring the turbidity of the solution and the amount of material precipitated.

Recently, the charge on proteins has been determined by titration with synthetic polyelectrolytes. Generally poly(diallyldimethyl-ammonium chloride) (PDDA) and PVBMA are used as the polycationic standard reagents and potassium poly(vinyl alcohol) sulfate (KPVS) is used as a polyanionic standard reagent. Figure 14A shows the relationships between the titre of PDDA or KPVS and the weight of human serum albumin (HSA) (expressed as mg/50 ml of solution).[79] When pH > pI (pI is the isoelectric point of HSA), the

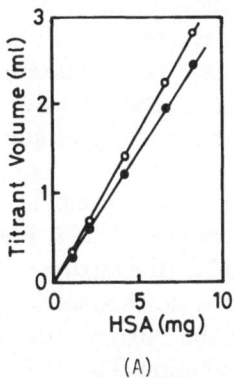
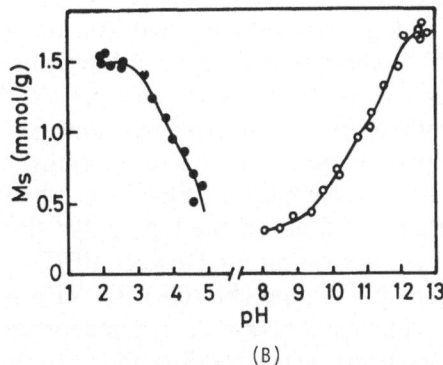

(A)   (B)

FIG. 14. Determination of the number of ionizable groups on proteins by the method of colloid titration. (A) Relationships between the weight of human serum albumin (HSA) and titrant volumes of poly(diallyldimethylammonium chloride) (○) at pH 12·85 and potassium poly(vinyl alcohol sulfate) (●) at pH 2·25. (B) Relationship between the mole number (M_s) of ionic groups of the standard polyelectrolytes and pH.

acidic groups on HSA can be titrated by PDDA. When pH < pI, the basic groups on HSA can be titrated by KPVS. The titre is proportional to the pH of the solution and the slope of this straight line gives the mole number (M_s) of ionic groups on HSA assuming stoichiometric PEC formation. The dependence of M_s on pH is depicted in Fig. 14B. In the regions of pH > 12·5 and pH < 2·5, M_s is saturated, reaching values of 1·71 and 1·48 (mmol/g), respectively. These values agree approximately with the numbers of acidic and basic groups on HSA, which are calculated on the basis of the amino acid sequence of HSA to be 1·76 and 1·51, respectively. In a similar manner, the number of acidic groups on a PEC of carboxyl haemoglobin (Hb) with KPVS was determined by titration with PDDA.[81] It was found that at pH = 7·7, all free carboxylic acid groups of Hb in the PEC are completely titrated by PDDA, but that when pH > 12, PDDA destroys some of the salt linkages between KPVS and Hb, causing an over-estimate in the determination of acid groups.

Proposals have been made to apply metachromatic reagents to colloid titrations.[80] The protein solutions are incubated with an excess of a polyelectrolyte, and then the residual polyelectrolyte is back-titrated by an oppositely charged polyelectrolyte using toluidine blue as a metachromatic reagent. The end-point is detected by measurements of the relative changes of absorbance at 635 nm—where free toluidine blue absorbs—with respect to that of the isosbestic point at 550 nm. In the region of 2 < pH < 9, the sensitivity in this system is 10^{-10} charge equivalents, and for amounts between 5×10^{-9} and 50×10^{-9} charge equivalents of polyelectrolytes, the coefficient of variation is 1–2%. This type of colloid titration is also effective for determining the charge densities and the binding capacities of anionic or cationic impurities in the suspensions.[68]

A non-stoichiometric PEC was found first by Tsuchida and coworkers,[144] in an ionene-type polycation system (3X)–NaSS. Figure 15 shows the relationship between PEC yields and Z (the mixing ratio). Poly(vinylpyridinium chloride) (QPVP)–NaSS is a stoichiometric PEC whose composition ($r = 1:1$) is independent of the mixing order and Z. By contrast, when NaSS is added to the 3X solution (Fig. 15B, $\bigcirc$), the yield increases linearly to about 100% at $Z = [\text{NaSS}]/[3X] = 1$, in the same way as in the QPVP–NaSS system, but further addition of NaSS results in a linear decrease of the yield and the formation of a water-soluble non-stoichiometric PEC where the ratio of [3X]/[NaSS] is 1:3. In that report, the cause of these phenomena was considered to

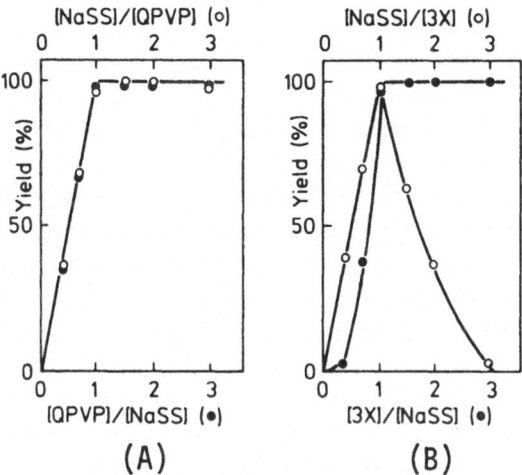

FIG. 15. Formation profiles of stoichiometric (A) and non-stoichiometric (B) polyelectrolyte complexes. QPVP, Quaternized poly(4-vinylpyridine); NaSS, poly(sodium styrenesulfonate); O, NaSS solution was added to polycation solutions; ●, polycation solutions were added to NaSS solution.

be the difference in chemical structures of the polycation, i.e. whether they are of the pendant or integral type. However, since then, there have been many reports about non-stoichiometric PECs,[3] and the conclusion must be drawn that the most essential factors for the formation of non-stoichiometric PECs are not the chemical structures, but the differences of chain length of the component polyelectrolytes and Z. As in the case depicted in Fig. 15, NaSS and QPVP have nearly the same chain length but 3X is much shorter. Non-stoichiometric PECs are obtained when Z, the ratio [lower molar mass polyelectrolyte (LPE)]/[higher molar mass polyelectrolyte (HPE)] is less than unity. The compositions of the PECs obtained correspond exactly to Z. When Z is larger than 1, a stoichiometric PEC is formed. That is, changing the degree of neutralization transforms non-stoichiometric PECs to stoichiometric PECs. These non-stoichiometric PECs are prepared by four methods,[131] one of which is by the gradual addition of a solution of HPE to a solution of LPE; another is the reverse addition as shown in Fig. 15B. When HPE is a weak polyelectrolyte and LPE is a strong one, these polymers are mixed at $\bar{\alpha} = 0$ and $Z > r$ under conditions where, firstly, the precipitate of stoichiometric PECs is formed with the composition corresponding to r. Then, on changing the pH of the solution, in accordance with the dissociation of excess free ionizable groups on HPE, water-soluble non-stoichiometric PECs are formed. When HPE and LPE are weak polyelectrolytes, the nature of the complexes alters with pH changes of the solution as follows: no complex→ non-stoichiometric PEC→ stoichiometric PEC→ non-stoichiometric PEC→ no complex (see Fig. 4).

Another interesting topic is whether the distribution of the component polyelectrolytes in the intermacromolecular complex is of the 'all-or-none' or 'random' type. As shown in Fig. 15, the yield of a stoichiometric PEC is proportional to Z and the r value of the PEC precipitates at $Z \neq 1$ is always unity. It follows that a completely neutralized PEC (a stoichiometric PEC) coexists with an excess of completely free polyelectrolytes in solution. This behaviour represents the all-or-none type complexation, and is depicted in eqn (19). This

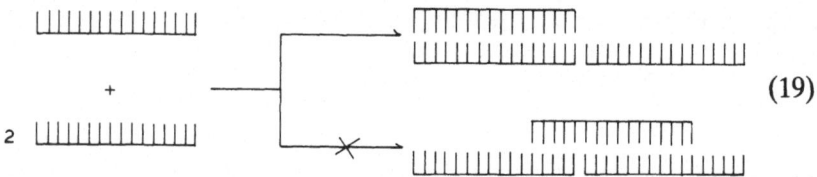

$$\text{(19)}$$

phenomenon implies that the reactivity of a polyelectrolyte complexed with a complementary one may be considered to be higher than that of free polyelectrolyte, probably owing to the changes of conformation, dissociation and micro-environment of the polyelectrolyte chains; in other words, cooperative and concerted interactions occur. Similar results were obtained by Kabanov et al.[170] for hydrogen-bonded interpolymer complexes, using an ultracentrifuge technique to construct sedimentation diagrams.

Under the conditions required to form the non-stoichiometric PECs, a structure was proposed for the PEC with $r = Z$, which contained hydrophilic blocks (non-complexed free sequences of HPE) and intra-ladder-like hydrophobic blocks (complexed sequences of HPE with LPE), connected to each other as shown in scheme (20). Thus a

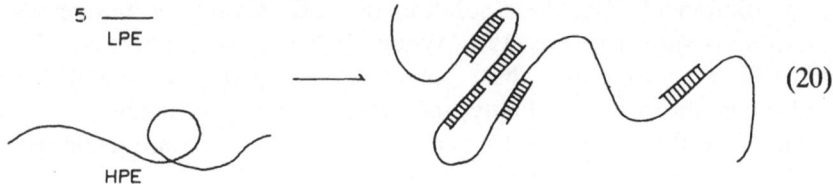

$$\text{(20)}$$

non-stoichiometric PEC behaves like a block copolymer. Hydrophobic sequences segregate and hydrodynamic behaviour, mainly due to the excluded volume of PEC, is considered to be dependent on the free HPE sequences located on the exterior of the PEC. For example, the second virial coefficient (A_2) is proportional to $(1 - r)^2$. A similar conclusion was drawn from determinations of the mass-average molar mass, the mean-square radius of gyration, $(\bar{R}_g^2)^{1/2}$, and the sedimentation coefficient, S, of non-stoichiometric PEC solutions. The conformation of the non-stoichiometric PEC is strongly affected by the ratio of molar masses of HPE and LPE, $r(Z)$, the charge densities of HPE and LPE, and the ionic strength; for example, there is the case where the stoichiometric PEC precipitate and the soluble HPE coexists in the system at high ionic strength.

The next consideration is the compatibility of polymers. In recent years, attention has been drawn to polymer blends and polymer alloys. Polymer blends are divided into three classes, i.e. compatible, partly compatible and incompatible systems, dependent on molecular, morphological and thermodynamic parameters. Almost all pairs of homopolymers are immiscible because the enthalpy of mixing cannot compensate for the entropy of mixing, but many efforts have been tried to make polymers miscible. General methods of increasing compatibility have been based on the incorporation of complementary active groups into the two individual polymers, for example through electrostatic, hydrogen bonding, dipole–dipole and donor–acceptor interactions[171] or by the addition of, as third component, a copolymer[172] whose monomer units are selected from the monomers of the two polymers to be mixed. Table 8 shows the glass transition temperatures, T_g, of component polymers including positively and negatively charged groups and their blends.[59] Unmodified polystyrene/polyisoprene and polystyrene/poly(ethyl acrylate) are incompatible or only slightly compatible, as detected by the coexistence of two glass transition temperatures, due to each component polymer. However, when more than 5% of ionizable groups are incorporated into the polymers, only one glass transition temperature, usually midway between those of the component polymers, is observed. These blends are transparent. From these results, it is clear that the incorporation of only about 5% of ionic linkages is sufficient to achieve compatibility in these systems. Similar phenomena were observed in the polystyrene/poly(butyl methacrylate) system[59] and in the styrene–vinylpyridine copolymer/styrene–methacrylic acid copolymer system or in the same system where methyl methacrylate replaces styrene.[60] The results of

TABLE 8

THE EFFECT OF IONIC LINKAGES ON COMPATIBILITY IN IMMISCIBLE POLYMER BLEND SYSTEMS

System		Sulfonation PA (mol%)	T_g of sulfonated PA (°C)	Q4VP in PC copolymer (mol %)	T_g of PC copolymer with Q4VP (°C)	T_g of PA/PC polymer blends (°C)
PA	PC					
PIP	PSt	0	−61	0	100	−61,100
		1·83	0	2·10	115	0,114
		5·25	65	5·06	130	85
		9·31	—	10·01	145	142
PSt	PEA	0	101	0	−13	−7,106
		1·75	115	1·65	—	−3,89
		5·1	124	5·4	4	72
		9·9	139	10·5	19	90

Abbreviations: PA, polyanion; PC, polycation; PIP, polyisoprene; PSt, polystyrene; PEA, poly(ethyl acrylate); Q4VP, 4-vinylpyridine quaternized with methyl bromide.

these investigations show that 5–10% of ionic sites are required to achieve compatibility through electrostatic interaction.

4.3 Organization of Membranes

The general properties of PEC membranes have already been discussed in Section 2. Figure 16 shows the relationship between the equilibrium gel water content of an equimolar PVBMA–NaSS neutral complex and water permeability, oxygen permeability, ionic conductivity and dialytic permeability for polysaccharides with a molar mass of 1000.[173] The magnitude of all these properties increases with the increase of water content. At the same water contents, the water permeability of a PEC membrane is, in general, more than ten times greater than that of reconstituted cellophane.[9] The permeability of PEC membranes towards low molar mass solutes is also greater than that of commercial membranes, but PEC membranes are effectively impermeable towards solutes with relatively high molar mass.[9] This type of PEC is stable in acids, weak bases and general organic solvents and compatible with olefins, nylons and most other thermoplastics.[173] For example, they are stable when immersed in 35% sulfuric and phosphoric acids at 400°C, and in 50–80% acids at room temperature.

Figure 17 shows the selective and active transport of Na$^+$ and K$^+$

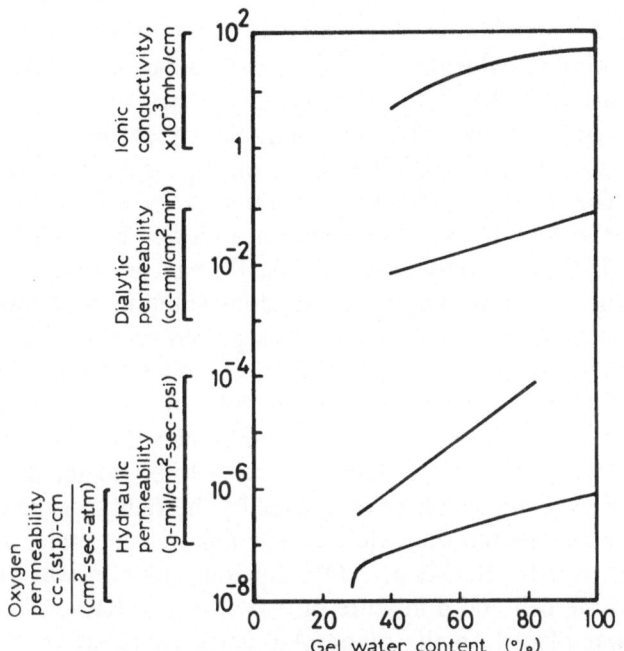

FIG. 16. Some physical properties of PEC hydrogels of poly(sodium styrene-sulfonate) with poly(4-vinylbenzyltrimethylammonium chloride) (1:1) as a function of water content in gel.

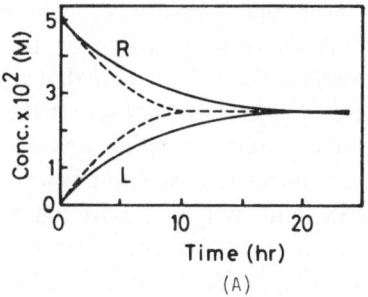

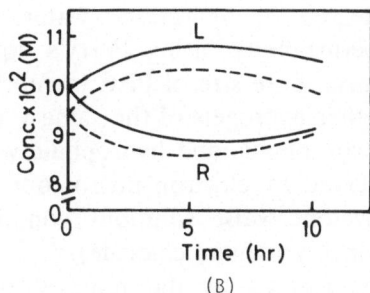

FIG. 17. Selective permeability (A) and active transport (B) of Na$^+$ (solid line) and K$^+$ (dotted line) through PEC membrane composed of 2-diethylaminoethyldextran, carboxymethyldextran and poly(potassium vinyl alcohol sulfate). L, left side cell; R, right side cell.

through PEC membranes composed of 2-diethylaminoethyldextran (EA), carboxymethyldextran (CMD) and potassium poly(vinyl alcohol) sulfate (KPVS).[113] In this system, the structure, especially the surface structure of the PEC, is strongly dependent upon preparation conditions such as pH, ionic strength, mixing order and purification methods. This PEC membrane was prepared by mixing these three components in 4% HCl aqueous solution, followed by casting the PEC from a HCl/dioxane/water solution (48:47:5 in per cent by volume) and washing with pure water. An aqueous solution containing equal concentrations of NaOH and KOH was placed on one side of a cell divided by the PEC membrane and an equal concentration of HCl was placed on the other side. Almost all the Na^+ and K^+ ions were found to have permeated through the PEC membrane within ca 12 h but the velocity of permeation of K^+ ions was found to be higher than that of Na^+ ions. This result seems to be caused by the radius of the hydrated Na^+ ion being greater than that of K^+ ions. In a similar transport experiment, a 0·1M NaOH or KOH aqueous solution was placed on one side of the cell and a mixture of 0·1M NaCl or KCl and 0·1M HCl solutions was placed on the other. An active transport of metal ions occurred and the rate of ion transport was proportional to proton concentration. Protons were found to be transported in the reverse direction to that of the metal ions, since the changes of dissociation of ionizable groups and the number of hydrogen bonds within the PEC membrane are reflected in the alkaline and acidic solutions on each side. This is a 'proton pump' mechanism, a term that signifies that it is the transport of protons that drives the transport of metal ions.

Mean pore sizes of PEC membranes were calculated from the degree of hydration (water content) and the velocity of water permeability, using Ferry's equation.[174] As shown in Table 9, the mean pore size of PEC membranes is found to be larger than that of other hydrogels of the same water content.[152] The morphology of the hydrophobic and hydrophilic water-containing microdomains was observed by electron microscopy of PEC membranes cast from water/acetone/NaBr solutions containing NaSS and poly(1,2-dimethyl-5-vinylpyridinium chloride).[175] Such domain structures are considered to be formed by the removal of acetone and NaBr from the PEC membranes on immersion in water. Therefore, on increasing the temperature of the water used in this process, the hydrophilic domains are made larger,[3] resulting in an increase of water permeability. The permeabilities of PEC membranes can also be controlled by the pH of

TABLE 9
AVERAGE PORE SIZE AND DIFFUSION CONSTANT OF PEC MEMBRANES

Hydrogel[a]	Water content (%)	Pore size (nm)	Water permeability (cm^{-1})	Diffusion constant (cm^2/s)
Ioplex	66·6	3·64	$8·18 \times 10^{-15}$	255·75
PHEMA	38·7	0·40	$0·08 \times 10^{-15}$	3·08
PPGMA	62·6	0·63	$0·28 \times 10^{-15}$	7·76
PGMA	80·3	2·17	$4·65 \times 10^{-15}$	88·59
PGMA	88·8	3·70	$14·79 \times 10^{-15}$	265·00

[a] Ioplex, equimolar PEC of poly(sodium styrenesulfonate) and poly(vinylbenzyltrimethylammonium chloride); PHEMA, poly(hydroxyethylmethacrylate); PPGMA, poly(propyleneglycol monoacrylate); PGMA, poly(glycerol methacrylate).

the solution, which changes the ionization of ionizable groups. The permeabilities of water and micro-salts through a PEC membrane composed of glycolchitosan (GC) and chondroitin sulfate C (CSC) are reduced at the PEC isoelectric point (pH = 6·0) as shown in Fig. 18.[147] This result seems to be caused by the morphological changes of the PEC membrane caused by the dissociation of ionizable groups in it. This phenomenon is very interesting and it is attractive to suggest a mechanochemical reaction, i.e. a conversion of chemical energy into mechanical energy. Such phenomena are not observed in·other ionic membrane systems.

As stated above, PECs have been studied with keen interest as new functional polymer materials, but in order to make them suitable for

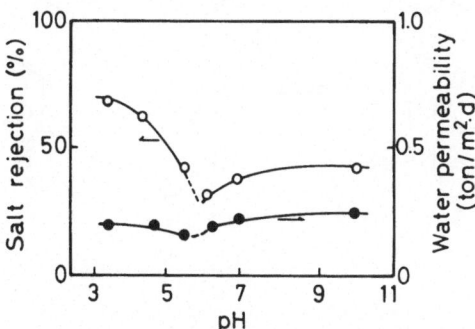

FIG. 18. The relationship between pH and salt (NaCl) rejection (○) or water permeability (●) in the membrane of polyelectrolyte complex composed of glycolchitosan and chondroitin sulfate C (83:17).

practical use there are some problems associated with their lack of mechanical strength, durability and stability towards chemicals. Many efforts have been made to resolve these issues, especially that of low mechanical strength. PEC membranes have been prepared by means of electrodeposition: a direct voltage is applied to an electrode that has been soaked in the aqueous solutions of both component polyelectrolytes.[176] These membranes are rapidly formed and are thin and uniform. Additionally, PEC membranes have been strengthened by post-crosslinking reactions,[177] blending[9] and matrix polymerization.

5 PHYSICAL CHEMISTRY OF MICRODOMAINS IN POLYELECTROLYTE COMPLEXES

5.1 Electrochemistry of Polyelectrolyte Complexes Coated on Electrodes

In recent years, various dipyridinium (viologen) compounds have been viewed as redox compounds, electron-transfer agents and electrochromic devices.[178] Also, there is currently considerable interest in chemically modified electrodes, especially polymer-coated ones.[179] In this context, the past few years have seen intensive studies of PECs of viologen polymers with various polyanions. Figure 19 shows typical cyclic voltammograms of viologen-coated basal plane pyrolitic graphite (BPG) electrodes.[47] The redox reaction of viologen units of polyviologens in PEC with NaSS is carried out chemically, electrically and photochemically as depicted in eqn (21).

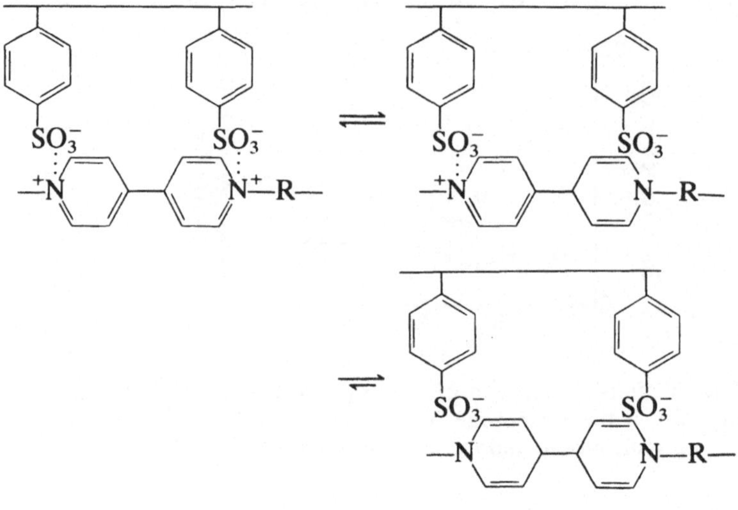

(21)

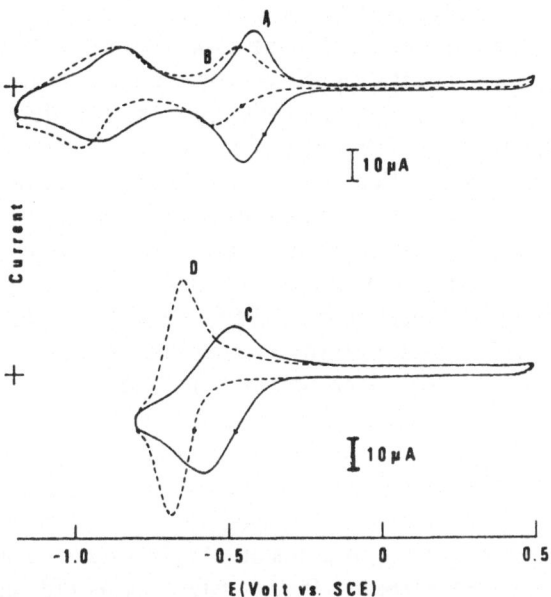

Fig. 19. Typical cyclic voltammograms of basal plane pyrolitic graphite (BPG) electrode coated with viologen compounds. (A) Poly(xylylviologen) (PXV). (B) PXV–poly(sodium styrenesulfonate). (C) PXV–Nafion. (D) Methyl viologen–Nafion. Scan rate, 100 mV/s.

The first step of the redox reaction, with a reduction potential of about −0·50 to −0·60 V against a saturated calomel electrode (SCE), is essentially reversible, but the second step with a reduction potential of about −0·90 to −1·0 V is partially irreversible.[47] Therefore, for the reversible redox reaction to be carried out efficiently, the potential sweep must be set to range from +0·8 to −0·8 V. The reduction potentials of methyl viologen (low molar mass model compound of polyviologen) are −0·68 V and −1·12 V vs SCE, respectively. The positive shift of reduction potential observed for a poly(xylyl viologen) (PXV)-coated electrode may be due to an intramolecular interaction of redox sites. However, the first and second redox potentials of the NaSS–PXV complex are shifted toward the negative by about 60 and 20 mV, respectively, and a peak broadening is observed. Such negative shifts by complexation are also observed in poly(alkyl viologen) (PAV)–poly(carboxylic acid)s systems but their absolute values are, of course, strongly dependent on the chemical structures of the component polyelectrolytes.[42] These results suggest that the electron

transfer is considerably restricted, probably because of changes of the
chemical environment, such as the introduction of a hydrophobic field
around viologen units of PXV produced by complexation, the decrease
of intrachain interactions of viologen units, or the decrease of the
migration of counterions and polymer chains.

In order to clarify the electrochemical characteristics of the domain
formed by PECs, the electron-transfer and ion (mass)-transfer processes were studied using a ferricyanide anion ($[Fe(CN_6)]^{3-}$) as the
redox substrate, having a standard reduction potential of $+0.15$ V vs
SCE.[47] The reaction mechanisms of PEC membrane-coated electrode
redox systems are divided into two classes as shown in Fig. 20.[180] The
kinetics of such coated electrode systems must be analyzed in a
stationary state by means of rotating disk voltammetry. The limiting
current (i_L) obtained experimentally is expressed as a function of i_E,
i_A, i_S and i_K[181] (where i_E is inherent current owing to electron carrier,
i_A is inherent current owing to diffusion of substrate in bulk solution,
i_S is inherent current owing to diffusion of substrate in membrane, and
i_K is kinetic inherent current). If D_E (where D_E is diffusion constant
of electron carrier in membrane) and i_E are very large, the redox

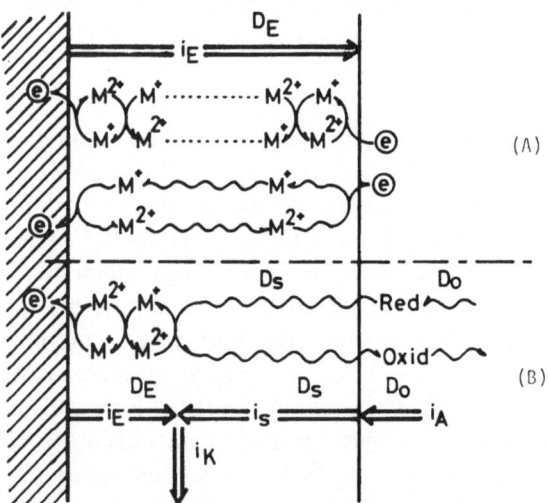

FIG. 20. Schematic diagram to illustrate the function of a polymer-coated
electrode. (A) Electron-transfer process. (B) Interfacial-reaction process.
(Where D_0 is diffusion constant of substrate in bulk solution, and D_s is
diffusion constant of substrate in membrane.)

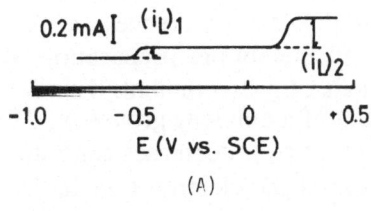

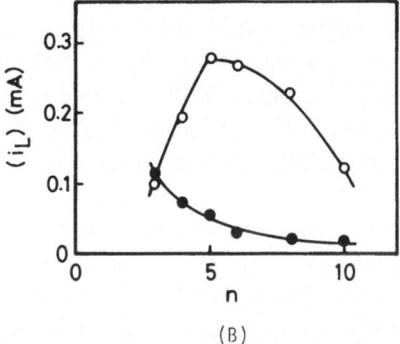

FIG. 21. Redox reaction through membrane of poly(sodium styrenesulfonate)–poly(alkylviologen) (PAV) coated on graphite electrode. (A) Typical rotating disk voltammogram for the reduction of ferricyanide. (B) Effect of charge density of PAV on $(i_L)_1$(●) and $(i_L)_2$ (○).

reaction takes place at the interface of the membrane and the solution as a result of an electron-transfer process. However, if these parameters are relatively small, then non-uniform interface reactions take place. Figure 21 shows a typical rotating disk voltammogram of the graphite electrode coated with NaSS–PAV (molar ratio 2:1) in a potential sweep range from +0·5 to −0·8 V.[47] Two i_L values are observed; $(i_L)_2$ corresponds to the direct reduction of ferricyanide at the surface of the electrode, indicating an ion-transfer process through the domain of PEC, and $(i_L)_1$ corresponds to the reduction via the redox cycle of viologen units in PEC, suggesting the importance of the electron-transfer process. The i_L values are given by:

$$(i_L)_1^{-1} = i_A^{-1} + (i_S i_K)^{-1/2}[\tanh (i_K/i_S)^{1/2}]^{-1}$$

and

$$(i_L)_2^{-1} = i_A^{-1} + (i_S i_K)^{-1/2}[\tanh (i_K/i_S)^{1/2}]$$

However, only one peak potential assigned to $(i_L)_1$ was observed by means of cyclic voltammetry using platinum electrodes coated with NaSS–PXV.[45] This difference has not yet been resolved, but can probably be attributed to the differences between polyviologen structures, thicknesses and membrane preparation methods, and to basically different methods of measurement. The dependence of $(i_L)_1$ and $(i_L)_2$ upon the number of methylene groups on PAV(n) is also shown in Fig. 21(B). The fact that $(i_L)_1$ decreases continuously with increasing n suggests a suppression of the electron transfer between adjacent redox sites in the PEC domain. By contrast, $(i_L)_2$ reaches a maximum at $n = 5$–6. This result suggests that the cavity of the microdomain in such PEC membranes is controlled by a suitable balance of the hydrophobicity, which increases with the increase of n, and charge density, which decreases with the increase of n. Also, the characteristics of the domain, especially hydrophobicity, control the ion (mass)-transfer.

The electrocatalytic reduction of oxygen by a graphite electrode coated with PXV–NaSS complex was reported by Oyama et al.[47] The half-wave potential of O_2 reduction at -0.40 V vs SCE is independent of the pH of the solution, in contrast to the pH dependence of a bare electrode. Additionally, the reduction occurs at the potential more positive by 350 mV at the maximum than that of a bare electrode. These results suggest that the hydrogen ion does not participate in the rate-determining step of the O_2 reduction process and that in this process only a small percentage of the viologen units are changed to monocation radicals. Therefore, the mechanism of the irreversible reduction of O_2 to H_2O_2 in the PEC membrane is considered to be represented by eqns (22) and (23):

$$MV^+ + O_2 \xrightarrow{k_1} MV^{2+} + O_2^- \tag{22}$$

$$O_2^- + MV^+ \xrightarrow{k_2} O_2^{2-} + MV^{2+} \tag{23}$$

where MV represents a viologen unit, and k_1 and k_2 denote the second-order rate constants. Selective recovery of some complex anions and cations by the control of a given potential was performed by PXV- or PXV–NaSS-coated electrodes.[48] When a PXV-coated electrode is soaked in a mixed solution containing equal amounts of $[Fe(CN)_6]^{4-}$ and $[Mo(CN)_8]^{4-}$, the latter species is predominantly adsorbed on the PXV membrane matrix. This selectivity is attributable to the differences of their ionic radii or hydrophobicities. Such adsorbed anions are desorbed by a decrease of the cationic charges of PXV when the given potential is -0.70 V (monocation radical) or

$-1{\cdot}20$ V (non-ion). On the other hand, using PXV–NaSS-coated electrodes, the charge of the membrane can be changed reversibly from non-ionic to anionic, so that a complex cation such as $[Ru(NH_3)_6]^{3+}$ can be adsorbed and desorbed reversibly merely by controlling the membrane potential.

The PECs of viologen polymers with NaSS represent new materials which can be used for electrochromic display devices. Whilst viologens show an absorption maxima (λ_{max}) at about 260–300 nm, their monocation radicals generally absorb visible light at around 550 nm (ε is of the order of 10^4). Therefore, the colour can be readily changed by merely adjusting the potential of the PEC-coated electrode. In the PXV–NaSS system, the total charge is about $2{\cdot}0$ mC/cm^2, and can be completely reduced at $-0{\cdot}8$ V (monocation radical). The optical density at 560 nm is proportional to the charge density and when completely reduced the colour changes to red–purple. Figure 22 shows the responses of the current and the optical density at 560 nm to a square-wave electrode potential alternating from $+0{\cdot}5$ to $-0{\cdot}8$ V.[45] Akahoshi et al.[45] found that the response time is less than 1 s and that the absorption decreases to less than 5% after 100 cycles. This seems

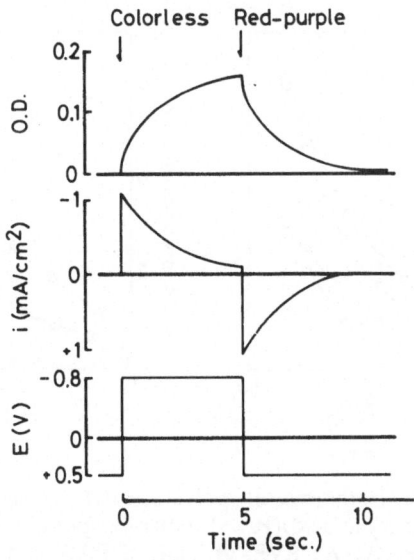

FIG. 22. The responses of the current and the optical density at 560 nm to square-wave electrode potential.

to be caused by the lack of crystallization of monocation radicals of viologen units. Moreover, Nomura *et al.*[46] found that the response time and the efficiency of the redox reaction was improved by the incorporation of ITO—a mixture of oxidized indium and oxidized tin—in the PEC membrane.

5.2 Novel Techniques for Separation of Proteins

PECs show specific adsorption properties for various compounds. Figure 23 shows the adsorption isotherms of gaseous water[24] and organic solvents on PECs.[182] An equimolar PEC of PAA–PVP shows a greater affinity for water than the component polyelectrolytes over almost the entire relative vapour pressure range. At relatively low vapour pressures, the adsorption isotherms of the PEC and PAA can be analyzed by the BET model, suggesting that they have particular adsorption sites for water. On the other hand, in the PVBMA–NaSS system, the amount of adsorbed water was found to increase proportionally with the increase of relative humidity up to 80%, and then to increase drastically, probably because of the capillary condensation of water into microvoids in the PEC.[134] The adsorptivity of some gaseous solvents on PECs of NaSS–PDDA, NaSS–PVBMA, NaSS–

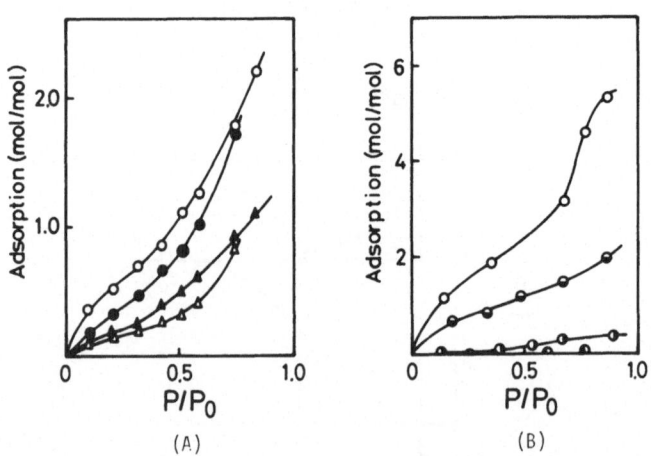

FIG. 23. Adsorption isotherms of polyelectrolyte complexes. (A) Water vapour on the poly(acrylic acid)(PAA)–poly(vinylpyridine)(PVP) (1:1) complex: ○, PEC; △, PAA; ▲, PVP; ●, sum of PAA and PVP. (B) Organic vapours on the poly(sodium styrenesulfonate)–poly(diallyldimethylammonium chloride) (1:1) complex: ○, water; ◐, methanol; ◑, acetone; ◐, benzene.

ionene, and PAA–poly(amine sulfonate) (PAS) was found to increase in the order water > methanol > acetone > benzene. This order is in accordance with the order of the polarity of these adsorbates, and suggests that these compounds are adsorbed at the sites of ionic linkages, probably through dipole interactions. The absolute values for adsorbed molecules, however, is dependent on the balance of ionic and hydrophobic domain structures in the PEC, and reflects the chemical structures of component polyelectrolytes. For example, at 50% relative humidity, the number of water molecules adsorbed at equilibrium per ionic site is 2·4 for NaSS–PDDA, 1·3 for NaSS–PVBMA, 3·4 for NaSS–ionene and 0·7 for PAA–PAS. Moreover, Yano and Wada[183] reported that the adsorptivity of water vapour by the NaSS–poly(vinylphenyltriethylammonium bromide) complex was affected by slight differences of structure of the polycation, i.e. whether it was m- or p-substituted. From these results, it is clear that the variations between adsorption isotherms may be attributed to the chemical structures of ionic sites and the environment around the ionic sites.

Snake-cage resins are interesting as they have both cationic and anionic ion-exchange abilities over a wide range of pH.[184] As one can see in Fig. 24, not only non-electrolytes, such as sucrose and glycerol, can be separated from an electrolyte such as NaCl, but electrolytes such as NaOH and NaCl can also be separated. This resin is stable and characterized by adsorbing both positive and negative ions at the same time over almost the entire pH region.

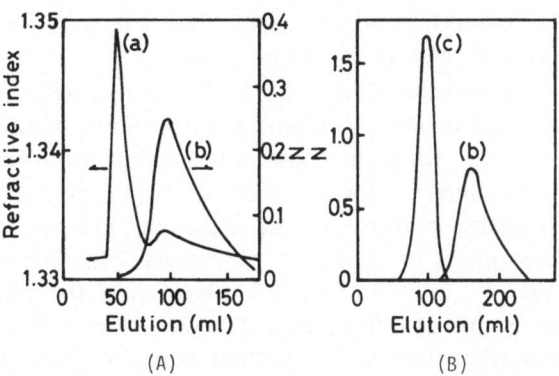

FIG. 24. Separation and purification of microsolutes by snake-cage resin. (A) Saccharose (a), and NaCl (b). (B) NaCl (b), and NaOH (c).

Separation and purification of blood plasma components, utilizing an electrostatic interaction, have been performed by ion-exchange chromatography.[185] Since proteins are polyampholytes, their net charge, charge distribution, dissociation constants of individual ionizable groups, solubility and higher ordered structures are strongly dependent on pH, ionic strength, temperature and solvent. Therefore, one can separate proteins by changing these factors so as to control the strength of interactions between proteins and ion-exchange resins. In general, anion-exchange resins are suitable for adsorption of acidic and weakly basic proteins, and cation-exchange resins are suitable for adsorption of non-ionic and basic proteins. Adsorbed proteins are eluted by gradient elution and step-wise elution methods. In fact, albumin, globulin and blood clotting factor IX have been separated and purified from blood plasma on an extensive scale by using anion-exchange Sephadex columns.[186] On the other hand, blood clotting factors II, VII, VIII, IX and X, von Willebrand factor (vWF), albumin and globulin were separated by using a cation-exchange resin derived from a copolymer of ethylene and maleic anhydride.[187] This resin was found to be able to adsorb selectively B-type hepatitis antigen from solutions containing other immunoglobulins. Albumin and globulin have been separated with high efficiency by using a copolymer of styrene and maleic anhydride, crosslinked by hexamethylenediamine.[188]

Both Tomono et al.[188] and Ohno et al.[188] have found that the adsorption of albumin is influenced by the balance of hydrophilicity and hydrophobicity of the resin. However, globulin is adsorbed both by hydrophobic and also by hydrophilic resins. Thus, by using a resin with 16% of crosslinking, almost all globulin can be adsorbed within half an hour at 20°C with only a little albumin.

Doenecke[189] established that NaSS could interact with the chromosomal chains of nucleosomes by dissociating the ionic linkages between DNA and histones through cooperative substitution reactions of macromolecular chains to form a PEC of NaSS–histone. Figure 25 shows the centrifuge patterns of the isokinetic gradient of rat liver nucleosome monomers titrated by NaSS. It was established that when the [NaSS]/[DNA] ratio reaches 0·8, about half the population of nucleosome monomer is left intact and another half sediments as free DNA, and that when the ratio is greater than 1·6, only free DNA is observed. The absence of intermediate peaks suggests an all-or-none type cooperative substitution reaction. However, when nucleosome

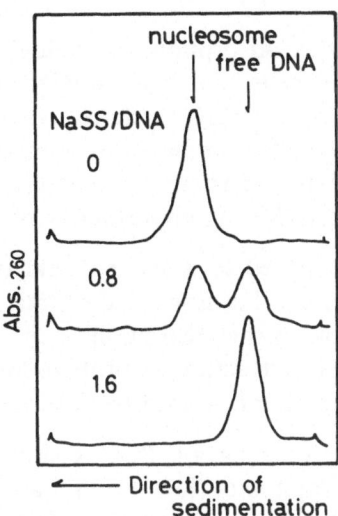

FIG. 25. Centrifuge patterns of isokinetic gradient of rat liver nucleosome monomers titrated with poly(sodium styrenesulfonate) (NaSS) at various mixing ratios.

dimers containing histone octamer are used, an intermediate peak corresponding to a partial removal of histone octamer is observed. Probably this arises because the DNA–histone complex is stabilized by a hydrophobic interaction as well as an electrostatic interaction.

In order to design biomedical polymers, it is necessary first to elucidate the molecular interaction with living organisms or the biological species of living organisms, in heterogeneous and homogeneous systems. The polymer characteristics controlling the interactions between polymer surfaces and biological species are charge balance, charge distribution, affinity to water (hydrophilicity and hydrophobicity), surface free energy, micro-phase separation, chemical and physical structures and dynamic fluctuation of molecular structure, e.g. micro-Brownian motion. In this context, many efforts have been made to develop biomedical polymers, e.g. by block copolymers with micro-phase separation,[190] by incorporation of heparin and anti-clotting factors into the polymer surface[151] and by forming polymer surfaces precoated with biological components.[191] PECs have advantages over these copolymers in what their characteristics can be easily changed. PECs with moderate negative charges possess highly significant non-thrombogenic activity, superior to that displayed by conventional low surface energy polymers.[150]

In recent years, the adsorption of biological components, e.g. proteins, platelets and cells, has been studied intensively by Akaike and his coworkers.[153] In PEC systems composed of NaSS and poly(triethyl-4-vinylphenethylammonium bromide) (P4EVP) or the copolymer of triethyl-4- and triethyl-3-vinylphenethylammonium bromide (P3, 4EVP), the following important facts were elucidated.

(1) Proteins are adsorbed to a lesser extent on neutral PECs than on charged PECs because electrostatic interactions play a predominant role in the adsorption.
(2) The higher ordered structures of proteins are destabilized with the increase in the amount of adsorbed proteins.

By controlling the annealing of PEC coatings, hydrophilic (high wettability) and hydrophobic (low wettability) PEC surfaces were prepared, probably by the reorientation of ionic sites necessary to complete the neutralization. Figure 26 represents the adsorption and conformational change profiles of albumin (Alb) and globulin (Glb) adsorbed on these hydrophilic and hydrophobic PECs.[192] On hydrophobic surfaces, free ionic sites are reduced, so the effect of charge balance of the PEC on the adsorption is reduced and the conformational changes of proteins are small. One notes, however, that the β-sheet content of Glb adsorbed on the hydrophobic PEC is higher

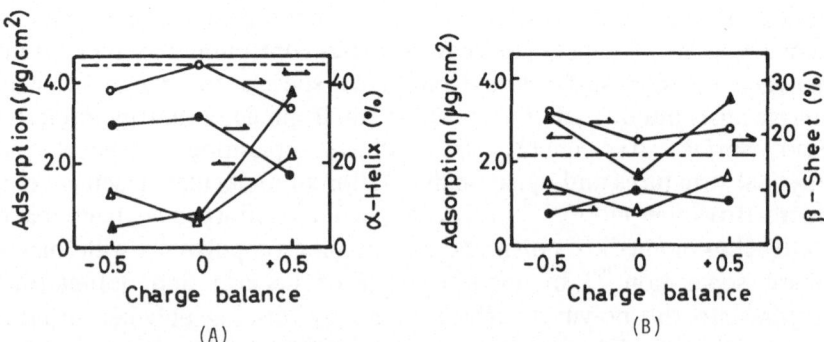

FIG. 26. Adsorption and conformational changes of albumin (A) and globulin (B) on poly(sodium styrenesulfonate)–poly(triethyl-4-vinylphenethyl-ammonium bromide) complex with various compositions: △, ▲, adsorption; ○, ●, conformation; ▲, ●, hydrophilic surface; △, ○, hydrophobic surface; — · —, conformations of native proteins.

than that of native Glb. Similar phenomena were observed when Glb was adsorbed on hydrophobic polymer surfaces such as poly(methyl methacrylate) and polystyrene.[193] These results show that the adsorbed Glb may take up different orientations in accordance with the surface character. Adhesion, aggregation, destruction and formation of pseudopods of the blood platelets were found to be dependent upon the net charge and chemical structure of PECs.[153] Kataoka et al.[153] considered that NaSS–P4EVP with moderate negative charge was superior to all other PEC structures in respect of compatibility with blood.

Kataoka et al.[193] proposed the use of PECs as agents for cell separation. Cells barely adhered on a non-ionic hydrogel such as poly(hydroxyethyl methacrylate) but adhered both on wettable glass and non-wettable polystyrene. These results suggest that cell adhesion may be more closely correlated with an interface free energy than with surface free energy. In the case of PECs, 30–40% of polymorphonuclear leukocytes and more than 80% of canine peripheral lymphocytes are adsorbed. This result indicates that separation of these cells may be possible.

The blood compatibilities of PECs composed of polysaccharides such as diethylaminoethyldextran, sodium dextran sulfate (DS) and NaSS were studied by Kikuchi et al.[76] In comparison with glass and poly(vinyl chloride), such PECs have good compatibility. It has been found that clot formation on PECs prepared at low concentrations of H^+ ions is higher than that on PECs prepared at high H^+ ion concentrations, because the latter PECs contain a large number of the polyanions, NaSS and especially DS, which is more active in suppressing blood clotting than NaSS. The differences in surface structures of the PECs, controlled by the order of mixing, H^+ ion concentration and other preparation conditions, may play an important role in determining the compatibility. It is well known that heparinized surfaces, through electrostatic interactions or covalent bonds, show a good blood compatibility. The interactions of a series of ionenes with heparins have been studied by Rembaum et al.,[194] who established that an almost stoichiometric reaction takes place. A complex coacervate is formed which is generally insoluble in an aqueous solution, even at high ionic strength. In vivo tests showed that a polyurethane tube coated with a PEC consisting of heparin and a polyurethane containing quaternary ammonium groups in the side chains exhibited very low thrombus formation activity.[195]

5.3 Coacervation and Immobilization

As mentioned previously, PECs contain both hydrophobic and electrostatic domains. Complex coacervates, a PEC-concentrated liquid
phase in droplets, have been identified as prebiological systems,[196]
since a characteristic interface exists around the domains. Such
interfaces exhibit unique properties such as the formation of a stable
local environment, adsorption and concentration of various compounds, specific transport processes and a chemical potential. Enzymatic reactions in coacervates were studied as a model system of
chemical evolution by Oparine et al.[197]

Ohno and Tsuchida[198] studied the change of mobilities of the
polymer chains and bound water with the formation of PECs. Figure
27 shows the pH dependence of the degree of fluorescence polarization (P), reflecting the mobility of the polymer chains of PECs
composed of polycations and PMAA*, with covalently bound 8-
anilino-1-naphthalenesulfonate as a fluorescent probe. Since no PEC is
formed at pH 2, the P value is very low and similar to that of PMAA*
alone. At pH > 6, where the PEC is completely formed, the mobility
of PMAA* is strongly depressed by complexation, detected by an
increase of the P value. On lowering the pH, the mobility is increased

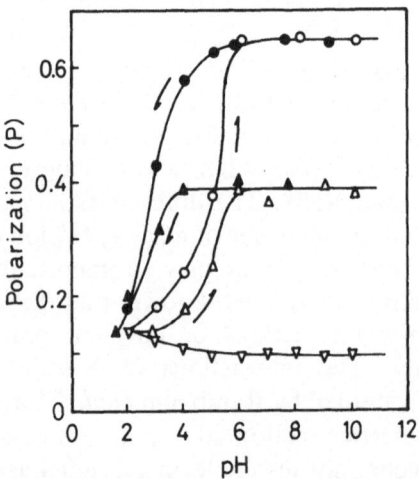

FIG. 27. Hysteresis phenomena in the formation of polyelectrolyte complexes:
changes of the mobilities of the polymer chains by complex formation. ∇,
Poly(methacrylic acid) with covalently bound 8–anilino-1-naphthalenesulfonic
acid (PMAA*); ○, ●, PMAA*–2X; △, ▲, PMAA*–4,4-ionene.

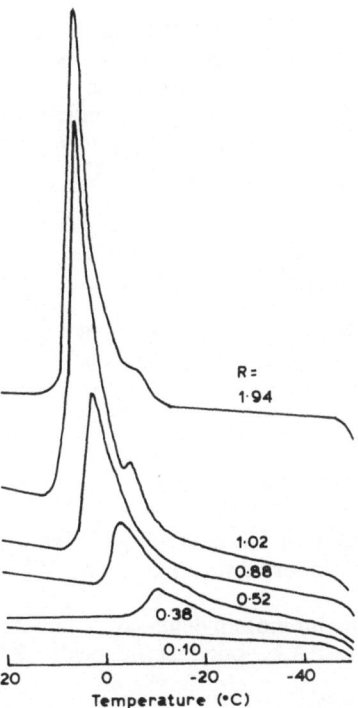

FIG. 28. DSC profiles of melting endotherms of water bound to poly(acrylic acid)–2X complex with various water contents, $R(gH_2O/g$ dry PEC).

again but a hysteresis loop is observed. This phenomenon suggests cooperative interaction. The degrees of depression and cooperativity are of course dependent on the hydrophobicity and rigidity of the polycation. Figure 28 shows the differential scanning calorimetric (DSC) melting endotherms of water in hydrated PAA–2X PEC with various water contents, R. Bound water molecules can be classified into three groups:

Type I: Non-freezing water, water molecules the most strongly bound to ionic sites; freezes below $-50°C$ and constitutes about 0.38 g of water/g dry PEC.

Type II: bound water molecules with a constant freezing point in a range from $-50°C$ to $0°C$.

Type III: bound water molecules with a variable freezing point.

However, this latter type of water cannot be distinguished from normal water molecules in PECs at higher water contents. The

TABLE 10

OXYGEN ADDUCTS OF THE HAEM DERIVATIVES INCORPORATED INTO COMPLEX
COACERVATES

[Haem derivative] = 0.08 mmol/cm^3; [ionic moiety of the added polymer] = 5 mmol/dm^3; [buffer solution (pH = 10)]/[ethylene glycol] = $1:1$ by vol.; temperature $-30°C$.

Haem derivatives[a]	Added polymer[b]	λ_{max} (nm)		Lifetime (min)
		Deoxy	O_2 adduct	
1	None	422,529,558	(Oxidized)	—
	2X or PAA	422,528,558	(Oxidized)	—
	2X and PAA	422,527,558	410,535,570	7
	2X and PMAA	422,527,558	410,535,572	19
2	None	426,558	(Oxidized)	—
	2X and PAA		402,540,580	2

[a] 1, Iron(II) protoporphyrin IX mono-N-[3-(1-imidazolyl)propyl]amide chloride; 2, iron(II) protoporphyrin IX mono-N-[5-(2-methyl-1-imidazolyl)pentyl]amide chloride.
[b] 2X, Poly[(N,N-dimethylimino)ethylene-(N,N-dimethylimino)methylene-1,4-phenylene dichloride]; PAA, poly(acrylic acid); PMAA, poly(methacrylic acid).

number of type I water molecules is decreased by PEC formation because of the desolvation of some ionic sites.

Okihana et al.[54] reported that complex coacervates could absorb low molar mass organic compounds and protect them from photochemical decomposition. The complex coacervates formed from KPVS and partially aminoacetalized PVA (PVA-A) contains twice as much glycine and diglycine as the surrounding equilibrium liquid. Their decomposition by UV light appears to be a first-order reaction. While the kinetic constants in the absence of coacervate droplets are 0.44 h^{-1} for glycine and 3.5 h^{-1} for diglycine, in the presence of coacervates they decrease to 0.093 and 1.81 h^{-1}, respectively. On the other hand, the reversible oxygen adducts of a haem (iron(II)protoporphyrin IX) in an aprotic solvent (water/ethylene glycol, $1:1$) cooled to $-30°C$ were stabilized by incorporating them into the complex coacervates of 2X and PAA or PMAA.[33] Table 10 shows the UV and visible absorption maxima (λ_{max}) of the haem derivatives and the life of oxygen adducts. Haem derivatives are incorporated into coacervates to a considerable extent as is shown by a value of 150 dm^3/litre for the equilibrium constant, K. Only haem derivatives in the coacervate droplets show the spectra of oxygen adducts and these adducts have a life of a few minutes. This result may be caused by retardation of the formation of binuclear complexes and by the hydrophobic environment around the haem in the coacervate.

The preparation and properties of immobilized enzymes in PECs were reported by Margolin *et al.*[28] Penicillin amidase (PA) is covalently bound to the polycation QPVP. This polycation forms a water-soluble non-stoichiometric PEC with PMAA where the molar ratio of QPVP/PMMA is $1:3$. The activity of the enzyme (E) is assayed with benzylpenicillin as a substrate (S) under the condition $[S]_0 \gg [E]_0$, assuming a Michaelis–Menten type reaction. The catalytic constant (k_{cat}) decreases from 50 to $33\,s^{-1}$ when the native enzyme is linked to QPVP and to $20\,s^{-1}$ when subsequently incorporated in the PEC. This phenomenon may be explained by the effect of charged groups in the micro-environment around the enzyme active centre. K_m, however, scarcely changes from the value for PA immobilized in polyacrylamide gel, which suggests the absence of impediments to diffusion of S in PEC systems. This PA-immobilized PEC is phase-separated as shown in Fig. 29. Under alkaline conditions the

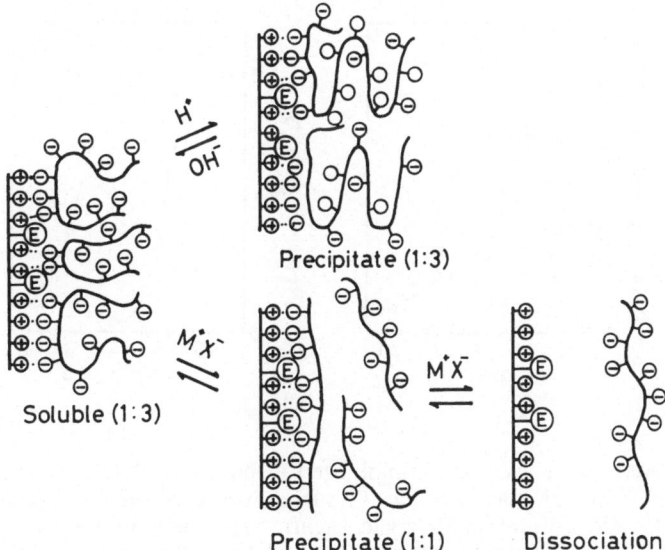

FIG. 29. Schematic representation of phase separation of quaternized poly-(vinylpyridine)–poly(methacrylic acid) complex containing penicillin amidase.

PEC is highly negatively charged and is, therefore, soluble in water, but under acid conditions the charge density decreases and precipitation occurs. Therefore the optimum pH for the PA reaction lies in the alkaline region and the constant for competitive inhibition by the reaction product (phenylacetic acid) decreases by two orders of magnitude. On the other hand, increasing the ionic strength to about 0·35M NaCl causes the stoichiometric PEC to precipitate and so reduces enzymatic activity. When the ionic strength is further increased above 0·4M, the PEC goes into solution again. It follows that the activity of PA immobilized in the PEC is easily controlled by the reversible conversion from the soluble to the insoluble state and the transition can be achieved by slightly adjusting the pH and ionic strength of the solution. At a given ionic strength, the self-regulating enzymatic system is controlled according to the following sequence: accumulation of ionic product→ change in ionic strength→ change in solubility of PEC→ change in activity.

The hydrolysis activity of invertase immobilized in PECs was reported by Osada et al.[21] Figure 30 shows the pH dependence and the protective effect against inhibitor, such as Hg^{2+}, on the immobilized enzyme using a saccharose substrate. Whilst invertase in the

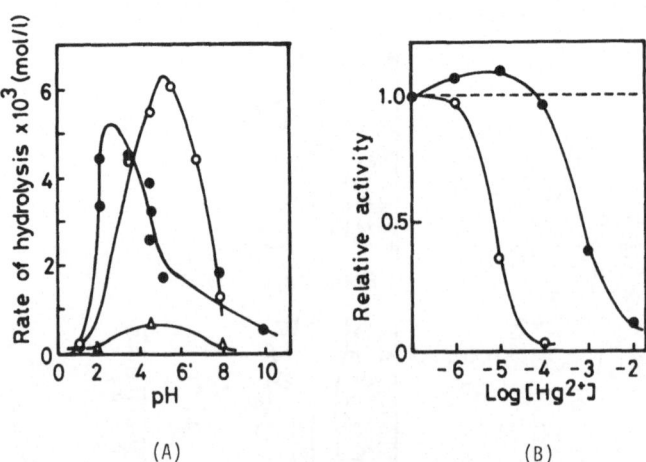

(A)   (B)

FIG. 30. Activity of invertase immobilized in polyelectrolyte complexes. (A) pH dependence of the activity: ○, free invertase; ●, immobilized in poly(ethyleneimine)-poly(methacrylic acid); △, immobilized in poly(2-acrylamido-2-methylpropanesulfonic acid)–poly(4-vinylbenzyltrimethylammonium chloride). (B) Protective effect against an inhibitor, Hg^{2+}: ○, free invertase; ●, immobilized in poly(ethyleneimine)–poly(methacrylic acid).

poly(ethyleneimine) (PEI)–poly(methylacrylic acid) (PMAA) PEC exhibits optimum relative activity of about 80% compared with free enzyme (at pH 2·0), in PVBMA–poly(2-acrylamido-2-methylpropane-sulfonic acid) the enzyme exhibits very low activity. It is very interesting that no activity is observed at 10^{-4}M Hg^{2+} for the free enzyme but the immobilized enzyme in PEI–PMAA retains 60% of its activity under the same conditions. Thus immobilization in PEC is found to enhance stability by about two orders against the inhibitory effect of Hg^{2+}.

The stabilization and immobilization of enzymes by direct complexation with synthetic polyelectrolyes have also been studied. For example, both formate and alcohol dehydrogenases containing —SH groups are stabilized by complexation with a polycation, QPVP,[199] against oxidation catalyzed by metal ions. This stabilization may result from the electrostatic repulsion between coated polycation and metal ions. Sweet potato β-amylase was found to be activated 1·3- to 3·0-fold by complexation with polycations, but polyanions strongly inhibited the enzyme activity. These activators are classified as mixed or partially non-competitive types.[99] Papain can be immobilized without the reduction of the enzyme activity by stoichiometric complexation with KPVS.[94] The redox reactions of mammalian cytochrome c are affected by complexation with polyanions such as phosphitin, the storage protein from egg yolk, and heparin and dextran sulfate.[200] Such complexation strongly inhibits the reaction of cytochrome c with negatively charged reagents, but affects the reactions with uncharged reagents only slightly. On the contrary, the oxidation of the reduced cytochrome c by positively charged reagents is greatly enhanced. These results can be explained by supposing the positive charges on the free cytochrome c are replaced by a net negative charge of polyanions. Any steric effects on polyanion bindings can be seen to be small in comparison with such electrostatic effects.

Interesting applications of PECs for the immobilization of cells and micro-organisms were studied by Kokufuta et al.[74] Cells from Escherichia coli and Nitrosomonas europaea were immobilized by an excess of trimethylammonium glycol chitosan (TGCI) in the cell suspension, followed by the addition of KPVS. Cells were observed to be situated in the pores and on the surface of the PEC matrix, mainly as the result of electrostatic interactions between polyelectrolytes and ionizable groups located on the cell surfaces. The number of grown cells released in the presence of the PEC is smaller than in the absence of the PEC, but their glucose consumptions are exactly the same. These results indicate that glucose in immobilized cell systems is consumed not only by the cells released from the PEC but also by entrapped cells which grow in the PEC. Hence, there is almost the same glucose oxidizing activity for entrapped cells as for the native cells. When cells from *Nitrosomonas europaea* are used, the ammonia-oxidizing activity of the entrapped cells is found to be retained even after 2000 h incubation in a batch process, and no degradation of the PEC is observed.

So far we have seen that PECs possess specific characteristics, originating mainly from the coexistence of hydrophilic microdomains, which contain ionic sites with high charge densities and bound water, and hydrophobic microdomains. In addition to these properties other functions of PECs have been demonstrated. Matrix polymerizations in PEC systems are carried out by polymerization of ionic monomers in the presence of oppositely charged polyelectrolytes.[201] It is to be expected that the polyelectrolytes will affect the kinetics of polymerization. Some structural factors of the freshly polymerized materials will reflect those of the parent polyelectrolytes, such as molar mass and its distribution, tacticities and optical isomerism, and the structures of the PECs that are obtained. Radical polymerization of methacrylic acid (MAA) in the presence of atactic P2VP was studied by Smid et al.,[27] who established that there are two types of polymerizations: (1) at relatively low concentrations of P2VP, the sequences of adsorbed MAA on P2VP are sufficiently long for a propagation to occur along the P2VP chains corresponding to a zipping up mechanism; and (2) at relatively high concentrations of P2VP, the sequences are shorter owing to the distribution of MAA on an excess of P2VP chains so that the rate-determining step may be the transport of MAA from bulk solution to the matrix domains.

The condensation polymerization of urea and formaldehyde in water in the presence of PAA was studied by Papisov et al.[44] They found that these complexes, consisting of a stoichiometric PEC together with an excess of one of the component polyelectrolytes, stabilized through hydrogen bonding and electrostatic interactions and formed fibrillized structures. PMAA, polymerized in the presence of chitosan, was found to be rich in syndiotacticity.[201] Acridine orange, adsorbed on this PMAA, showed an induced CD spectrum below pH 6·48. This result demonstrated that this PMAA is optically active, due to the conformational asymmetry of its main chain.

Knowledge of the interactions of synthetic macromolecules with cell membranes provides much useful information for a number of applications, for example cell protection, cell dispersive support, fusogens, bactericides, cell separators, cell-cultured material and interferon inducers. A polycation such as PLL is bound to the negatively charged surface of chloroplast thylakoid membranes and interferes with photosystem I activity and the cation-dependent rearrangement of chlorophyll proteins.[108] On adding micro-anions such as $[Fe(CN)_6]^{4-}$, SO_4^{2-} or Cl^-, the inhibition effects of PLL were found to be prevented or reversed without decreasing the interaction between PLL and the membranes, by screening the positive charges of the PLL-coated membrane surfaces. From these results it is clear that the effects of PLL are not caused by a direct binding of PLL to the membrane proteins, but by the changes of the electric charges on the exterior membrane surface, followed by the change of transmembrane electrical potential.

Patel and his coworkers[202] found that the heparin could lyze the multilammelar liposomes, irrespective of the net charges of the liposomes. Heparin also shows a lytic activity to red blood cells, probably due to the interaction between heparin and phospholipid components of the cell membrane. Cell fusions have been applied to the creation of new intermediate hybrid cells, especially of higher plants, and to the micro-injection of some substrates, e.g. drugs, proteins and nucleic acids. It is well known that PEG is a most useful fusogen but the mechanisms of the cell fusion have not yet been clarified. The interactions between water-soluble polymers and lipo-somes made of soybean and egg-yolk lecithins were studied by Sakamoto et al. and Abe et al.[203] Polycations exhibit similar effects as fusogens to those of PEG even at very low concentration—10^{-3}–10^{-2} (mass %). However, in such conditions, polycations interact with liposome membranes through electrostatic interactions so strongly that they perturb and finally lyze the membrane. Thus, a good fusogenic result is achieved by using a mixture of PEG and a small amount of PLL for human erythrocytes.[204] By contrast, liposomes of egg-yolk lecithin, which are less negative and more stable than those of soybean lecithin, aggregated without the perturbation of the membrane.[203]

In another type of application, PEC formation has been used for the selective recovery of organic and metallic ions. For example, a polymer containing crown ethers such as poly(vinylbenzo-[18]-*crown*-6) with K^+ ions was found to behave like a polycation and could be interacted with a polyanion such as PAA to form a PEC.[205] Three-component PECs containing poly(carboxylic acid)s, P4VP and Cu^{2+} have been described by Kabanov et al.[206] Such PECs, with coordinated metal ions, are expected to be developed as new polymer catalysts, redox systems and membranes with selective permeability as well as selective adsorption resins, because the microdomains in PECs provide spatially fixed ligands and hydrophobic and electrostatic fields.

In spite of the importance of the specific and unique characteristics of PECs and their many useful applications, there are still very few new developments and a paucity of fundamental and systematic work. PECs with various specific functions can be easily obtained because numerous combinations of polycations and polyanions are possible and there are many factors that can be used to control properties. Conversely, this very variety makes the formulation of a system philosophically difficult. However, the authors believe that, in the near future, active work in this field must be undertaken to develop new tailor-made materials for special uses.

REFERENCES

1. MICHAELS, A. S., *Ind. Eng. Chem.*, 1965, **57**, 32.
2. TSUCHIDA, E., ABE, K. and HONMA, M., *Macromolecules*, 1976, **9**, 112.
3. GULYAWVA, Z. G., POLETAEVA, O. A., KALACHEV, A. A., KASAIKIN, V. A. and ZEZIN, A. B., *Vysokomol. Soedin.*, *Ser. A*, 1976, **18**, 2800; KHARENKO, O. A., KHARENKO, A. V., KALYUZHNAYA, R. I., IZUMURDOV, V. A., KASAIKIN, V. A., ZEZIN, A. B. and KABANOV, V. A., *Vysokomol. Soedin.*, *Ser. A*, 1979, **21**, 2719; KASAIKIN, V. A., KHARENKO, O. A., KHARENKO, A. V., ZEZIN, A. B. and KABANOV, V. A., *Vysokomol. Soedin.*, *Ser. B*, 1979, **21**, 84; GULYAEVA, Z. G., ZEZIN, A. B., RAZVODOVSKII, E. F. and BERESTETSKAYA, T. Z., *Vysokomol. Soedin.*, *Ser. A*, 1974, **16**, 1852; IZUMURDOV, V. A., KASAIKIN, V. A., ERMAKOVA, L. N. and ZEZIN, A. B., *Vysokomol. Soedin.*, *Ser. A*, 1978, **20**, 400.
4. TSUCHIDA, E., OSADA, Y. and ABE, K., *J. Polym. Sci.*, *Polym. Chem. Ed.*, 1976, **14**, 767.
5. ABE, K., KOIDE, M. and TSUCHIDA, E., *Macromolecules*, 1977, **10**, 1259.
6. TSUCHIDA, E. and ABE, K., *Adv. Polym. Sci.*, 1982, **45**, 1.
7. TSUCHIDA, E., HORIE, K. and ABE, K., (Eds) *Intermacromolecular Complexes*, 1983, Gakkai Shuppan Center, Tokyo.
8. TSUCHIDA, E., *Functional Polymers*, 1974, Kyoritsu Shuppan, Tokyo, Ch. 8.
9. BIXLER, H. J. and MICHAELS, A. S., *Encycl. Polym. Sci. Technol.*, 1974, **10**, 765; BEKTUROV, E. A. and BAKAUOVA, Z. K., *Vestn. Akad.*

Nauk Kaz. SSR, 1984, (3), 11; STROEM, G., *Sven. Papperstidn.*, 1984, **87**, 14, 19.

10. KABANOV, V. A. and PAPISOV, I. M., *Vysokomol. Soedin.*, *Ser. A*, 1979, **21**, 243.
11. BEKTUROV, E. A. and BIMENDINA, L. A., *Adv. Polym. Sci.*, 1981, **41**, 99.
12. TSUCHIDA, E., OSADA, Y. and OHNO, H., *J. Macromol. Sci.—Phys.*, 1980, **B17**, 683.
13. NAGASAWA, M., IZUMI, M. and KAGAWA, I., *J. Polym. Sci.*, 1959, **37**, 375.
14. MORAWETZ, H., *Polyelectrolyte Solution*, 1961, Academic Press, New York; OOSAWA, F., *Polyelectrolytes*, 1971, Marcel Dekker, New York; REMBAUM, A. and SELEGNY, E., *Polyelectrolytes and Their Applications*. 1975, D. Reidel, Dordrecht.
15. GUERON, M., *Biopolymers*, 1980, **19**, 353.
16. KATCHALSKY, A. and SPITNIK, P., *J. Polym. Sci.*, 1947, **2**, 432.
17. TSUCHIDA, E., OSADA, Y. and ABE, K., *Makromol. Chem.*, 1974, **175**, 583.
18. PHILIPP, B., HONG, L. T., LINOW, K. J. and COWIE, J. M. G., *Eur. Polym. J.*, 1981, **17**, 615; SHARPE, Jr, A. J. and WINDHAGER, R. H., *Tappi*, 1982, **65**, 123; SATO, T., ITO, O., MORI, S. and OTSU, T., *J. Polym. Sci.*, *Polym. Chem. Ed.*, 1984, **22**, 1661.
19. VENGEROVA, N. A., RUDMAN, A. R., EL'TSEFAN, B. S., SNEGIREVA, N. S., SHILOKHVOST, V. P., TSIVINSKAYA, L. K., KALYUZHNAYA, R. I., SIDOROVA, L. P. and RESHETILOVA, T. I., *Vysokomol. Soedin.*, *Ser. A*, 1983, **25**, 1245.
20. TAKAGISHI, T., KOZUKA, H. and KUROKI, N., *J. Polym. Sci.*, *Polym. Chem. Ed.*, 1981, **19**, 3237; TAKAGISHI, T., KOZUKA, H., KIM, G. J. and KUROKI, N., *J. Polym. Sci.*, *Polym. Chem. Ed.*, 1982, **20**, 2231; TAKAGISHI, T., KOZUKA, H. and KUROKI, N., *J. Polym. Sci.*, *Polym. Chem. Ed.*, 1983, **21**, 447.
21. OSADA, Y., IINO, Y. and NUMAJIRI, Y., *Chem. Lett.*, 1982, 559.
22. KABANOV, V. A., ZEZIN, A. B., ROGACHEVA, V. B., IZUMURDOV, V. A. and RYZHIKOV, S. V., *Dokl. Akad. Nauk SSSR*, 1982, **262**, 1419; KABANOV, V. A., ZEZIN, A. B., ROGACHEVA, V. B. and RYZHIKOV, S. V., *Dokl. Akad. Nauk SSSR*, 1982, **267**, 862; KAIMINS, I., OZOLINA, G. and ZEZIN, A. B., *Vysokomol. Soedin.*, *Ser. B*, 1983, **25**, 179; ROGACHEVA, V. B., GRISHINA, N. G., ZEZIN, A. B. and KABANOV, V. A., *Vysokomol. Soedin.*, *Ser. A*, 1983, **25**, 1530; KABANOV, V. A., ZEZIN, A. B., KALYUZHNAYA, R. I. and SHALBAEVA, G. B., *Dokl. Akad. Nauk SSSR*, 1984, **277**, 1166; SHALBAEVA, G. B., NIKOLAEVA, T. V., MIL'CHENKO, E. N., KALYUZHNAYA, R. I. and ZEZIN, A. B., *Vysokomol. Soedin.*, *Ser. A*, 1984, **26**, 1270.
23. PAL, M. K., BOEHMER, V., KERN, W. and NATH, J., *Makromol. Chem.*, 1981, **182**, 1649.
24. OYAMA, H. T. and NAKAJIMA, T., *J. Polym. Sci.*, *Polym. Chem. Ed.*, 1983, **21**, 2987; OYAMA, H. T. and NAKAJIMA, T., *J. Appl. Polym. Sci.*, 1984, **29**, 2143.

25. ABE, K., HASEGAWA, M. and SENOH, S., *Polym. Prep., Jpn.*, 1983, **32,** 492.
26. PLATA, N. A., ALIEVA, E. D. and KALACHEV, A. A., *Vysokomol. Soedin., Ser. A*, 1981, **23,** 640; KABANOV, V. A., KARGINA, O. V., UL'YANOVA, M. V. and LITVINOV, I. A., *Vysokomol. Soedin., Ser. B,* 1982, **24,** 17: KABANOV, V. A., MUSTAFAEV, M. I., NEKRASOV, A. V., NORIMOV, A. S., PETROV, R. V., KHAITOV, R. M. and KHAUSTOVA, L. I., *Dokl. Akad. Nauk SSSR*, 1984, **274,** 998; ZUBOV, V. P., LUTSENKO, V. V. and KORNEV, A. P., *Vysokomol. Soedin., Ser. A,* 1984, **26,** 1484.
27. SMID, J., TAN, Y. Y. and CHALLA, G., *Eur. Polym. J.,* 1983, **19,** 853; SMID, J., TAN, Y. Y. and CHALLA, G., *Eur. Polym. J.,* 1984, **20,** 887; FUJIMORI, K., TRAINOR, G. T. and COSTIGAN, M. J., *J. Polym. Sci., Polym. Chem. Ed.,* 1984, **22,** 2479.
28. MARGOLIN, A. L., IZUMURDOV, V. A., SVEDAS, V., ZEZIN, A. B., KABANOV, V. A. and BEREZIN, I. V., *Biochim. Biophys. Acta,* 1981, **660,** 359; MARGOLIN, A. L., IZUMURDOV, V. A., SVEDAS, V. K. and ZEZIN, A. B., *Biotechnol. Bioeng.,* 1982, **24,** 237; IZUMURDOV, V. A., SAVITSKII, A. P. and KABANOV, V. A., *Dokl. Akad. Nauk SSSR*, 1983, **272,** 1408; IZUMURDOV, V. A., SAVITSKII, A. P., ZEZIN, A. B. and KABANOV, V. A., *Vysokomol. Soedin., Ser. B,* 1983, **25,** 805; IZUMURDOV, V. A., MARGOLIN, A. L., SHERTYUK, S. F., SVODAS, V., ZEZIN, A. B. and KABANOV, V. A., *Dokl. Akad. Nauk SSSR,* 1983, **269,** 631.
29. SKORODINSKAYA, A. M., KEMENOVA, V. A., CHERNAVA, O. V., EFIMOV, V. S., LAKIN, K. M., ZEZIN, A. B. and KABANOV, V. A., *Khim.-Farm. Zh.,* 1983, **17,** 1463; NURGALIEVA, F. F., SAGDIEVA, E. G., TASHMUKHAMEDOV, S. A., TILLAEV, R. S. and KHAMIDOVA, G. R., *Dokl. Akad. Nauk UzSSR 1983,* (8), 38.
30. IZUMURDOV, V. A., ZEZIN, A. B. and KABANOV, V. A., *Vysokomol. Soedin., Ser. A*, 1983, **25,** 1972.
31. BRONICH, T. K., *Deposited Doc., VINITI,* 1981, **575,** 79.
32. Nitto Electric Industrial Co. Ltd, Jpn. Kokai Tokkyo Koho, 1982, 82 91 707
33. TSUCHIDA, E., NISHIDA, H., TAGUCHI and MACHIDA, K., *Makromol. Chem., Rapid Commun.,* 1982, **3,** 161; OHNO, H., SHIBAYAMA, M. and TSUCHIDA, E., *Makromol. Chem.,* 1983, **184,** 1017.
34. GULYAEVA, Zh. G., ZANSOKHOVA, M. F., RAZVODOVSKII, E. F., EFINOV, V. S., ZEZIN, A. B. and KABANOV, V. A., *Vysokomol. Soedin., Ser. A*, 1983, **25,** 1283.
35. TSUCHIDA, E., Jpn. Kokai Tokkyo Koho, 1983, 58 36633.
36. CRUMP, R. A. and HOPKINS, J., 1981, British Patent 1570723.
37. BEKTUROV, E. A., LEGKUNETS, R. E., SHAYAKHMETOV, S. S., and KUSAIBERGENOV, S. E., *Makromol. Chem., Rapid Commun.,* 1981, **2,** 761; LEGKUNETS, R. L., IZOTOV, M. Z., BEKTUROV, E. A. and SHAYAKHMETOV, S. S., *Izv. Akad. Nauk Kaz. SSR, Ser. Khim.,* 1981, (3), 42; SHAYAKHMETOV, S. S., LEGKUNETS, R. E., IZOTOV, M. Z. and BEKTUROV, E. A., *Izv. Akad. Nauk. Kaz. SSR, Ser. Khim.,* 1981, (3), 47.
38. RUDMAN, A. R., KALYUZHNAYA, R. I., VENGEROVA, N. A.,

EL'STEFON, B. S., RAZVODOVSKII, E. F. and ZEZIN, A. B., *Vysokomol. Soedin., Ser. A*, 1983, **25**, 2405.

39. SADOVA, A. N., KUZNETSOVA, L. E. and BALAKIREVA, R. S., *Deposited Doc., SPSTL*, 1981, **13**, 107; OKUBO, T., HONGYO, K. and ENOKIDA, A., *J. Chem. Soc., Faraday Trans. I*, 1984, **80**, 2087.

40. MARGOLIN, A. L., IZUMURDOV, V. A., SHERSTYUK, S. F., ZEZIN, A. B. and SVEDAS, V., *Mol. Biol. (Moscow)*, 1983, **17**, 1001.

41. DZHAGIPAROVA, A. T. and KUDAIBERGENOV, S. E., *Deposited Doc., VINITI*, 1981, **575**, 118; LEGKUNETS, R. E., DZHAGIPAROVA, A. T., KUDAIBERGENOV, S. E., BEKTUROV, E. A. and SHAYAKHMETOV, S. S., *Izv. Akad. Nauk Kaz. SSR, Ser. Khim.*, 1982, (6), 54.

42. WAN, G. X., OHNO, H. and TSUCHIDA, E., *Makromol. Chem., Rapid Commun.*, 1983, **4**, 87.

43. POBEDIMSKAYA, T. G., KRUPIN, S. V., KARATAEVA, L. V. and BARABANOV, V. P., *Izv. Vyssh. Uchebn. Zaved., Khim. Khim. Teknol.*, 1981, **24**, 1411; SHIBALOVISH, V. G., PERINA, G. P., SHOKINA, L. V., BONDARENKO, V. M. and NIKOLAEV, A. F., *Deposited Doc., SPSTL*, 1981, **28**, 77.

44. PAPISOV, I. M., KUZOVLEVA, O. E., MARKOV, S. V. and LITMANOVICH, A. A., *Eur. Polym. J.*, 1984, **20**, 195.

45. AKAHOSHI, H., TOSHIMA, S. and TAYA, K., *J. Phys. Chem.*, 1981, **85**, 818.

46. NOMURA, K., MASUMI, T., NISHIOKA, K., DEGUCHI, H. and OHNO, H., *Mitsubishi Denki Giho*, 1982, **56**, 149.

47. OHNO, H., HOSODA, N. and TSUCHIDA, E., *Polym. Prep. Jpn.*, 1982, **31**, 522; TSUCHIDA, E. and OHNO, H., *Polym. Prep.*, 1982, **23**, 66; OYAMA, N., OKI, N., OHNO, H., MATSUDA, H. and TSUCHIDA, E., *J. Phys. Chem.*, 1983, **87**, 3642; KATO, M., OKI, N., OHNO, H. and TSUCHIDA, E., *Polymer*, 1983, **24**, 846; OHNO, H., HOSODA, N. and TSUCHIDA, E., *Makromol. Chem.*, 1983, **184**, 1061.

48. OYAMA, N., OHSAKA, T., SATO, K. and YAMAMOTO, H., *Anal. Chem.*, 1983, **55**, 1429.

49. WANG, S., XI, F. and LI, Z., *Kao Fen Tzu Turg Hsun*, 1981, (3), 224; WANG, S., XI, F. and LI, Z., *Gaofenzi Tongxun*, 1982, (4), 241.

50. PHILIPP, B., HONG, L. T., DAWYDOFF, W., LINOW, K. J. and SCHUELKE, U., *Z. Anorg. Allg. Chem.*, 1981, **479**, 219.

51. REINERT, K. E. W., *Bioelectrochem. Bioeng.*, 1981, **8**, 301.

52. KABANOV, V. A., KARGINA, O. V., MISHUTINA, L. A., LUBANOV, S. YU., KATUZYUSKI, K. and PENCZEKE, S., *Makromol. Chem., Rapid Commun.*, 1981, **2**, 343.

53. SCHEMPP, W. and TRAN, H. T., *Wochenbl. Papierfabr.*, 1981, **109**, 726; ONABE, F., *Mokuzai Gakkaishi*, 1982, **28**, 437; ONABE, F., *Mokuzai Gakkaishi*, 1982, **28**, 445.

54. CHO, C. S., *Bull. Korean Chem. Soc.*, 1981, **2**, 76; OKIHANA, H. and PONNAMPRUMA, C., *Origins Life*, 1982, **12**, 347.

55. ERMAKOVA, L. N., *Deposited Doc., VINITI*, 1981, **3167**, 194; ERMAKOVA, L. N., NUSS, P. V., KASAIKIN, V. A., ZEZIN, A. B. and KABANOV, V. A., *Vysokomol. Soedin., Ser. A*, 1983, **25**, 1391.

56. DESBRIERES, J. and RINAUDO, M., *Eur. Polym. J.*, 1981, **17**, 1265; FURUSAWA, K., KANESAKA, M. and YAMASHITA, S., *J. Colloid Interface Sci.*, 1984, **99**, 341.
57. NAGATA, I. and MORAWETZ, H., *Macromolecules*, 1981, **14**, 87; FOWKES, F. M., TISCHLER, D. O., WOLFE, J. A., LANNIGAN, L. A., ADEMU-JOHN, C. M. and HALLIWELL, M. J., *J. Polym. Sci., Polym. Chem. Ed.*, 1984, **22**, 547.
58. DAUTZENBERG, H., LINOW, K. J. and PHILIPP, B., *Acta Polym.*, 1982, **33**, 619; ONABE, F., YAMAZAKI, A., USUDA, M. and KADOYA, T., *Mokuzai Gakkaishi*, 1983, **29**, 60; MUSABEKOV, K. B., OMAROVA, K. I. and IZIMOV, A. I., *Acta Phys. Chem.*, 1983, **29**, 89.
59. EISENBERG, A., SMITH, P. and ZHOU, Z. L., *Polym. Eng. Sci.*, 1982, **22**, 1117; ZHOU, Z. L. and EISENBERG, A., *J. Polym. Sci., Polym. Phys. Ed.*, 1983, **21**, 595; SMITH, P. and EISENBERG, A., *J. Polym. Sci., Polym. Lett. Ed.*, 1983, **21**, 223; RUTKOWSKA, M. and EISENBERG, A., *Macromolecules*, 1984, **17**, 821.
60. BILIEHENKO, V. N., EZHOVA, T. G., USKOVA, E. T., RAEVSKII, U. S. and USKOV, I. A., *Ukr. Khim. Zh. (Russ. Ed.)*, 1982, **48**, 877; DJADOUN, S., *Polym. Bull. (Berlin)*, 1983, **9**, 313; HIGASHITANI, K., HOSOKAWA, G., AIMOTO, H. and MATSUNO, Y., *Kagaku Kogaku Ronbunshu*, 1983, **9**, 543; HO, C. H. and HOWARD, G. J., *Colloids Surf.*, 1983, **7**, 265; XIAO, H. X., FRISCH, K. C. and FRISCH, H. L., *J. Polym. Sci., Polym. Chem. Ed.*, 1984, **22**, 1035.
61. POBEDIMSKAYA, T. G., KRUPIN, S. V., GUBAIDULLIN, F. A. and BARABANOV, V. P., *Neft. Khoz.*, 1982, (9), 32; POBEDIMSKAYA, T. G., KRUPIN, S. V. and BARABANOV, V. P., *Vysokomol. Soedin., Ser. B*, 1983, **25**, 488; IKKAKU, Y., OKUBO, M. and MATSUMOTO, T., *Nippon Setchaku Kyokaishi*, 1983, **19**, 52.
62. BLEIBERG, I., FABIAN, I. and ARONSON, M., *Biochim. Biophys. Acta*, 1981, **674**, 345; WEILER, I. M., *Immunopharmacology*, 1983, **6**, 245; TADOLINI, B. and CABRINI, L., *Appl. Biochem. Biotechnol.*, 1984, **9**, 143.
63. FERRACINI, E., FERRERO, A. and RIVA, F., *Colloid Polym. Sci.*, 1981, **259**, 602; CASU, B., TORRI, G., LEGRAMANDI, M. and RIGHETTI, P. G., *Carbohydr. Res.*, 1982, **104**, 299; CUNDALL, R. B., JONES, G. R. and MURRAY, D., *Makromol. Chem.*, 1982, **183**, 849.
64. COTTONARO, C. N., ROOHK, H. V., BARTLETT, R. H., SERVAS, F. M. and SPERLING, D. R., *Trans.-Am. Soc. Artif. Intern. Organs*, 1982, **28**, 478.
65. GROEBE, V., LUU, T. H., PHILIPP, B. and SCHWARZ, H. H., *Acta Polym.*, 1981, **32**, 488; PHILIPP, B., LINOW, K. J. and SCHLEICHER, H., *Papier (Darmstadt)*, 1981, **35**, 570; SCHEMPP, W., HESS, P. and KRAUSE, T., *Papier (Darmstadt)*, 1982, **36**, 41.
66. TOEI, K. and ZAITSU, T., *Bunseki Kagaku*, 1982, **31**, 543.
67. ONABE, F., *Mokuzai Gakkasishi*, 1983, **29**, 459; ONABE, F., *Mokuzai Gakkaishi*, 1983, **29**, 467; ONABE, F., *Mokuzai Gakkaishi*, 1983, **29**, 513.
68. STROEM, G. and STENIUS, P., *Colloids Surf.*, 1981, **2**, 357; BARLA, P.,

STROEM, G. and STENIUS, P., *Ekman-Days 1981, Int. Symp. Wood Pulping Chem.*, 1981, **5**, 103.
69. SULGA, G., MOZHIKO, L. N., KALYUZHNAYA, R. I., ZEZIN, A. B. and KABANOV, V. A., *Khim. Drev.*, 1982, (1), 87; SULGA, G., ZEZIN, A. B., KALYUZHNAYA, R. I., MOZHEIKO, L. and REKNER, F., *Khim. Drev.*, 1981, (2) 63; SHUL'GA, G. M., KALYUZHNAYA, R. I., MOZHEIKO, L. N., REKNER, F. V., ZEZIN, A. B. and KABANOV, V. A., *Vysokomol. Soedin., Ser. A*, 1982, **24**, 1516; SHUL'GA, G. M., KALYUZHNAYA, R. I., ZEZIN, A. B., KABANOV, V. A., MOZHEIKO, L. N. and REKNER, F. V., *Vysokomol. Soedin., Ser. A*, 1984, **26**, 291; PILIPENKO, A. T., TARASEVICH, Y. T., SOLOMENTSEVA, I. M., ZNAMENSKAYA, M. V., ZUL'FIGAROV, O. S., BARAN, A. A. and ALESHKINA, T. P., *Khim. Tekhnol. Vody*, 1984, **6**, 333.
70. NEDELCHEVA, M. and STOILKOV, G., *Chem. Prum.*, 1983, **33**, 424; LINDSTROEM, T. and WAGBERG, L., *Tappi J.*, 1983, **66**, 83; HAARS, A., BAUER, A. and HUETTERMANN, A., *Holzforschung*, 1984, **38**, 171.
71. DOMARD, A. and RINAUDO, M., *Macromolecules*, 1981, **14**, 620.
72. KENNEDY, J. F., BARKER, S. A. and BRADSHOW, I. J., *Biochem. Soc. Trans.*, 1982, **10**, 136.
73. SUHAILA, M. and SALLEH, A. B., *Biotechnol. Lett.*, 1982, **4**, 611; Kaimins, I. and OZOLINA, G., *Latv. PSR Zinat. Akad. Vestis, Khim. Ser.*, 1984, (3), 366.
74. KOKUFUTA, E., MATSUMOTO, W. and NAKAMURA, I., *Biotechnol. Bioeng.*, 1982, **24**, 1591; KOKUFUTA, E., MATSUMOTO, W. and NAKAMURA, I., *J. Appl. Polym. Sci.*, 1982, **27**, 2503; KOKUFUTA, E., FUJII, S. and NAKAMURA, I., *Makromol. Chem.*, 1982, **183**, 1233.
75. KATAOKA, S. and ANDO, T., *Polym. Commun.*, 1984, **25**, 24; EZAKI, B., SAITO, M., IMAMURA, T. and KINA, K., *Bunseki Kagaku*, 1984, **33**, 108; MUTJE PUJOL, P., LOPEZ, A. L. T. and FRONT, C. C., *Ing. Quim. (Madrid)*, 1984, **16**, 75.
76. KIKUCHI, Y. and SAKAI, K., *Nippon Kagaku Kaishi*, 1981, 574; KIKUCHI, Y. and SASAYAMA, S., *Makromol. Chem.*, 1982, **183**, 2153; KIKUCHI, Y. and KASEDA, K., *Nippon Kagaku Kaishi*, 1982, 842.
77. KIKUCHI, Y. and HORI, K., *Nippon Kagaku Kaishi*, 1982, 847; KIKUCHI, Y. and TAKABAYASHI, T., *Bull. Chem. Soc., Jpn.*, 1982, **55**, 2307.
78. HARROP, R., PHILLIPS, G. O., ROBB, I. D. and WILLIAMS, P. A., *Prog. Food Nutr. Sci.*, 1982, **6**, 331.
79. KOKUFUTA, E., SHIMIZU, H. and NAKAMURA, I., *Macromolecules*, 1982, **15**, 1618.
80. HORN, D. and HEUCK, C. C., *J. Biol. Chem.*, 1983, **258**, 1665; ICHIRO, H., SUEHIRO, T., NAGASAWA, J., YAMAUCHI, A. and SAGASAKA, M., *Sen'i Gakkaishi*, 1983, **39**, T532.
81. KOKUFUTA, E., SHIMIZU, H. and NAKAMURA, I., *Macromolecules*, 1981, **14**, 1178.
82. SAKAMOTO, M., KURAMOTO, N., KOMIYAMA, J. and IIJIMA, T., *Int. J. Biol. Macromol.*, 1982, **4**, 207; SUZAWA, T., SHIRAHAMA, H. and FUJIMOTO, T., *J. Colloid Interface Sci.*, 1983, **93**, 498; KURAMOTO, N.,

SAKAMOTO, M., KOMIYAMA, J. and IIJIMA, T., *Makromol. Chem.*, 1984, **185**, 1419.

83. ORLIEVSKAYA, O. V., KOSTSREVA, I. A., PONOMAREVA, R. B., PAPUKOVA, K. P. and SAMSONOV, G. V., *Vysokomol. Soedin.*, *Ser. B*, 1982, **24**, 63; KUZNETSOVA, N. P., GUDKIN, L. R. and SAMSONOV, G. V., *Vysokomol. Soedin.*, *Ser. A*, 1983, **25**, 2580; KUZNETSOVA, N. P., MISHAEVA, R. N., GUDKIN, L. R. and SAMSONOV, G. V., *Vysokomol. Soedin.*, *Ser. B*, 1983, **25**, 750; LONGO, W. E., IWATA, H., LINDHEIMER, T. and GOLDBERG, E. P., *Polym. Prep.*, 1983, **24**, 56.

84. IZUMURDOV, V. A., KASAIKIN, V. A., ERMAKOVA, L. N., MUSTAFAEV, M. I., ZEZIN, A. B. and KABANOV, V. A., *Vysokomol. Soedin.*, *Ser. A*, 1981, **23**, 1365; KABANOV, V. A., MUSTAFAEV, M. I. and GONCHAROV, V. V., *Vysokomol. Soedin.*, *Ser. A*, 1981, **23**, 2611; KABANOV, V. A. and MUSTAFAEV, M. I., *Vysokomol. Soedin.*, *Ser. A*, 1981, **23**, 255; VINOGRADOV, I. V., KABANOV, V. A., MUSTAFAEV, M. I., NORIMOV, A. Sh., PETROV, R. V. and KHAITOV, R. M., *Dokl. Akad. Nauk SSSR*, 1982, **263**, 228; IZUMURDOV, V. A., ZEZIN, A. B. and KABANOV, V. A., *Dokl. Akad. Nauk SSSR*, 1984, **275**, 1120.

85. BAGCHI, P. and BIRUBAUM, S. W., *J. Colloid Interface Sci.*, 1981, **83**, 460; KHAITOV, R. M., NORIMOV, A. Sh., VINOGRADOV, I. V., KABANOV, V. A. and MUSTAFAEV., M. I., *Immunologiya (Moscow)*, 1982, (6), 52.

86. SIMPSON, H. S., BALLANTYNE, F. C., PACKARD, C. J., MORGAN, H. G. and SPHEPHERD, J., *Clin. Chem.*, 1982, **28**, 2040.

87. POTEMPA, L. A., SIEGEL, J. N. and GEWURZ, H., *J. Immunol.*, 1981, **127**, 1509; POTEMPA, L. A. and GEWURZ, H., *Mol. Immunol.*, 1983, **20**, 501.

88. KAZATCHKINE, M. D., MAILLET, F., FISCHER, E. and GLOTZ, D., *Agents Actions*, 1981, **11**, 645; HOBBS, R. N., LEA, D. J., PHUA, K. K. and JOHNSON, P. M., *Ann. Rheum. Dis.*, 1983, **42**, 435; MAILLET, F. and KAZATCHKINE, M. D., *Immunology*, 1983, **50**, 27.

89. PAUTOV, V. D., KUZNETSOVA, N. P., MISHAEVA, R. N. and ANUFRIEVA, E. V., *Vysokomol. Soedin.*, *Ser. A*, 1983, **25**, 1599; KASTAREVA, I. A., ORLIEVSKAYA, O. V., YURCHENKO, V. S., PONOMAREVA, R. B. and SAMSONOV, G. V., *Vysokomol. Soedin.*, *Ser. B*, 1983, **25**, 415.

90. HEGARDT, F. G., GIL, G. and CALVET, V. E., *J. Lipid Res.*, 1983, **24**, 821; CAYGILL, J. C., MOORE, D. J. and KANAGASABAPATHY., L., *Enzyme Microb. Technol.*, 1983, **5**, 365.

91. LIEBEL, M. A. and WHITE, A. A., *Biochem. Biophys. Res. Commun.*, 1982, **104**, 957; ROUSTAN, C., FATTOUM, A., JEANNEAU, R., PRADEL, L. A., SCHUHMANN, D. and VANEL, P., *Biophys. Chem.*, 1982, **15**, 169.

92. ROSZKOWSKA, W. and WOROWSKI, K., *Rocz. Akad. Med. im. Julizna Marchlewskiego Bialymstoku*, 1982, **27**, 187.

93. RAMESH, V. and SINGH, C., *J. Mol. Catal.*, 1981, **10**, 341; ALLEN, R. A., *Thromb. Haemostatics*, 1982, **47**, 41; RAZUMAS, V., SAMALIUS, A. and KULYS, J., *J. Electroanal. Chem. Interfacial Electrochem.*, 1984, **164**, 195; KLEIN, J. and LUCKHAM, P. F., *Colloids Surf.*, 1984, **10**, 65.

94. KOKUFUTA, E., TAKAHASHI, K. and NAKAMURA, I., *Polym. Prep.*, *Jpn.*, 1983, **32**, 269.

95. MULIMANI, V. H. and DAY, R. A., *Curr. Sci.*, 1981, **50**, 629.

96. TRAORE, F. ISKHAKOVA, K. M. and AKHMEDOV, K. S., *Dokl. Akad. Nauk UzSSR* 1983, (6), 38: TRAORE, F. ISKHAKOVA, K. M. and AKHMEDOV, K. S., *Dokl. Akad. Nauk UzSSR*, 1983, (6), 40.

97. SHPILEVSKAYA, I. N., APAZID, A. I. and PETROVA, T. N., *Sb. Nauchn. Tr.-Tashk. Gos. Univ. im. V. I. Lenina*, 1982, **679**, 36; SHPILEVSKAYA, I. N., AZAPID, A. I. and PETROVA, T. N., *Sb. Nauchn. Tr.-Tashk. Goc. Univ. im. V.I. Lenina*, 1982, **679**, 40.

98. ELGINDY, N. A. and ELEGAKEY, M. A., *Drug Dev. Ind. Pharm.*, 1981, **7**, 587; ELGINDY, N. A. and ELEGAKEY, M. A., *Drug Dev. Ind. Pharm.*, 1981, **7**, 739; ELEGAKEY, M. A. and ELGINDY, N. A., *Drug Dev. Ind. Pharm.*, 1983, **9**, 895.

99. MULIMANI, V. H. and DAY, R. A., *Indian J. Biochem. Biophys.*, 1981, **18**, 157; MULIMANI, V. H., ROTH, A. C. and DAY, R. A., *J. Biosci.*, 1982, **4**, 127; MULIMANI, V. H. and DAY, R. A., *Indian J. Biochem. biophys.*, 1982, **19**, 292; MULIMANI, V. H., MADAIAH, M. and DAY, R. A., *Curr. Sci.*, 1983, **52**, 407; MULIMANI, V. H. and DAY, R. A., *Indian J. Biochem. Biophys.*, 1983, **20**, 263.

100. IGARASHI, K., SAKAMOTO, I., GOTO, N., KASHIWAGI, K., HONMA, R. and HIROSE, S., *Arch. Biochem. Biophys.*, 1982, **219**, 438; YEN, W. S., RHEE, K. W. and WARE, B. R., *J. Phys. Chem.*, 1983, **87**, 2148.

101. AVDYUKOVA, N. V., NIKIFOROVA, N. V. and RADINA, L. B., *Mol. Biol.* (*Moscow*), 1982, **16**, 619; NOVOSELER, M. A., *Biofizika*, 1983, **28**, 570.

102. MARX, R. and DOENECKE, D., *Z. Naturforsch., C: Biosci.*, 1981, **36C**, 149; LAIGLE, A. and LACOMBE, C., *Stud. Biophys.*, 1982, **89**, 11; NICOTRA, J., DeBARI, V. A. and NEEDLE, M. A., *Immunol. Lett.*, 1982, **4**, 249.

103. KLEVAN, L. and SCHUMAKER, V. N., *Nucleic Acids Res.*, 1982, **10**, 6809.

104. HARTUNG, K., SCHLICK, E., STEVENSON, H. C. and CHIRIGOS, M. A., *J. Immunopharmacol.*, 1983, **5**, 129; KOL'TSOV, V. D., KRASNOVA, L. V., BALANDIN, I. G. and MESHKOVA, E. N., *Byull. Eksp. Biol. Med.*, 1983, **95**, 74; JOYCE, G. F., VISSER, G. M., VAN BOECKEL, C. A. A., VAN BOOM, J. H., ORGEL, L. E. and VAN WESTRENE, J., *Nature* (*London*), 1984, **310**, 602.

105. GLADILIN, K. L., ORLOVSKII, A. F., KIRPOTIN, D. B. and VORONTSOVA, V. YA., *Vysokomol. Soedin., Ser. A*, 1983, **25**, 168; DAVIES, M. E. and FIELD, A. K., *J. Interferon Res.*, 1983, **3**, 89; TAKAGIWA, T., *Osaka Kogyo Daigaku Kiyo, Rikohen*, 1984, **28**, 129; TAKEHARA, M., KAKUDA, T. and NAKATA, M., *ICMR Ann.*, 1983, **3**, 81.

106. PETROV, R. V., KHAITOV, R. M. and ATAULLAKHANOV, R. I., *Immunologiya* (*Moscow*), 1982, (5), 28; FORMCHENKO, V. M., AZHERMACHEV, A. K., CHUGUNOV, V. A. and BABAEV, P. V., *Kolloidn. Zh.*, 1983, **45**, 273; OTTENBRITE, R. M., KUUS, K. and KAPLAN, A. M., *Polym. Prep.*, 1983, **24**, 25; KHAITOV, R. M., PETROV,

R. V., ABDULLAEV, D. M. and ATAULLAKHANOV, F. I., *Byull. Eksp. Biol. Med.*, 1984, **97**, 588; PETROV, R. V., KHAITOV, R. M. and ATAULLAKHANOV, R. I., *Biol. Membr.*, 1984, **1**, 599.

107. VAARA, M. and VILJANEN, P., *FEMS Microbiol. Lett.*, 1983, **19**, 253; VAARA, M. and VAARA, T., *Antimicrob. Agents Chemother.*, 1983, **24**, 107; VAARA, M. and VAARA, T., *Antimicrob. Agents Chemother.*, 1983, **24**, 114.

108. MIEHAUX, M., PAQUOT, M., BAIJOT, B. and THONART, P., *Biotechnol. Bioeng. Symp.*, 1982, **12**, 475; RICHTER, M. L. and HOMANN, P. H., *Arch. Biochem. Biophys.*, 1983, **222**, 67; YOUNG, D. H. and KAUSS, H., *Plant Physiol.*, 1983, **73**, 698; BARNES, J. L. and VENKATACHALAM, M. A., *Pathog. Cationic Proteins, Symp. 1982*, 1983, 281; BARNES, J. L., RADNIK, R. A., GILCHRIST, E. P. and VENKATACHALAM, M. A. *Kidney Int.*, 1984, **25**, 11; KUBOI, T. and FUJII, K., *Soil Sci. Plant Nutr. (Tokyo)*, 1984, **30**, 311; BRYSIN, V. G., ZAUROV, D. D., GUDKOVA, L. A. and TOPCHIEV, D. A., *Lab. Delo*, 1984, (2), 75.

109. SARAGA, L. T. M., *Polym. Prep.*, 1982, **23**, 54; KAWABATA, N., HAYASHI, T. and MATSUMOTO, T., *Appl. Environ. Microbiol.*, 1983, **46**, 203; ALIEV, K. V., RINGSDORF, H., SCHLARB, B. and LEISTER, K. H., *Makromol. Chem. Rapid Commun.*, 1984, **5**, 345; IKEDA, T., TAZUKE, S. and SUZUKI, Y., *Makromol. Chem.*, 1984, **185**, 869.

110. HASKINS, K., DONOSO, J. A. and HIMES, R. H., *J. Cell Sci.*, 1981, **47**, 237; MIYATAKA, K. and FLAVIN, M., *Int. J. Biochem.*, 1983, **15**, 1305.

111. KAJITA, S. and MATSUI, C., *Uirusu*, 1981, **31**, 33; NAKAJIMA, T., TERAOKA, T., SHIGEMATSU, T. and KASUGAI, H., *Nippon Noyaku Gakkaishi*, 1983, **8**, 499.

112. SENO, S., TSUJI, T. and ONO, T., *Biorheology*, 1983, **20**, 653.

113. KIKUCHI, Y., KUBOTA, N. and GOTO, N., *Polym. Prep., Jpn.*, 1983, **32**, 365; KIKUCHI, Y. and SHIMIZU, K., *Bull. Chem. Soc., Jpn.*, 1981, **54**, 2549.

114. SRINIVASAN, R. and KAMALAM, R., *Biopolymers*, 1982, **21**, 251; SRINIVASAN, R. and KAMALAM, R., *Biopolymers*, 1982, **21**, 265; SRINIVASAN, R. and VISWANATHAN, P., *Proc. Indian Acad. Sci., Chem. Sci.*, 1983, **92**, 551.

115. SCHMITT, E., HOLTZ, M., ESTHER, G., COURTNEY, J. M. and KLINKMANN, H., *Z. Urol. Nephrol.*, 1983, **76**, 99.

116. PLATE, N. A. and VALUEV, L. I., *Thromb. Res.*, 1982, **27**, 131; PLATE, N. A., VALUEV, L. I., GUMIROVA, F. Kh., ZIMIN, N. K. and ROSENFEL'D, M. A., *Vopr. Med. Khim.*, 1982, **28**, 97.

117. CRESCENZI, V., RIZZO, R. and MANZINI, G., *Period. Biol.*, 1981, **83**, 31; BOGRACHEVA, T. Y., GRINBERG, V. Y. and TOLSTOGUZOV, V. B., *Carbohydr. Polym.*, 1982, **2**, 163; MITTERER, A., EIGNER, W. D., SCHULZ, J., JUERGENS, G. and HOLASEK, A., *Int. J. Biol. Macromol.*, 1982, **4**, 227; BLOMHOFF, H. K. and CHRISTENSEN, T. B., *Biochim. Biophys. Acta*, 1983, **743**, 401; TEISNER, B., DAVEY, M. W. and GRUDZINSKAS, J. G., *Clin. Chim. Acta*, 1983, **127**, 413.

118. CHERRY, J. P., *Food Carbohydr., Pap. Symp.*, 1982, **1981**, 375; ZARIN,

N. A., *Vopr. Med. Khim.*, 1983, **29**, 41; HOLBROOK, J. J., LUSCOMBE, M., MARSHALL, S. E. and PEPPER, D. S., *Pathog. Cationic Proteins, Symp. 1982*, 1983, 173; LENDERS, J. P. and CRICHTON, R. R., *Biotechnol. Bioeng.*, 1984, **26**, 1343.

119. LEVY, D. E., HORNER, A. A. and SOLOMOU, A., *J. Exp. Med.*, 1981, **153**, 883; CARDIN, A. D., WITT, K. R. and JACKSON, R. L., *Anal. Biochem.*, 1984, **137**, 368; BEUGELING, T., VAN DER DOES, L., SEDEREL, L. C., VAN DUIJL, J. F. and BANTJES, A., *Adv. Biomater.*, 1984, **5**, 331; HATTON, M. W. C., ROLLASON, G. and SEFTON, M. V., *Thromb. Haemostasis*, 1983, **50**, 873; HENNINK, W. E., KIM, S. W. and FEIJEN, J., *J. Biomed. Mater. Res.*, 1984, **18**, 911; HENNINK, W. E., DOST, L., FEIJEN, J. and KIM, S. W., *Trans. Am. Soc. Artif. Intern. Organs*, 1983, **29**, 200.

120. SKRYZYDLEWSKI, Z., WOROWSKI, K., WISNIEWSKI, L. and ROSZKOWSKA, W., *Patol. Pol.*, 1981, **32**, 121.

121. SATO, H., NAKANISHI, E. and NAKAJIMA, A., *Int. J. Biol. Macromol.*, 1981, **3**, 66; KUDRYASHOV, B. A., UL'YNOV, A. M. and ZHITNIKOVA, E. S., *Vestn. Mosk. Univ., Ser. 16: Biol.*, 1981, (3), 26.

122. KOVACS, J., FREY, A. and SEIFART, K. H., *Biochem. Int.*, 1981, **3**, 645; LARIONOVA, N. I., UNKSOVA, L. E., MIRONOV, V. A., SAKHAROV, I. YU., KAZANSKAYA, N. F. and BEREZIN, I. V., *Vysokomol. Soedin., Ser. A*, 1981, **23**, 1823; SAKHAROV, I. YU., LARIONOVA, N. I., KAZANSKAYA, N. F. and BEREZIN, I. V., *Enzyme Microb. Technol.*, 1984, **6**, 27.

123. MCMULLEN, J. N., NEWTON, D. W. and BECKER, C. H., *J. Pharm. Sci.*, 1982, **71**, 628; WATASE, M. and NISHINARI, K., *Biorheology*, 1983, **20**, 495; COMP, P. C., *Nouv. Rev. Fr. Hematol.*, 1984, **26**, 239; COMPER, W. D., MACDONALD, P. M. and PRESTON, B. N., *J. Phys. Chem.*, 1984, **88**, 6031; LABBE, J. P., BERTRAND, R., AUDEMARD, E., KASSAB, R., WALZTHOENY, D. and WALLIMANN, T., *Eur. J. Biochem.*, 1984, **143**, 315.

124. TSYGANKOV, A. YU. and GLADILIN, K. L., *Dokl. Akad. Nauk SSSR*, 1983, **268**, 1003; YOUNG, B. R., LAMBRECHT, L. K., ALBRECHT, R. M., MOSHER, D. F. and COOPER, S. L., *Trans. Am. Soc. Artif. Intern. Organs*, 1983, **29**, 442; DOI, H., IDENO, S., KUO, F. H., IBUKI, F. and KANAMORI, M., *J. Nutr. Sci. Vitaminol.*, 1983, **29**, 679.

125. COOKE, R., DURAND, R., TEISSERE, M., PENON, P. and RICARD, J., *Biochem. Biophys. Res. Commun.*, 1981, **98**, 36; GEORGHIOU, S., *Mod. Fluoresc. Spectosc.*, 1981, **3**, 193; WATANABE, F. and SCHWARZ, G., *J. Mol. Biol.*, 1983, **163**, 485; KUMAR, N. V. and GOVIL, G., *Biopolymers*, 1984, **23**, 1979; LIU, N., ZHANG, Y. and YANG, K., *Weishengwu Xuebao*, 1983, **24**, 50.

126. OOSAWA, F., *Biopolymers*, 1968, **6**, 1633.

127. OKIHANA, H. and NAKAJIMA, A., *Bull. Inst. Chem. Res., Kyoto Univ.*, 1976, **54**, 63.

128. POLDERMAN, A., *Biopolymers*, 1975, **14**, 218.

129. ABE, K., *Intermacromolecular Complexes*, 1983, Gakkai Shuppan Center, Tokyo, p. 111.

130. ABE, K., OHNO, H. and TSUCHIDA, E., *Makromol. Chem.*, 1979, **178**, 2285.
131. KABANOV, V. A. and ZEZIN, A. B., *Sov. Sci. Rev.*, *Sect. B*, 1982, **4**, 207; KABANOV, V. A. and ZEZIN, A. B., *Macromol. Chem. Phys.*, *Suppl.* 1984, **6**, 259; KABANOV, V. A. and ZEZIN, A. B., *Pure Appl. Chem.*, 1984, **56**, 343; SKORODINSKAYA, A. M., KEMENOVA, V. A., EFIMOV, V. S., MUSTAFAEV, M. I., KASAIKIN, V. A., ZEZIN, A. B. and KABANOV, V. A., *Khim.-Farm. Zh.*, 1984, **18**, 283; ROGACHEVA, V. B., RYZHIKOV, S. V., ZEZIN, A. B. and KABANOV, V. A., *Vysokomol. Soedin, Ser. A*, 1984, **26**, 1674; ZEZIN, A. B., KASAIKIN, V. A., KABANOV, N. M., KHARENKO, O. A. and KABANOV, V. A., *Vysokomol. Soedin.*, *Ser. A*, 1984, **26**, 1519; IZUMURDOV, V. A., SAVITSKII, A. P., ZEZIN, A. B. and KABANOV, V. A., *Vysokomol. Soedin.*, *Ser. A*, 1984, **26**, 1724; DIKOV, M. M., OSIPOV, A. P., EGOROV, A. M., BEREZIN, I. V., MUSTAFAEV, M. I. and KABANOV, V. A., *Biokhimiya (Moscow)*, 1984, **49**, 1300.
132. OSADA, Y., ABE, K. and TSUCHIDA, E., *Nippon Kagaku Kaishi*, 1973, 2219; OSADA, Y., ABE, K. and TSUCHIDA, E., *Nippon Kagaku Kaishi*, 1973, 2222.
133. TSUCHIDA, E., OSADA, Y. and ABE, K., *Makromol. Chem.*, 1974, **175**, 583.
134. MICHAELS, A. S. and MIEKKA, R. G., *J. Phys. Chem.*, 1961, **65**, 1765; MICHAELS, A. S., MIR, L. and SCHNEIDER, N. S., *J. Phys. Chem.*, 1965, **69**, 1447; MICHAELS, A. S., FALKENSTEIN, G. L. and SCHNEIDER, N. S., *J. Phys. Chem.*, 1965, **69**, 1456.
135. SAITO, M., ABE, K., OSADA, Y. and TSUCHIDA, E., *Nippon Kagaku Kaishi*, 1974, 977; ABE, K. and TSUCHIDA, E., *Makromol. Chem.*, 1975, **176**, 803.
136. SHINODA, K. and NAKAJIMA, A., *Bull. Inst. Chem. Res., Kyoto Univ.*, 1975, **53**, 392; SHINODA, K. and NAKAJIMA, A., *Bull. Inst. Chem. Res., Kyoto Univ.*, 1975, **53**, 400; NAKAJIMA, A. and SHINODA, K., *J. Colloid Interface Sci.*, 1976, **55**, 126.
137. ZEZIN, A. B., LUTSENKO, V. V., ROGACHEVA, V. B., ALEKSINA, O. A., KALYUZHNAYA, R. I., KABANOV, V. A. and KARGIN, V. A., *Vysokomol. Soedin.*, *Ser. A*, 1972, **14**, 772; ABE, K., KOIDE, M. and TSUCHIDA, E., *Polym. J.*, 1977, **9**, 73; ABE, K. and TSUCHIDA, E., *Polym. J.*, 1977, **9**, 79.
138. SHINODA, K., SAKAI, K., HAYASHI, T. and NAKAJIMA, A., *Polym. J.*, 1976, **8**, 208.
139. HAMMES, G. G. and SCHULLERY, S. E., *Biochemistry*, 1968, **7**, 3882; NAKAJIMA, A., SHINODA, K., HAYASHI, T. and SATO, H., *Polym. J.*, 1975, **7**, 550; SATO, H., HAYASHI, T. and NAKAJIMA, A., *Polym. J.*, 1976, **8**, 517; MITA, K., ICHIMURA, S. and ZAMA, M., *Biopolymers*, 1978, **17**, 2783.
140. OLINS, D. E., OLINS, A. L. and VAN HIPPEL, P. H., *J. Mol. Biol.*, 1968, **33**, 265; PINKSTON, M. F., RITTER, A. H. and LI, H. J., *Biochemistry*, 1976, **15**, 1676; MANDEL, R. and FASMAN, G. D., *Biochemistry*, 1976,

15, 3122; CONSTANTINO, P., VERDINI, S., DE SANTIS, P., RIZZO, R. and SAVINO, M., *Biopolymers,* 1979, **18,** 9; SEIPKE, G., ARFMANN, H. A. and WAGNER, K. G., *Int. J. Biol. Macromol.,* 1980, **2,** 268.

141. TAZAWA, T. and TS'O, P. O. P., *J. Mol. Biol.,* 1972, **66,** 115; SPRINGGATE, M. W. and POLAND, D., *Biopolymers,* 1973, **12,** 2241; SUURKUUSK, J., ALVARES, J., FREIE, E. and BILTONEN, R., *Biopolymers,* 1977, **16,** 264; SIANO, D. B., *Biopolymers,* 1978, **17,** 2897; BABA, Y., FUJIOKA, K. and KAGEMOTO, A., *Polym. J.,* 1978, **10,** 241; HERNANDEZ, L. A., VIETA, R. S. and JAO, T. C., *Biopolymers,* 1980, **19,** 1715.

142. GELMAN, R. A. and BLACKWELL, J., *Biopolymers,* 1974, **13,** 139; CUNDALL, R. B., LANTON, J. B. and MURRAY, D., *Makromol. Chem.,* 1979, **180,** 2913.

143. NIELSEN, C. E., *J. Appl. Phys.,* 1970, **41,** 4626.

144. TSUCHIDA, E., OSADA, Y. and SANADA, K., *J. Polym. Sci., Part A-1,* 1972, **10,** 3397.

145. VOGEL, M. K., CROSS, R. A., BIXLER, H. J. and GUZMAN, R. J., *J. Macromol. Sci.—Chem.,* 1970, **A4,** 675.

146. KALYUZHNAYA, R. I., VOLYNSKII, A. L., RUDMAN, A. R., VENGEROVA, N. A., RAZVODOVSKII, YE. F., EL'STON, B. S. and ZEZIN, A. B., *Vysokomol. Soedin., Ser. A,* 1976, **18,** 71.

147. HOSONO, M., SUGII, S., KITAMURA, R., HONG, Y. M. and TSUJI, W., *J. Appl. Polym. Sci.,* 1977, **21,** 2125; HOSONO, M., SUGII, S. and TSUJI, W., *Kobunshi Ronbunshu,* 1977, **34,** 843; SATO, H., MAEDA, M. and NAKAJIMA, A., *J. Appl. Polym. Sci.,* 1979, **23,** 1759.

148. CRAY, C. A., Thesis, Massachusetts Institute of Technology, 1965.

149. SMOLEN, V. F. and HAHMAN, D. E., *J. Colloid Interface Sci.,* 1973, **42,** 70.

150. MARKLEY, L. L., BIXLER, H. J. and CROSS, R. A., *J. Biomed. Mater. Res.,* 1968, **2,** 145.

151. HALBERT, S. P., *J. Biomed. Mater. Res.,* 1970, **4,** 549; REMBAUM, A., *J. Macromol. Sci., Chem.,* 1970, **A4,** 715; YEN, S. P. S. and REMBAUM, A., *J. Biomed. Mater. Res.,* 1971, **5,** 83; BRUCK, S. D., *J. Biomed. Mater. Res.,* 1971, **5,** 139; FUKUDA, H. and KIKUCHI, H., *J. Biomed. Mater. Res.,* 1978, **12,** 531; KIKUCHI, Y. and HORI, K., *Nippon Kagaku Kaishi,* 1980, 1157; JAQUES, L. B., *Pharmacol. Rev.,* 1980, **31,** 99.

152. REFOJO, M. F., *J. Appl. Polym. Sci.,* 1967, **11,** 1991; DUBY, J. and PRIJOT, E., *Arch. Opthal. (Patis),* 1969, **29,** 393; REFOJO, M. F., *J. Biomed. Mater. Res.,* 1969, **3,** 333.

153. KATAOKA, K., AKAIKE, T., SAKURAI, Y. and TSURUTA, T., *Makromol. Chem.,* 1978, **179,** 1121; KATAOKA, K., TSURUTA, T., AKAIKE, T. and SAKURAI, Y., *Makromol. Chem.,* 1980, **181,** 1363.

154. TSUCHIDA, E. and OSADA, Y., *Makromol. Chem.,* 1974, **175,** 593.

155. IKAWA, T., ABE, K., HONDA, K. and TSUCHIDA, E., *J. Polym. Sci., Polym. Chem. Ed.,* 1975, **13,** 1505.

156. ABE, K., Thesis, Waseda University, 1978.

157. TSUCHIDA, E. and MATSUDA, M., unpublished data.

158. BLUMSTEIN, A., KAKIVAYA, S. R. and SALAMONE, J. C., *J. Polym. Sci.,*
     Polym. Lett. Ed., 1974, **12,** 651; BLUMSTEIN, A., KAKIVAYA, S. R.,
     BLUMSTEIN, R. and SUZUKI, T., *Macromolecules,* 1975, **8,** 435.
159. KABANOV, V. A., KARGINA, O. V. and ULYANOVA, M. V., *Vysokomol.*
     Soedin., Ser. B, 1974, **16,** 759; KABANOV, V. A., KARGINA, O. V. and
     ULYANOVA, M. V., *J. Polym. Sci., Polym. Chem. Ed.,* 1976, **14,** 2351.
160. SALAMONE, J. C., POULIN, S., WATTERSON, A. C. and OLSON, A. P.,
     Polymer, 1979, **20,** 611.
161. ISE, N. and OKUBO, T., *Acc. Chem. Res.,* 1980, **13,** 303; ISE, N., *Kagaku*
     (Tokyo), 1983, **52,** 501; ISE, N., *Kagaku Sosetsu,* 1983, **40,** 151.
162. DEBYE, P. and HÜCKEL, E., *Physik. Z.,* 1923, **24,** 185.
163. FUOSS, R. M. and EDELSON, D., *J. Polym. Sci.,* 1951, **6,** 767; DOTY, P.
     and STEINER, R. F., *J. Chem. Phys.,* 1952, **20,** 85; VEIS, A. and
     EGGENBERGER, D. N., *J. Am. Chem. Soc.,* 1954, **76,** 1560.
164. LIN, S. C., LEE, W. I. and SHURR, J. M., *Biopolymers,* 1978, **17,** 1041;
     PATOWSKI, A., GULARI, E. and CHU, B., *J. Chem. Phys.,* 1980, **73,**
     4178; GIORDANO, R., MAISANO, G., MALLAMACE, F., MICALI, N. and
     WANDERLINGH, F., *J. Chem. Phys.,* 1981, **75,** 4770.
165. DE GENNES, P. G., PINCUS, P. and VELASCO, R. M., *J. Phys.,* 1976, **37,**
     1461; NIERLICH, M., WILLIAMS, C. E., BOUE, F., COTTON, J. P.,
     DAUD, M., FARROUX, B., JANNINK, G., PICOT, C., MOAN, M., WOLFF,
     C., RINAUDO, M. and DE GENNES, P. G., *J. Phys.,* 1979, **40,** 701.
166. PLESTIL, J., MIKES, J. and DUSEK, K., *Acta Polym.,* 1979, **30,** 29; ISE,
     N., OKUBO, T., HIRAGI, Y., KAWAI, H., HASHIMOTO, T., FUJIMURA,
     M., NAKAJIMA, A. and HAYASHI, H., *J. Am. Chem. Soc.,* 1979, **101,**
     5836.
167. ISE, N., OKUBO, T., YAMAMOTO, K., MATSUOKA, H., KAWAI, H.,
     HASHIMOTO, T. and FUJIMURA, M., *J. Chem. Phys.,* 1983, **78,** 541.
168. ISE, N., OKUBO, T., YAMAMOTO, K., KAWAI, H., HASHIMOTO, T.,
     FUJIMURA, M. and HIRAGI, Y., *J. Am. Chem. Soc.,* 1980, **102,** 7901.
169. TOEI, K. and KOHARA, T., *Anal. Chim. Acta,* 1976, **83,** 59.
170. PAPISOV, I. M., BARAMOVSKII, V. YU, and KABANOV, V. A.,
     Vysokomol. Soedin., Ser. A, 1975, **17,** 2104.
171. HAMMER, C. F., *Macromolecules,* 1971, **4,** 69; ZAKREZEWSKI, G. A.,
     Polymer, 1973, **14,** 348: SCHURER, J. W., DE BOER, A. and CHALLA,
     G., *Polymer,* 1975, **16,** 201; OHNO, N. and KUMANOTANI, J., *Polym. J.,*
     1979, **11,** 947; TING, S. P., BULKIN, B. J. and PEARCE, E. M., *J. Polym.*
     Sci., Polym. Chem. Ed., 1981, **19,** 1451.
172. OGINO, I., *Polymer Alloy,* 1981, Tokyo Kagaku Dojin, p. 111.
173. LYSAGHT, M. J., *Ionic Polymers,* 1975, Applied Science Publishers,
     London, p. 281.
174. FERRY, J. D., *Chem. Rev.* 1936, **18,** 373.
175. KUZNETSOVA, N. L., BOGINO, N. A., TSIVINSKAYA, L. K. and
     YAVORSKAYA, Y. S., *Vysokomol. Soedin., Ser. A,* 1974, **16,** 2435.
176. TOGAWA, K., NAKAMOTO, M. and KUWATA, K., Jpn. Kokai Tokkyo
     Koho, 1977, 52 52877.
177. KIYATA, T., ASAMI, S. and SHIMIZU, A., Jpn. Kokai Tokkyo Koho,
     1977, 52 78291.

178. MEYER, T. J., *Acc. Chem. Res.*, 1978, **11**, 94; WHITTEN, D. G., *Acc. Chem. Res.*, 1980, **13**, 82; FENDLER, J. H., *Acc. Chem. Res.*, 1980, **15**, 7; BRUGGER, P. A. and GRÄTZEL, M. J., *J. Am. Chem. Soc.*, 1980, **102**, 2461; MATSUO, T., SAKAMOTO, T., TAKUMA, K., SAKURA, K. and OHSAKA, T., *J. Phys. Chem.*, 1981, **85**, 1277; HENGLEIN, A., *J. Phys. Chem.*, 1982, **86**, 2291.

179. BOLTS, J. M., BOCARSLY, A. B., PALAZZOTO, M. C., WALTON, E. G., LEWIS, N. S. and WRIGHTON, M. S., *J. Am. Chem. Soc.*, 1979, **101**, 1378; OYAMA, N. and ANSON, F. C., *Anal. Chem.*, 1980, **52**, 1192; DEGRAND, C. and MILLER, L. L., *J. Am. Chem. Soc.*, 1980, **102**, 5728; DOBLHOFER, K. and DÜRR, W., *J. Electroanal. Chem.*, 1980, **127**, 1041; ABRUNA, H. D., DENISEVICH, P., UMANA, M., MEYER, T. J. and MURRAY, R. M., *J. Electrochem. Soc.*, 1981, **103**, 1.

180. SHIGEHARA, K. and TSUCHIDA, E., *Kagaku (Kyoto)*, 1984, **39**, 352.

181. ANDRIEUX, C. P., BOUCHIAT, J. M. and SAVEANT, J. M., *J. Electroanal. Chem.*, 1982, **131**, 1; ANDRIEUX, C. P., BOUCHIAT, J. M. and SAVEANT, J. M., *J. Electroanal. Chem.*, 1982, **134**, 163; SATO, K., YAMAGUCHI, S., MATSUO, M., OHSAKA, T. and OYAMA, N., *Bull. Chem. Soc., Jpn.*, 1983, **56**, 2004: ANSON, F. C., SAVEANT, J. M. and SHIGEHARA, K., *J. Am. Chem. Soc.*, 1983, **105**, 1096.

182. KUROKAWA, Y., SHIRAKAWA, N., TERADA, M. and YUI, N., *J. Appl. Polym. Sci.*, 1980, **25**, 1645.

183. YANO, O. and WADA, Y., *J. Appl. Polym. Sci.*, 1980, **25**, 1723.

184. HATCH, M. J., DILLON, J. A. and SMITH, H. B., *Ind. Eng. Chem.*, 1957, **49**, 1812; LESSLAUER, W. and LÄUGER, P., *J. Phys. Chem.*, 1967, **71**, 2544; MICHAELI, I. and BEJERANO, T., *J. Polym. Sci., Part C*, 1969, **22**, 909.

185. PETERSON, E. A. and SOBER, H. A., *J. Am. Chem. Soc.*, 1956, **78**, 751; SOBER, H. A., GUTTEN, F. J., WYCKOFF, M. M. and PETERSON, E. A., *J. Am. Chem. Soc.*, 1956, **78**, 756.

186. HEYSTEK, J., BRUMMELHUIS, H. G. J. and KRIJNEN, H. W., *Vox Sang.*, 1973, **25**, 113; CURLING, J. M., *Methods of Plasma Protein Fractionation*, 1980, Academic Press, London, p. 77; SUOMELA, H., *Methods of Plasma Protein Fractionation*, 1980, Academic Press, London, p. 107.

187. JOHNSON, A. J., MACDONALD, V. E., SEMEAR, M., FIELDS, J. E., SCHUK, J., LEWIS, C. and BRIND, J., *J. Lab. Clin. Med.*, 1978, **92**, 194; HARRIS, R. B., JOHNSON, A. J., SEMEAR, M., DELENTE, J. and FIELDS, J. E., *Vox Sang.*, 1979, **35**, 129.

188. TOMONO, T., YOSHIDA, S. and TOKUNAGA, E., *J. Polym. Sci., Polym. Lett. Ed.*, 1979, **17**, 335; OHNO, H., NII, A., ABE, K. and TSUCHIDA, E., *Makromol. Chem.*, 1981, **182**, 1407.

189. DOENECKE, D., *Biochem. Internal.*, 1980, **1**, 237.

190. LYMAN, D. J., *Trans. Am. Soc. Artif. Internal. Organs*, 1975, **21**, 49; FURUSAWA, K., IMAIKE, T., OHTA, Y., SUDA, Y. and TSUDA, K., *Kobunshi Ronbunshu*, 1979, **36**, 273; OKANO, T., NISHIYAMA, S., SHINOHARA, I., AKAIKE, T., SAKURAI, Y., KATAOKA, K. and TSURUTA, T., *J. Biomed. Mater. Res.*, 1981, **15**, 393; SHIMADA, M.,

UNOKI, M., INABA, N., TAHARA, H., SHINOHARA, I., OKANO, T., KATAOKA, K. and SAKURAI, Y., *Kobunshi Ronbunshu*, 1982, **39**, 257.
191. NOISSHIKI, Y., *Kagaku no Ryoiki, Zokan*, 1982, **135**, 64.
192. AKAIKE, T., TSURUTA, T., KOSUGE, K., MIYATA, S., KATAOKA, K. and SAKURAI, Y., *Bull. Heart Inst., Jpn.*, 1980, **21**, 4.
193. KATAOKA, K., MAEDA, M., NISHIMURA, T., NITADORI, Y., TSURUTA, T., AKAIKE, T. and SAKURAI, Y., *J. Biomed. Mater. Res.*, 1980, **14**, 817.
194. REMBAUM, A., *Appl. Polym. Symp.*, 1973, **22**, 299; PAL, M. K. and GHOSH, A. K., *Makromol. Chem.*, 1973, **169**, 273.
195. REMBAUM, A., YEN, S. P. S., LANDEL, R. F. and SHEN, M., *J. Macromol. Sci.—Chem.*, 1970, **A4**, 715; YEN, S. P. S. and REMBAUM, A., *J. Biomed. Mater. Res., Symp.*, 1971, **1**, 83; REMBAUM, A., YEN, S. P. S., CHEONG, E., WALLACE, S., MOLDY, R. S. and DREYER, W. J., *Macromolecules*, 1976, **9**, 328.
196. BUNGENBERG DE JONG, H. G. and KRUYT, H. R., *Kolloid Z.*, 1930, **50**, 39; EVREINOVA, T. N., *Concentration of Matter and Action of Enzymes on Coacervate*, 1966, Nauka, Moscow.
197. OPARINE, A., *Dokl. Akad. Nauk SSSR*, 1958, **122**, 661; CEPERPOVKAYA, K. B., *Dokl. Akad. Nauk SSSR*, 1960, **135**, 1532; EVREINOVA, T. N., *Biofizika*, 1964, **29**, 1035.
198. OHNO, H. and TSUCHIDA, E., *Makromol. Chem., Rapid Commun.*, 1980, **1**, 585.
199. DIKOV, M. M., OPISOV, A. P., EGOROV, A. E., KARULIN, A. Y., MUSTAFAYEV, M. I. and KABANOV, V. A., *J. Solid-Phase Biochem.*, 1980, **5**, 1.
200. PETERSON, L. C. and COX, R. P., *Biochem. J.*, 1980, **192**, 687.
201. TAKAGISHI, T., KOZUKA, H. and KUROKI, N., *J. Polym. Sci., Polym. Chem. Ed.*, 1981, **19**, 3237.
202. PATEL, H. M., PARVEZ, N., FIELD, J. and RYMAN, B. E., *Biosci. Rep.*, 1983, **3**, 39.
203. SAKAMOTO, H., ABE, K., ITOH, Y. and SENOH, S., *Polym. Prep., Jpn.*, 1984, **33**, 366; ABE, K., OKAMOTO, S. and SENOH, S., *Polym. Prep., Jpn.*, 1984, **33**, 367.
204. HONDA, K. and TSUCHIDA, E., *Kagaku no Ryoiki, Zokan*, 1982, **134**, 123.
205. VARMA, A. J., MAJEWICZ, T. and SMID, J., *J. Polym. Sci., Polym. Chem. Ed.*, 1979, **17**, 1573.
206. ZEZIN, A. B., KABANOV, N. M., KOKORIN, A. I. and ROGACHEVA, V. B., *Vysokomol. Soedin., Ser. A*, 1977, **19**, 118; KABANOV, N. M., KOKORIN, A. I., ROGACHEVA, V. B. and ZEZIN, A. B., *Vysokomol. Soedin., Ser. A*, 1979, **21**, 209; BERE, A. and HELENCE, C., *Biopolymers*, 1979, **18**, 2659.

Chapter 6

IONIC POLYMER MEMBRANES

PAULINE J. BROOKMAN and JOHN W. NICHOLSON

Laboratory of the Government Chemist, Department of Trade and Industry, London, UK

1 INTRODUCTION

The use of membranes to fractionate solutions and mixtures of gases is well established. Dialysis was first used in 1854[1] but was not commercially exploited at that time. In 1870, the first synthetic membrane for ultrafiltration using copper ferrocyanide was prepared. Much later, however, with the development of polymer chemistry, a wide variety of synthetic materials for membranes became available. Many of these membranes were made from nitrocellulose, by a process first developed by William J. Elford.

Dialysis is defined as the selective transport of solutes through a membrane to give a difference in concentration between the two liquids either side of the membrane. No external pressure is applied. The technique is generally used for the purification of solutions. The major large-scale application of dialysis is for patients suffering from kidney failure who therefore require the separation of toxic substances such as urea from their blood stream by artificial means. The development of a suitable haemodialysis membrane material is a difficult task, since the membrane is required to be highly selective so that other vitally important components of the blood are not coextracted with the toxic urea. Ionic polymer membranes, based on a polyacrylate-type material, have been used for this application.

Dialysis represents only a part of total membrane usage. As it is controlled by differences in chemical potential alone, without requiring the use of external forces such as pressure or elevated temperature, it is a cheap process. However, in order to have a large transfer of material, the surface area of the membrane solution interface must be large, and such membranes are expensive to produce.

Recently, the manufacture of 'hollow fibre membranes' has become commonplace; these fibres are used either singly or in bundles to form the membrane, each fibre usually having a diameter of less than 1 mm. The fibres not only have excellent mass-transfer properties but are also self-supporting, obviating the need for supports for fragile thin conventional membranes. Hollow fibre membranes have been used in blood fractionation and desalination. There are two classes: (1) open, functioning similarly to conventional membranes, allowing fluid to be exchanged over the large surface of the fibre; (2) loaded, allowing the permeant to be modified chemically as it diffuses through the fibre; this class of membrane is used for controlled release of biomaterials.[3] The disadvantage of these fibre membranes is their susceptibility to fouling and blocking by particulate matter.[4]

For many separations, therefore, ultrafiltration is a suitable alternative. In this technique, separation is a form of 'molecular filtration', i.e. solutes are fractionated according to molecular diameter, usually in the region 2–20 nm.

Ionic polymers can be used to form membranes which find diverse applications in a number of fields. As such they represent a branch of membrane technology.

Reverse osmosis is the application of pressure in excess of the osmotic pressure to reverse the direction of solvent flow. The technique is principally used for purification of solvents, the major industrial application being desalination.[5] Reverse osmosis is favoured over other forms of purification on energy grounds: the main advantage is that heating is not required to effect a phase change in the resulting product.

Unlike ultrafiltration, reverse osmosis is not size separation by sieve filtration; there is no correlation between the membrane porosity and the size of the permeant solvent molecules. The mechanism involves the absorption of solvent into the membrane. In aqueous conditions the water held in the membrane matrix is said to be in an ice-like configuration.[6] When pressure is applied, water molecules from the high-pressure side are forced into the membrane structure while, at the opposite face of the membrane, water is released. Other materials are not transported through the membrane, as the process involves specific clustering of water molecules; the only exceptions are those compounds capable of forming hydrogen bonds to water, such as urea and methanol.

2 STRUCTURAL AND TRANSPORT PROPERTIES OF IONOMER MEMBRANES

2.1 Structural Characteristics

Although ion-containing polymers had been known for some time, it was the classic work of Brown[7] on carboxylate elastomers, reported in 1957, that established the concept of 'ionic crosslinks'. This work was followed shortly afterwards in the early 1960s by DuPont's development of Surlyns,[7a] which are copolymers of ethylene and methacrylic acid neutralized with zinc or sodium salts. The novelty of these materials lies in the presence of a minority proportion of ionizable groups. The DuPont company coined the name ionomer to describe this special class of ionic polymers that contain a hydrocarbon or fluorocarbon chain and a small proportion (generally <10 mol %) of neutralized acid groups.[8] These materials have been extensively reviewed by Longworth[7a] in Part 1 of this *Developments Series*.

The presence of these ionic groups gives the polymer greater mechanical strength, i.e. improved tensile strength and impact resistance, together with increased chemical resistance. Other types of ionomers commonly encountered are copolymers of (a) styrene and acrylic acid, (b) ethyl acrylate and methacrylic acid and (c) styrene and vinylpyridine. DuPont workers also developed the perfluorosulphonate ionomer named Nafion, whose structure is given below where M represents either a metal cation in the neutralized form or hydrogen in the acid form. Equivalent weights of the various Nafions lie in the region 1000–2000.[9]

$$(CF_2CF_2)_x(CF_2-CF)_y$$
$$|$$
$$O-CF_2CF(CF_3)OCF_2CF_2SO_3M$$

These ionomers are prepared by copolymerizing the hydrocarbon–fluorocarbon with the appropriate acid or ester, followed by neutralization or hydrolysis with a base. These reactions may be carried out in solution or in melt.

Another new type of perfluorinated membrane named Flemion has

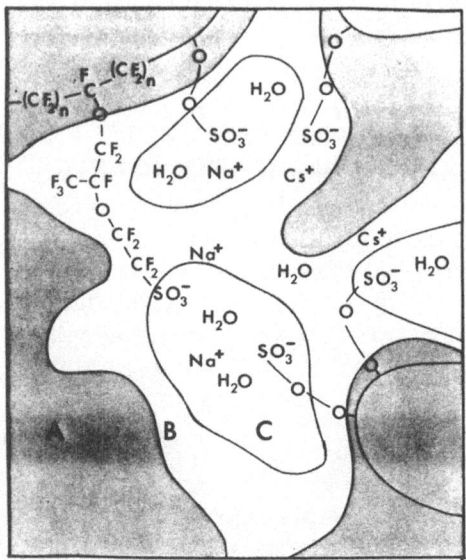

FIG. 1. Three-region structure model for Nafion. A, Fluorocarbon; B, interfacial zone; C, ionic clusters. Reprinted from ref. 39 with permission from the American Chemical Society.

been developed by the Asahi Glass Co. of Japan. This polymer contains carboxylic acid groups in place of the sulphonate groups.[10]

The unique and fundamental feature of all types of ionomer is their morphology. In all cases they are found to contain phase-separated clusters of ionic groups. Numerous workers have investigated the structure of ionomers and developed models to represent their findings. A variety of such models have been reviewed by Mauritz and Hopfinger.[11] Of the models discussed some, for example those of Eisenberg,[12] favour the situation where the cluster is a result of an agglomeration of multiplets. Ion multiplets are of a specific size and therefore contain a maximum number of ion-pairs; for example, Eisenberg and King[13] stipulate a maximum of eight pairs for a spherical multiplet. The formation of larger multiplets was discovered by Pettit and Bruckenstein,[14] although they are only present in solvents of low dielectric constant and always at low concentration.

The maximum radius (r_m) of the ion multiplet, assuming a spherical arrangement, is given by:

$$r_m = 3v_p/S_{ch}$$

where v_p is the volume of an ion-pair and S_{ch} is the area of the hydrocarbon chain in contact with the surface of the multiplet sphere.[15] If the multiplets are not spherical this restriction of size does not apply.

Cluster formation is not unique to these materials, as carboxylated rubbers were also found to undergo ionic aggregation.[16] The driving force in this case is the highly unfavourable thermodynamic state of the ions in a non-polar medium. Carboxylated rubbers cured with metal oxides are of high strength, as the vulcanization gives an 'internally relaxed network' where the clusters act as a reinforcing filler. Toughness due to clustering is also present in unneutralized carboxyl block copolymers by cluster formation; separation is encouraged by block formation.

The model for cluster formation from these multiplet units as proposed by Eisenberg[12] is that the multiplet containing the maximum number of ion-pairs is prevented from becoming any larger by the hydrocarbon 'skin'. It thus attracts other multiplets of varying sizes by electrostatic forces to a clustered arrangement, whilst maintaining individual multiplets, separated by a distance greater than the thickness of the hydrocarbon chain. The size of each cluster is then a balance between the electrostatic forces attracting multiplets and elastic forces from repulsion of the polymer backbone. Therefore, at a given temperature the opposing forces will be in equilibrium.

One of the strongest and most frequently used techniques to identify the morphology of ionomers is that of small-angle X-ray scattering (SAXS). The small-angle data of the scattering pattern give the electron density fluctuations of low frequency, and therefore it follows from Bragg's Law that the information is on the larger microphase morphology, in the range of approximately 1–100 nm.[18] This contrasts with the more widely encountered wide-angle X-ray diffraction, where the fluctuations are at the atom-to-atom bond distance. The pattern is complicated by thermal density fluctuations.

Longworth and Vaughan[19] were the first to use the SAXS technique for ionomers; they studied the effect of various counterions and concluded that the ionomer peak derived from an ordered spacing of hydrocarbon chains in ethylene ionomers. They argued that the fundamental structural unit is not the multiplet, as described earlier, but the coordinated metal ion. For example, the Group I metals would be surrounded by three carboxylate groups in an octahedral arrangement. Clusters are then formed by association of these coordinated metal ions. These clusters would be less tightly occupied than those discussed by Eisenberg, as each sub-group would be separated by the remainder of the polymer backbone and not the single layer thickness described earlier. In this model it is difficult to decide what forces are producing the structural repeat unit of size 20 Å.

One of the first papers to discuss microdispersed ionomers appeared in 1968.[20] In this paper Bonnotto and Bonner proposed a model for ethylene–acrylic acid ionomers, in which there are residual crystalline areas of short sequence from the polyethylene and amorphous regions containing the ionic species, i.e. ionized COO^-, pendant COO^-, and hydrogen-bonded and short-chain olefin units. These areas exhibit large cohesive forces as they are ionic.

Unlike the typical covalent crosslinks found in other organic polymers, the ionic forces are diffuse and non-directional and they act at greater distances. In a study of the effect of charge in the ionic domains, one would have assumed that different cations should behave differently according to the ratio of their charge intensity to their ionic radius (Q/A); this behaviour is observed in polyphosphate compounds where the change in counterion Q/A (84 for calcium and 41 for sodium) changes the T_g from 525 to 280 K.[17] However, in the ionomers investigated, monovalent and divalent counterions had similar properties at similar degrees of neutralization; hence it was concluded that the cations were not associated with any particular COO^- groups. Ward and Tobolosky[17] concluded that the most important parameter is the degree of neutralization and not the cation valency.

2.2 Spectroscopic Studies

2.2.1 Vibrational Spectroscopy

The motion of cations in ionic polymers is observed in the far-infrared region, where the reduced mass, force field and damping depend on the environment and therefore give differing absorption frequency.[21] The symmetric stretch of the sulphonate group was found by Lowry and Mauritz,[22] using Fourier Transform infrared (FTIR) spectroscopy, to be dependent on its chemical environment. Both the nature of the counterion and the extent of hydration affect the absorption frequency of the sulphonate group. At high degrees of hydration, this frequency is independent of both the water concentration and the nature of the counterion. The degree of hydration at which a frequency shift is seen is related to the hydration number and hydration energy of the particular counterion. Lowry and Mauritz[22] interpreted these effects as evidence for contact ion-pairs between the sulphonate group and the cation. This conclusion is supported by the infrared spectroscopic data summarized by Falk.[23] Lowry and Mauritz argued that the changes in frequency arise from changes in the polarization of the S—O dipole; the magnitude of these changes is at a maximum for the Li^+ ion and almost insignificant for Rb^+ and Cs^+ ions.

Hydration, quite obviously, shields the S—O bond from such interactions. Lowry and Mauritz, therefore, concluded that contact-pairs are present in Nafion at low degrees of hydration for Na^+, Li^+ and K^+ ions.

From these studies, structural information may be deduced. At low water contents (less than two H_2O molecules per sulphonic group), one quarter of all the hydroxyl groups are exposed to the fluorocarbon chain that is in small clusters. At higher water contents the fraction of H-bonded H_2O molecules increases. A more recent study of asymmetric stretching vibration modes in ethylene methacrylate ionomers has been reported by Brozoski et al.[24] They concluded that the effects observed in the carboxylate vibration region (coalescence to a single broad peak), with annealing at room temperature, were not a result of cluster formation by multiplets as proposed by Eisenberg,[15] but were due to differences in the degree of hydration only.[25]

Brozoski and coworkers[25] also performed a study, employing FTIR, on the carboxylate geometry around the cation using knowledge of the established coordinating nature of the cation. The sharp singlet and doublet peaks in the IR region observed in ionomers suggested the presence of multiplets, but these workers used symmetry group analysis to elucidate the multiplicity of the bands. Some ionomers were found to give singlets, e.g. those of K, Cs, and Zn, yet others gave doublets, e.g. those of Na, Ca, Sr and Ba. On the basis of ionic radii and known coordinating tendencies of Li and Na these cations tend to adopt a coordination number of six. The spectrum of each of these contains a doublet which would correspond satisfactorily with an octahedral structure which has two absorption modes in the infrared spectrum. The conclusions of this work are presented in Table 1.

2.2.2 Raman Spectroscopy

To obtain further structural information on ionomer systems Neppel et al.[26] have studied Raman spectra in the range 425–100 cm^{-1} and at

TABLE 1

SYMMETRY GROUPS AND INFRA RED SPECTRA OF METAL IONS IN IONOMERIC MATERIALS[a]

Stereochemistry	Coordination no.	Symmetry group	Stretching modes	No of IR-active bands
Square planar	4	D_2H	2	1
Tetrahedral	4	S_4	2 (degenerate)	1
Octahedral	6	D_3	3	2
Body centred cubic	8	D_4H	4	1

[a] Adapted from Ref. 25.

various temperatures (150–250°C) for a group of ethyl acrylate–sodium acrylate copolymers with various sodium acrylate contents from 100 to 0 mol %. Two absorption bands were assigned, one to multiplets and one to ion clusters. Both of these absorption bands were found to be absent in sodium acrylate and ethyl acrylate homopolymers. Since clustered ion-pairs would oscillate with greater reduced mass than multiplets, the absorption bands would be at a lower frequency, so it was possible to assign the absorption bands. At all temperatures the bands increase in intensity with the increase in sodium acrylate content up to 10 mol % for the multiplet band ($246 \, cm^{-1}$) and up to 35 mol % for the cluster band ($175 \, cm^{-1}$). With the increase in temperature, the multiplet band increases whilst the cluster band decreases for any composition. This effect of temperature is assumed to be indicative of breakdown of clusters to multiplets, but no critical temperature for this transition was observed; this observation conflicts with those of other workers[15] who presumed that such a critical temperature existed.

2.2.3 Mössbauer Spectroscopy

Mössbauer spectroscopy is a useful technique for the study of ionomers that contain either iron or tin. The numerous papers in this field have been reviewed by Rodmacq et al.[27] and by Meagher et al.[28] The latest work, which combines Mössbauer and electron probe spectroscopy, show that these ions are entirely located in a separate aqueous phase in Nafion that is not directly involved in the crystalline or amorphous phases of the fluorocarbon backbone. In the aqueous phase the iron occurs in a variety of different forms. The two techniques give useful information about the precipitation of iron hydroxide when the membranes are re-exchanged with other cations.

2.2.4 Nuclear Magnetic Resonance Studies of Ion Pairing

Since ionic polymer membranes such as Nafion are insoluble, conventional NMR studies, without the facility for Magic Angle spinning and cross-polarization, yield poorly resolved spectra. Despite these difficulties some information has been gained. Of particular significance are studies using the counterion as a probe e.g. ^{23}Na, ^{7}Li or ^{133}Cs. The ^{23}Na spectra show increasing linewidth and chemical shift with decreasing water content. The data from this work suggested that the three or four molecules around the sodium ions are in the first hydration sphere.[29] A four-state model originally proposed by Eigen,[30,31] depicted in Fig. 2, is based on these ion hydrate associations; the model is also discussed by Komoroski and Mauritz.[32]

2.3 Stress Relaxation Studies

Yeo and Eisenberg[9] found that the natural relaxation of the Nafion membranes invalidated time–temperature superposition procedures. This implies that there are two mechanisms of relaxation. These can be considered, firstly, as resulting from chain diffusion which is normal for organic polymers, and secondly, as a consequence of the cluster structure. This second mechanism has been ascribed to the movement of ion-terminated chains from one cluster to another. Above 180°C, the time–temperature superposition technique is again valid as the clusters become more mobile.

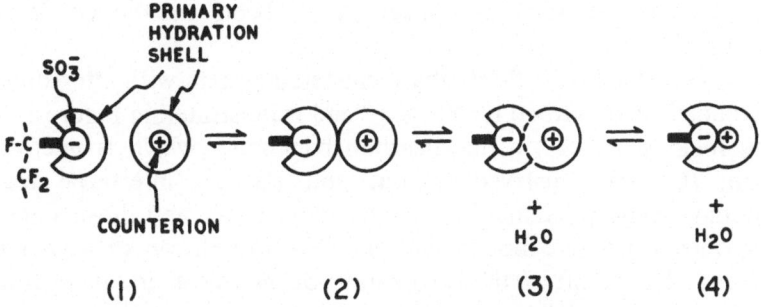

FIG. 2. Four-state model of hydration-mediated equilibrium between unbound and side-chain-associated counterions in ionomer membranes. Reprinted from ref. 39 with permission from the American Chemical Society.

Generally Nafions are similar to other ionomers including the formation of ionic clusters. However, they do exhibit some unique properties. One such property is the large reduction in electron density on ionization which is accompanied by an increase in the diffusion coefficient for water. Also, the glass transition temperature of the ionic regions in the presence of water is much lower than that of the rest of the matrix. The Nafion membranes consist of three microdispersed phases as described by Yeager[34] (see Section 2.4).

2.4   Transport Properties of Ionomer Membranes

Morphology differences discussed earlier clearly indicate that transport phenomena will be affected by clustered structures. Conventional ion-exchange membranes are composed of polystyrenesulphonate crosslinked by divinyl benzene; they would be expected to have differing transport properties from their ionomer counterparts, and from a comparison of these properties details of the nature of the clustered state may be gained.

Investigations have been carried out in dilute and in concentrated solutions. In dilute conditions the system is simpler, as the ionic polymer exhibits Donnan exclusion of anions which prevents absorption of electrolytes from solution, so that only the cationic exchange occurs. However, there is a need for experiments at higher concentrations and higher temperatures in conditions more closely modelling those of membranes used in the chlor-alkali industry.

Transport through a membrane is either along concentration gradients, i.e. diffusion, or along electric potential gradients, i.e. ionic migration and electro-osmosis. Many models have been based on spectroscopic evidence alone but some attempts based on ion transport data, to explain the mechanism have been mentioned in the literature.[33]

A model based on a three-phase clustered system with interconnecting channels is described by Yeager[34] and is illustrated in Figure 1. The three regions are (A) a fluorocarbon backbone, which is microcrystalline; (C) the clustered region; and (B) an interfacial region containing some pendant side chains, some water and those sulphonate groups which are not in clusters. The proportion of each phase present is dependent on the Q/A ratio[17] of the cation and its hydration energy.

In studies of diffusion rates, Na^+ ions were found to be transported at a greater rate than Cs^+ ions; using the triphasic model the assumption is made that both cations will diffuse at an equal rate through regions B and C but that the Cs$^+$ ion will experience a more tortuous route through region A.

In many of the transport models the membrane is not considered as a three-dimensional structure, but as containing cluster formations in only one phase. Since clusters are only of the order of 5 nm in diameter, they clearly do not alone account for transport through a membrane that may be up to a few millimetres thick. The concept termed the 'percolation mechanism',[35] put quite simply, uses a model based on the assumption that clusters are linked by conductive channels.

Figure 3(a) shows low concentrations; the clusters are therefore poorly connected as they are physically too far apart for macroscopic ion flow to occur. Successive diagrams (b), (c) and (d) show increasing cluster formation and hence greater ion flow. At low concentrations the material is an insulator. As the concentration is increased an

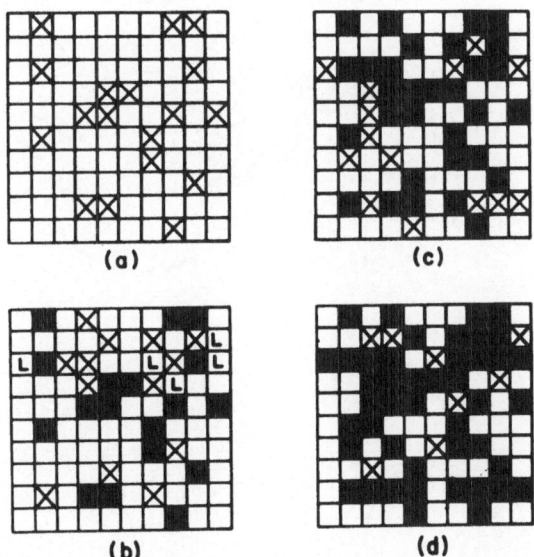

FIG. 3. Two-dimensional illustration of the concept of percolation. The shaded areas and crossed areas correspond respectively to sites that were previously occupied and sites that have just been occupied. Those marked L in (b) are empty sites that must be occupied before the onset of ion transport. The empty and occupied sites represent the fluorocarbon backbone and the electrolyte phase respectively. Reprinted from ref. 39 with permission from the American Chemical Society.

insulator-to-conductor phase transition occurs, and the average size of the channels becomes greater.

A simple power law[36] accounts for the threshold conductivity when this phase transition occurs;

$$\sigma = \sigma_0(c - c_0)^n$$

where σ_0 and n are constants. The index n is dependent only on the spatial dimensions and is applicable to any percolative system; it normally has a numerical value between 1·3 and 1·7. The threshold volume c_0 describes the way in which the components are dispersed. In this equation the dimensional and topological factors are controlled by the term $(c - c_0)^n$, whilst information on how ions move can be obtained empirically as the factor O_0 is separated out.

In dry Nafion polymer, the clusters (equivalent weight 1200) are

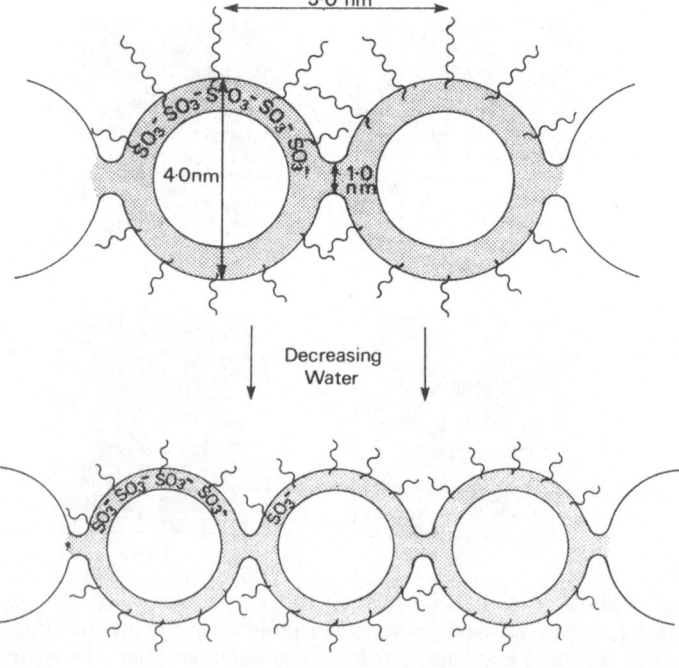

FIG. 4. Redistribution of ion-exchange sites that occurs on dehydration of Nafion polymer. Reprinted from ref. 39 with permission from the American Chemical Society.

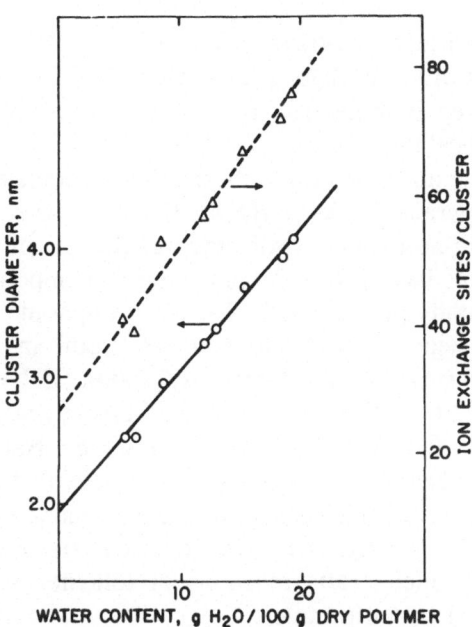

FIG. 5. The variation of cluster diameter (O) and ion exchange sites (△) per cluster with water content in Nafion EW 1200. Reprinted from ref. 39 with permission from the American Chemical Society.

about 1·9 nm in diameter and contain 26 exchange sites.[35] (As shown in Fig. 4.) As the polymer absorbs further amounts of water, the cluster diameter and number of exchange sites per cluster increase. This information is important as the clusters are not only growing by water absorption; as the number of exchange sites is also increased (as depicted in Fig. 5), there is a rearrangement process leading to fewer, larger clusters in the fully hydrated membrane.

As mentioned earlier, Nafion would be expected to exhibit different rates of cation exchange from other ion exchange resins; this is rationalized by two explanations.[34] Firstly, the lack of formal cross-links means that the structure is flexible, and controlled by the water content, which in turn is determined by the nature of the cation. Polystyrenesulphonate membranes conversely contain large quantities of absorbed water, and the structure is fixed by the crosslinks of divinylbenzene. Secondly, the perfluorinated backbone creates a much lower charge density because of electron withdrawal by the fluorine atoms away from the sulphonate group. This reduces the possibility of even solvent-separated ion-pairs. Thus, as a result, Nafion has less solvent enthalpy of ion exchange than other ion-exchange resins. This second parameter is reflected in the application of Nafion as a 'superacid' solid catalyst.

Acids that have up to 10^{12} times the acidity of common mineral acids are termed 'superacids'. As early as 1927,[37] it was observed that perchloric acid in non-aqueous solvents was capable of protonation of very weak organic bases, for example carbonyl compounds. A system containing antimony pentafluoride and fluorosulphonic acid (HSO_3F–SbF_3), called Magic Acid,[®] and is most useful in catalysis as it stabilizes species such as the tertiary butyl cation $(CH_3)_3C^+$. Attempts have been made to use these acids in a solid form by attaching them to materials such as fluorinated alumina.[38] However, Nafion used in its protonated form has shown higher catalytic activity than any of the other 'solid superacid catalysts', providing the reaction temperature is less than 200°C. Among the uses reported are isomerization of m-xylene and Friedel–Crafts reactions of toluene and phenol with oxalates and alkyl chloroformates.[38]

3   CHEMISTRY AND APPLICATIONS OF IONIC POLYMER MEMBRANES

3.1   Nafion Membranes

Ionic polymer membranes have been used in a variety of applications, such as reverse osmosis, dialysis and in membrane electrolytic cells for the chlor-alkali industry. Nafions are among those ionic polymers which have found commercial application as membranes and they have fairly recently been the subject of a substantial review;[39] what follows is necessarily a much briefer account.

The Nafions include a range of perfluorinated polymers containing small proportions of sulphonic or carboxylic functional groups. Being ionomers, rather than polyelectrolytes, they are not soluble in water. They also exhibit a remarkable degree of chemical resistance, notably to alkalis, and this, together with their ion-exchange properties, has led to their use as membranes in various electrolytic processes. One such process is chlor-alkali production. The term 'chlor-alkali' refers to products obtained from the electrolysis of sodium chloride, i.e.

® Registered trademark of Cationics Inc., Columbia, SC.

chlorine, sodium hydroxide and sodium carbonate. The first two are produced simultaneously during the electrolysis, while the latter is included because it shares many of the end-uses of sodium hydroxide, and it is also made from the process in small quantities.[40] Nafion membranes, which are permeable to sodium ions, but which exclude chloride, are clearly of interest for such an application.[41–43] The arrangement of a typical membrane cell which might be used for electrolytic chlor-alkali production is shown in Fig. 6.

In practice, although the sulphonated Nafion membranes were found to be impermeable to Cl⁻, they did allow the passage of hydroxyl ions, and this led to losses in current efficiency, particularly compared with the older diaphragm cells.[40] However, since the commercial introduction of the original Nafion membranes in 1969, the chlor-alkali industry has been actively engaged in efforts to improve these materials, not only by reducing their permeability towards hydroxyl ions, but also by improving their selectivity for sodium ions. Various methods have been tried, such as converting the

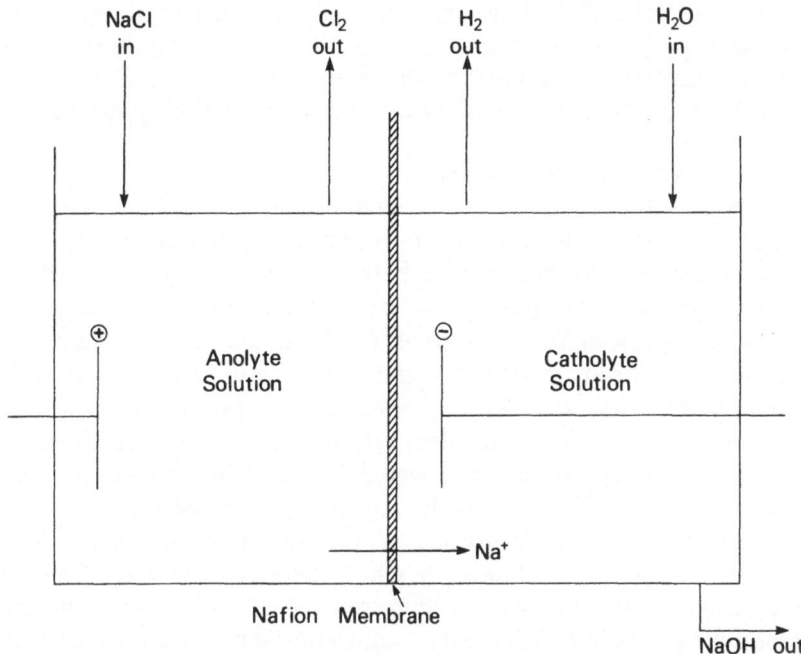

FIG. 6. Diagram of membrane cell for electrolytic chlor-alkali production.

sulphonate groups to sulphonamide or carboxylic functionality.[43] The latter approach proved to be the more successful, and present-day membranes tend to be of either the wholly carboxylic or the mixed carboxylic/sulphonic types, with the latter proving to be particularly effective in commercial chlor-alkali plants.[44]

Nafion membranes have also been considered for use in electrochemically regenerative hydrogen/halogen cells.[45] Such cells are of interest as large-scale energy stores for industrial use, since they are, in effect, reversible electrolytic cells. They have also been considered for use in fuel cells.[46,47] A fuel cell is a special case of a primary battery, which has high-energy reactants constantly fed into it, and low-energy products continuously removed.[48] Chemical energy is thus converted directly into electrical energy with no intermediate combustion cycle, and these cells, being free from Carnot cycle limitations, consequently have high thermal efficiency.[49] Typical fuels are hydrogen and carbon monoxide, while oxygen itself is used as oxidant. In such cells, the membrane acts effectively as a solid electrolyte, and plays the vital part of preventing the molecular form of the fuel from mixing with the oxidant and undergoing reaction directly.[50] Yeager and his coworkers[46,47] have investigated the transport and water absorption properties of Nafion membranes under conditions of high solution concentration typical of those encountered in fuel cells, and concluded that these materials showed promise for this application.

3.1.1 Ion Selectivity of Nafions

The fundamental feature of Nafion membranes, which underlies all of the applications mentioned so far, is that of ionic selectivity. These materials are able to discriminate between ions according to size and charge depending on their water content and on the form of the functional groups in the membrane (acid or neutralized). Steck and Yeager[51] measured selectivity coefficients for alkali-metal ions, alkaline-earth ions, and certain other metal ions, such as Ag^+ and Zn^{2+}, using Nafion 120 (equivalent weight 1200) both as received and when expanded by boiling with water for 5 h. They also determined water contents as a function of the nature of the metal ion.

Experimentally, concentrations of the salt solutions were determined by column ion-exchange, in which the metal ion was exchanged for hydrogen ion, and the concentration of the acid form was then measured by titration. Selectivity coefficients were calculated from the following equation:

$$K_A^B = \frac{\bar{X}_B^2 C_A^2 \gamma_A^2}{\bar{X}_A^2 C_B \gamma_B}$$

where K_A^B is the selectivity coefficient for B over A, X_A, X_B represent the equivalent ionic fractions of the ions A and B respectively in the membrane, and C_A, C_B are the solution molarities of A and B. The single-ion activity coefficients, γ_A and γ_B, were calculated using the extended Debye–Hückel equation and Kielland's individual ion size parameters.[51] The selectivity coefficients measured in this way are shown in Table 2.

Steck and Yeager[51] found that the metal-for-hydrogen exchange was always accompanied by the release of water, and by some polymer contraction, both of which processes increase entropy. Indeed, these workers concluded[51] that this increase in entropy is the controlling factor in determining the magnitudes of the selectivity coefficients.

3.1.2 Ion-selective Electrodes

The demonstrated selectivity of Nafion membranes implies that they have the potential for use in ion-selective electrodes, and they have indeed been used in this particular application. Polymer-membrane

TABLE 2
SELECTIVITY COEFFICIENTS (K_A^B) FOR METAL IONS IN NAFION 120 MEMBRANES[51]

Ion	K_A^B at 25°C (as received polymer)	K_A^B at 40°C (as received polymer)	K_A^B at 25°C (expanded form)
Li^+	0·579	0·555	0·586
Na^+	1·22	1·31	1·18
K^+	3·97		3·48
Ca^+	9·11	9·04	7·06
Tl^+	6·12		3·83
Ag^+	1·07		0·90
Mg^{2+}	2·30	2·36	2·15
Ca^{2+}	3·60		2·87
Ba^{2+}	5·55	5·27	4·61
Zn^{2+}	0·97		

ion-selective electrodes (PMISEs) are already established analytical tools, using plasticized PVC as the membrane.[52] The problem is that, under conditions of use, this plasticizer may either itself be lost, or it may release the electro-active species into the sample liquid,[53] thus limiting the lifetime of the PMISE. The use of Nafions circumvents this problem, since Nafions do not contain any plasticizer; the presence of ionic clusters provides a mechanism for ionic conductivity via diffusion of ions from cluster to cluster,[54] a process which does not rely on the presence of any other phase, such as plasticizer, in order to be operative. In the absence of any plasticizer, PMISEs based on these polymers would be expected to have effectively unlimited lifetimes.

In a study of Nafion-based PMISEs, Martin and Frieser[52] prepared electrodes sensitive to K^+, Cs^+ and tetrabutylammonium (TBA^+) ions. This was done by cutting disks of Nafion with a cork borer, and soaking them in 95% aqueous ethanol for 90 minutes, followed by successive equilibrations for 2 h each in sodium hydroxide solution, deionized water and hydrochloric acid. To convert the acid form of the Nafion to the appropriate salt, the disks were finally soaked for three days in one of the following solutions: $0.1M$ KCl, $0.1M$ TBA Br or $0.1M$ $CsNO_3$. The disks were then cemented to the ends of glass tubes using silicone rubber sealant, after which the appropriate internal reference solutions (Table 3) were poured into the tubes, and silver/silver chloride wires were inserted as internal reference electrodes. Under the right conditions, the response of all three PMISEs to their respective primary ions was found to be nearly Nernstian, i.e. the response of the cell containing the ion-selective electrode followed the form of the Nernst equation:

$$E = E^0 + \frac{RT}{zF} \ln A_X$$

where E^0 is the standard electrode potential of the cell, A_X is the

TABLE 3

INTERNAL   REFERENCE   SOLUTIONS   IN   NAFION-BASED
PMISESs[52]

Primary ion	Internal reference solution
K^+	$10^{-2}M$ KCl
TBA^+	$10^{-2}M$ TBA Cl
Cs^+	$5 \times 10^{-3}M$ KCl $+ 5 \times 10^{-3}M$ $CsNO_3$

TABLE 4
SELECTIVITY COEFFICIENTS OF NAFION-BASED PMISEs[52]

| Interferent | Selectivity Coefficients | |
	TBA^+	Cs^+
Ca^{2+}	10^{-4}	0·16
Na^+	10^{-4}	0·51
K^+	10^{-4}	0·84
H^+	—	0·73
TBA^+	(1)	767
Decyltrimethylammonium	325	—

activity of the ion X, and z is the number of charges on X. The sign is positive for cations and negative for anions.

In practice, the range of Nernstian response of any electrode is limited;[55] for the Nafion-based PMISEs, it extends over about 1·4 orders of magnitude,[52] which is much less than those electrodes with plasticized PVC membranes. By contrast, as shown by Table 4, the selectivity is good, and compares extremely well with that of a typical conventional PMISE. These electrodes showed particularly good selectivity towards large organic cations (Table 4). In addition, they also proved to be extremely long-lived, in comparison with other electrodes, most of the experimental membranes having retained their selectivity for over six months of use.

Already, Nafions have been shown to be extremely useful and versatile membrane materials, and there is every prospect that their range of application will continue to expand.

3.2 Poly(acrylic acid) Membranes

Poly(acrylic acid) is another polymer which has been found to have a range of applications in membranes, and has been used in five different ways:

(i) adsorbed onto an inorganic support, such as zirconia;[56,57]
(ii) as grafted side chains on other polymers, such as nylon 6[66] or fluorinated copolymers;[59]
(iii) as part of a conventional copolymer, with such comonomers as n-butyl methacrylate;[60]
(iv) crosslinked by polyvalent metal ions to form water-plasticized poly(acrylate) salts;[61,62]

(v) in the form of a neutralized polymer complex, such as poly(acrylic acid)–poly(ethylenepiperazine).[63]

Poly(acrylic acid) has a much higher density of functional groups than have the various Nafions; and it is a polyelectrolyte rather than a classical ionomer and needs to be made insoluble in some way in order to become an effective component of membranes for aqueous application. Each of the above methods provides a means of achieving the necessary insolubilization, and they will now be considered in turn in some detail.

3.2.1   Inorganic Supports for Poly(acrylic acid)

The use of inorganic supports, such as zirconia, to insolubilize poly(acrylic acid) has led to the preparation of membranes which are useful for reverse osmosis.[56,57,64] Such membranes may be formed dynamically, by exposing an oxide, such as ZrO_2, to an acidic solution containing poly(acrylic acid), and raising the pH to precipitate the polymer.[56] The resulting membranes may be used in the reverse osmosis of low-salinity water. Sachs and Lonsdale[64] found that when the inorganic supports used are sufficiently fine and porous, such membranes are capable of sustaining high water fluxes, and altogether, performed as well as conventional membranes made from modified cellulose acetate.

3.2.2   Graft Copolymers

Films containing poly(acrylic acid) side chains attached to water-insoluble polymer backbones have been prepared for use as membranes. Hegazy and his co-workers[59] prepared cation-exchange membranes by radiation grafting of acrylic acid onto hexafluoropropylene–tetrafluoroethylene copolymer film. They found that swelling behaviour, electrical conductivity and mechanical properties depend on the degree of grafting, but they did not subject these materials to detailed evaluation as membranes. A similar approach was made using nylon 6 as the support polymer for grafted poly(acrylic acid) side chains.[58] The resulting material, which contained about 60% by weight of acrylic acid, was treated in water containing paraformaldehyde, lactic acid and formic acid to give a non-crystalline film, and this membrane had a salt rejection rate of 96% when used in the reverse osmosis of aqueous potassium chloride.

3.2.3 Linear Copolymers

The next category of poly(acrylic acid) membranes are those based on copolymers with n-butyl methacrylate. These membranes, which are prepared with acrylic acid at various levels between 25 and 45%, have been used for haemodialysis and appear to have superior transfer properties for urea, creatine, uric acid and most other blood components, compared with cellulosic membranes, but they give a lower rate of permeation for glucose by comparison with conventional cellophane.[2] The permeability of the membranes was found to be related to acrylic acid content, those membranes which contained more acrylic acid being more permeable. There is the drawback, however, that increasing amounts of acrylic acid in the membrane leads to reduced mechanical sturdiness, and this limits the durability of the more permeable membranes. Treatment with ethylene oxide was found to improve the mechanical properties but, as with reduction of the acrylic acid content, this had the effect of reducing the permeability.[60] In another study,[65] the performance of anionic haemodialysis membranes of poly(acrylic acid-co-n-butyl methacrylate) was compared with that of a cationic membrane based on a copolymer of acrylonitrile and dimethylaminoethyl methacrylate. The nature of the polymer was found to have a marked influence on the dialysis properties of the resulting membranes, with the anionic membrane proving to be more permeable to creatine than urea, and the cationic one being permeable only to acidic solutes, such as uric acid and phosphates.[65]

3.2.4 Polyacrylate Salts

Ionically crosslinked poly(acrylic acid) has been used to prepare membranes for possible application in dialysis and reverse osmosis.[61,62] Two methods have been employed, a wet[61] and a dry[62] technique.

In the wet technique, a solution of poly(acrylic acid) and a solution of a crosslinking metal salt are prepared separately. The solution of polymer is cast onto a glass plate to form a film of predetermined drawdown thickness, after which the plate is carefully immersed into the crosslinking solution and is held there, at a particular temperature and for a certain length of time, until the membrane is completely crosslinked. After this synthesis, the membrane is soaked in deionized water before use.

This synthetic technique relies on two independent processes, namely diffusion of the metal ions from the solution into the polymer, and subsequent reaction of these ions with the carboxylic acid groups of the parent polymer. Habert and his coworkers[61] have developed what they describe as a 'qualitative rate model' in order to consider the various possibilities of such a synthesis, which may occur depending on the relative and absolute magnitudes of the rates of the two processes. Using this model they concluded that it is desirable to have the rate of membrane crosslinking controlled by both the rates of diffusion and reaction. Clearly, if diffusion were substantially the quicker of the two processes, the polymer would be able to dissolve before reaction proceeded to any significant extent. Conversely, if the rate of reaction exceeded that of diffusion, there is the possibility that a skin would form at the polymer–solution interface and prevent any further penetration by metal ions, leaving the bulk of the cast film of polymer uncrosslinked.

To overcome the problem of imbalance between diffusion and reaction rates, Habert et al.[62] devised an alternative synthetic procedure, their so-called dry technique. In this approach, the metal ions are incorporated into the casting solution, and crosslinking is effected by thermal treatment of the cast film. Membranes prepared in this way were conditioned by soaking overnight in acetone at room temperature, and then by soaking for a further 24 h in deionized water.

Membranes prepared by the wet technique have been crosslinked using a variety of metal cations.[61] The most successful are those based on aluminium, although, as shown in Table 5, other polyvalent ions have been used but with much less success. The various ions were

TABLE 5

MEMBRANE SYNTHESIS IN AQUEOUS SOLUTIONS OF SALTS OF POLYVALENT IONS[61]

Metal ion	Salt	Results
Zn^{2+}	Chloride	Rapid precipitation of white film.
Be^{2+}	Sulphate	Rubbery film.
UO_2^{2+}	Nitrate	Yellow leathery film formed at 50°C.
Zr^{4+}	Chloride	Weak gel.
	Nitrate	No crosslinking.
	Sulphate	No crosslinking.
Al^{3+}	Nitrate	Very satisfactory film formed.
Cr^{3+}	Chloride	Rubbery film formed with distinct skin.

introduced into the membrane by adding a salt of the relevant metal to the crosslinking bath, the salts being chlorides, sulphates or nitrates. In general, all of these salts give insoluble products, but some, such as those cross-linked by Be^{2+} ions, slowly disintegrated when soaked in water. The aluminium-containing membranes have proved to be the most satisfactory of all. They have also been prepared by the dry technique[62] and evaluated for permselectivity by comparing the permeabilities of various solutes under a concentration gradient typical of those used in dialysis. A permeability coefficient was estimated from the equation:

$$\ln\left(\frac{C_0}{C}\right) = \frac{PAt}{V}$$

Where C_0 and C represent the initial and final concentrations respectively, t represents the time and V represents the volume of the system. P, the permeability coefficient, was determined from the slope of the graph $\ln(C_0/C)$ versus time. Table 6 gives values for a typical membrane, in this case one having an aluminium content of 4·9%; it was prepared by curing for $3\frac{1}{2}$ h at 60°C, followed by a 2 h soak in acetone at room temperature. These results show a molecular-weight cut-off at about 200 for the organic compounds, as well as a significant exclusion of the ionic solute, NaCl. Such ionic exclusion suggests that the carboxylic groups in the polymer matrix exert a definite influence, one that has been used to advantage in reverse osmosis. Reverse osmosis has been carried out statically on sodium chloride solutions at room temperature using applied pressures of between 300 and 1000 psi. The experimental membrane made of poly(acrylic acid) crosslinked with aluminium gave results which compared extremely

TABLE 6

PERMEABILITY COEFFICIENTS FOR DIALYSIS OF SOLUTES THROUGH AN ALUMINIUM CROSSLINKED POLY(ACRYLIC ACID) MEMBRANE[62]

Solute	Molecular weight	$P\ (cm\,min^{-1}) \times 10^4$
Ethylene glycol	62	58
Diethylene glycol	106	21
Triethylene glycol	150	9·7
Tetraethylene glycol	194	0
NaCl	—	9·5

well with those obtained using commercial reverse osmosis membranes. In a typical separation with an experimental membrane, a 0·1% NaCl solution was subjected to a pressure of 600 psi and gave a flux of 1250 g/h m^2 and a solute rejection of 85%.[62]

In a further development of the approach of using ionically crosslinked poly(acrylic acid), Huang et al.[66] prepared composite membranes in which the polysalt was supported on a porous polysulphonate matrix. These membranes were prepared by coating the polysulphone support with a thin layer of dilute poly(acrylic acid) solution, drying this layer, and ionically crosslinking the product in solutions of different salts. This technique is similar in many ways to the wet technique previously used by Huang et al.[61]

The transport properties of the resulting membranes were determined using a high-pressure reverse osmosis cell at 25°C and 300 psi, with an effective membrane area of 18·1 cm^2 (4–8 cm in diameter). The feed solution consisted of 0·1% sodium chloride, and was circulated during the experiments. Results for the various membranes are given in Table 7. As can be seen, the salt rejection is generally poor, except for those membranes crosslinked by Zr^{4+} or UO_2^{2+}. Huang et al.,[66] therefore, concluded that the degree of crosslinking in most of these membranes is very low. In order to improve the crosslinking, the preparative technique was modified. Instead of using poly(acrylic acid), the porous polysulphone support was coated initially with a solution of sodium polyacrylate, a solution prepared by

TABLE 7

PREPARATION AND REVERSE OSMOSIS OF IONICALLY CROSSLINKED POLY(ACRYLIC ACID) COMPOSITE MEMBRANES[66]

Crosslinking solution metal salt, 0·25 M	pH	Crosslinking temperature(°C)	Crosslinking time(h)	Rejection (%)
$ZnCl_2$	5·6	60	8	16·0
$ZnSO_4$	5·3	70	4	10·0
$Ba(OH)_2$	13·3	70	2	17·4
$Cr(NO_3)_3$	1·5	70	4	3·9
$AlCl_3$	3·6	60	4	43·5
$AlK_3(SO_4)_3$	2·8	70	2	36·7
$ZrCl_4$	1·0	60	2	29·0
$ZrO(NO_3)_2$	1·2	70	6	32·4
$UO_2(NO_3)_2$	2·1	70	1	79·5

TABLE 8

PREPARATION AND REVERSE OSMOSIS OF IONICALLY CROSSLINKED POLY(ACRYLIC ACID) COMPOSITE MEMBRANES WITH NEUTRALIZED ACID, pH 7·0

Crosslinking solution metal salt, 0·25 M	pH	Crosslinking temperature(°C)	Crosslinking time (h)	Rejection (%)
AlCl₃	3·6	70	2	67·7
AlK₃(SO₄)₃	3·0	70	6	42·8
ZrCl₄	1·0	70	4	48·7
ZrO(NO₃)₂	1·2	70	4	34·8
UO₂(NO₃)₂	2·1	70	1	6·8

neutralizing the original dilute poly(acrylic acid) with sodium hydroxide. Crosslinking was then brought about, presumably by an ion-exchange mechanism, by soaking the supported membrane in identical solutions of polyvalent metal ions. As the results in Table 8 show, there is generally a substantial improvement in salt rejection.

These membranes also show a high degree of rejection of phenol in aqueous solution; Huang et al.[66] suggested that this was due to electrostatic repulsion between the adsorbed polyacrylate and either the polar phenol molecules or phenoxide ions of the solute.

3.2.5 Polyelectrolyte Complexes

Finally, poly(acrylic acid) has been used as one of the components of polyelectrolyte complex (PEC) membranes. These have been reviewed previously by Lysaght[67] and are also discussed at length by Tsohuda and Abe in Chapter 5 of this book.

The reaction between two oppositely charged strong polyelectrolytes has been much employed to prepare PEC membranes, notably for use in ultrafiltration. One such PEC membrane formed when poly(acrylic acid) was neutralized with poly(ethylenepiperazine) has been used for permeability studies on a range of uncharged metabolites, such as urea, creatine and vitamin B12.[63]

Another patent[68] describes the preparation of a desalination membrane consisting of layers of anionic and cationic polyelectrolyte, separated by a layer of polymer. Such membranes included poly(acrylic acid), poly(styrenesulphonic acid), and poly(vinylimidazoline bisulphate) supported on acrylonitrile–vinyl chloride copolymer. These membranes gave salt rejections for CaCl₂ of up to 93%.

3.3 Polyelectrolyte-modified Membranes

Another way of using ionic polymers in membrane applications, reported by Sata and Izuo,[69] has been to alter the permselectivity of natural cation-exchange materials by loading them with polyelectrolyte of opposite charge. This treatment has yielded membranes which, in electrodialysis, are preferentially permeable both to lower-valent ions and to large hydrated ions. The permselectivity is dependent on the type of polyelectrolyte, and its molecular weight, as well as the natural cation-exchange material.

In one experiment, where poly(ethyleneimine) was used as the polyelectrolyte, Sata and Izuo[69] investigated the effect of conformation of the adhered polymer on the transport properties of the resulting membrane. The parent cation-exchange membrane used was Neosepta Cl-25 a commercial material which is a sulphonic-based membrane manufactured by Tokuyama Soda Co. Ltd. The conformation of the poly(ethyleneimine) was altered by changing the composition of the aqueous solution in which the membrane was immersed; adding salt, adjusting the pH to high values or adding an acid having a multivalent anion cause the poly(ethyleneimine) to adopt contracted conformations, leading to large changes in transport properties. Sata and Izuo[69] concluded that the amount of poly(ethyleneimine) adsorbed is substantially more than would be expected on the assumption that adsorption would stop at a monolayer. Such a large amount of adsorbed polymer clearly causes a significant narrowing of the pathways along which ions migrate. However, this effect is not the whole explanation for the change in transport properties; Sata et al.[70] observed that when the pathways are made narrower by other means, such as the diffusion of neutral molecules of glucose into the membrane, this remarkable change in permselectivity did not occur. These workers noted that potassium, with a larger hydrated diameter than sodium, permeated more easily through the adsorbed polyelectrolyte membrane. These results led them to conclude that the polyelectrolyte layer changes permselectivity because of the intensity of the electrostatic repulsion between the cation and the cationically charged layer.[70]

3.4 Membranes Containing Phosphonic Acid Functional Groups

Lastly, the literature on ionic polymers also includes reference to membranes containing phosphonic acid functional groups having been prepared.[71] They belong to a relatively neglected class of ionic polymer. They have been prepared by the oxidative chlorophosphonylation of polyethylene, polypropylene or ethylene–propylene copolymers, followed by saponification with water. The overall chemistry of these syntheses is outlined by the Scheme.[72]

$$-CH_2- + PCl_3 + \tfrac{1}{2}O_2 \longrightarrow -\underset{\underset{POCl_2}{|}}{C}H- + HCl \xrightarrow{H_2O} -\underset{\underset{PO(OH)_2}{|}}{C}H-$$

Ionic polymers of this kind have been produced with 6·8–16·0% phosphorus in them. The products containing the larger amounts of phosphorus showed significant water solubility, and are thus polyelectrolytes, while those with lower amounts of phosphorus, having only a small proportion of functional groups within the polymer chain, are classical ionomers. For membrane applications, polymers having phosphorus contents of between 8 and 10% have been used, and they show ion-exchange properties similar to those of membranes containing carboxylic and sulphonic acid groups.[71]

4 IONIC POLYMERS AS MODELS OF BIOLOGICAL MEMBRANES

There have been a number of studies in which synthetic ionic polymers have been prepared for use as models of biological membranes. These models have been used in attempts to elucidate the mechanisms underlying membrane excitability, i.e. the ability of membranes to alter temporarily their permeability to ions in response to an external stimulus. Ionic polymers are used as models because such polymers possess electrical properties similar to those of biological membranes. These properties include persistent electrical polarization[73-76] and the ability to pass transient bursts of current at regular intervals across a membrane when subjected to an applied potential.[77,78] Such phenomena underlie the conduction of nerve impulses in Man and throughout the animal kingdom.

4.1 Neural Conduction
The generally accepted explanation for neural conduction is that put forward by Hodgkin and Huxley[79] at the conclusion of their fundamental studies in the late 1940s and early 1950s for which they won the Nobel Prize. They discovered that neural excitability is dependent

on the movement of sodium and potassium ions across the membrane. Briefly, the process of nerve impulse transmission is as follows: the excitable cell contains substantially more potassium ions than does the surrounding extra-cellular fluid, and substantially less sodium and chloride. At rest, the cell membrane is moderately permeable to potassium and chloride, while being almost impermeable to sodium. When an impulse arrives at a given point in the membrane, there is a large and specific increase in the permeability to sodium, as a result of which sodium becomes able to enter the cell. At the same time, there is a leakage from the cell of an approximately equivalent amount of potassium.[80] This increase in permeability leads to the passage of a transient current, carried mainly by the sodium ions.[79] Once this burst of depolarizing current has passed, the cell membrane reverts to its original state of impermeability towards sodium ions. The levels of sodium, potassium and chloride are then restored to those of the resting state, the sodium concentration being reduced by metabolic expulsion from the cell to the extra-cellular fluid.[80]

4.2   Model Systems

4.2.1   Transient Permeability Charges

The crucial phenomenon in this overall mechanism, then, is the brief change in the permeability of the membrane towards sodium ions. Despite its importance, it received but little attention in the original physiological studies.[79,80] Subsequently, however, it has aroused interest, and has been the subject of theoretical and experimental work. For example, to account for such behaviour, Wobschall[81] has developed a mathematical model in which a bilayer membrane is postulated whose structure, and hence permeability, is altered in an electric field. The model was successfully used to predict many aspects of the nerve action potential, but this success was dependent on an arbitrary choice of a number of parameters, so that these predictions did not, by themselves, constitute a sufficiently rigorous test. However, the model did show that changes in membrane structure induced by an electric field could provide a plausible explanation for the observed changes in permeability.

In another study, Shashoua[77] prepared synthetic multilayer membranes from ionic polymer components. Under the influence of a d.c. electric field, such membranes spontaneously generate transient bursts of current, with time constants and amplitudes analogous to the familiar spike potentials of naturally occurring neural membranes. Experimentally, the synthetic membranes were prepared in a Petri dish from a 10% solution of poly(acrylic acid/acrylamide) [80:20], in 0·15M sodium chloride, at pH 4·5. On top of this was poured a 2% polybase solution, also in 0·15M sodium chloride solution, the base being dimethylaminoethyl acrylate. An alternative membrane was prepared in which the polyacid component was the homopolymer of acrylic acid, with calcium hydroxide as the basic component of the crosslinking solution. These two procedures gave membranes which were of the order of 1000 Å thick. Despite their different chemical constitutions, both types of membrane have similar electrical properties. In particular, when subject to a gradually increasing direct current voltage, a threshold point is reached above which the membrane spontaneously begins to generate spontaneous spike potentials of varying amplitude (1–20 mV) and duration (2–10 ms). Shashoua also found that there was an upper critical value, above which the membrane simply passed a constant current.

These observations were explained as follows[82,83] The membranes, in particular those prepared from oppositely charged polyelectrolytes, are supposed to consist of three layers, each having different permselective properties. In order, the layers are (i) an upper basic one, (ii) a middle polyelectrolyte complex one, formed by interaction of the polyacid and polybase, and (iii) a lower polyanionic one containing an excess of carboxylic groups. When a potential is applied across the whole membrane, there is an accumulation of salt at the interface between two of the layers and this leads to a change in the configuration of the impermeable layer, and an increase in the permeability. Salt is then able to pass through the layer and away from the interfacial region, thereby reducing the high local concentration. As the salt concentration falls, there is a change in the electrostatic environment within the composite membrane, so that the original configuration of the impermeable layer is restored, causing salt to accumulate once more at the interface. The whole process can then begin again, and is in this way able to repeat itself in cyclic fashion as illustrated. This model predicted a frequency of spike potential close to that observed experimentally by Shashoua.[77]

In a similar study, Huang and Spangler[78] prepared membranes electrophoretically from calcium chloride and poly(glutamic acid) on a matrix of poly(sebacyl piperazine). These membranes were found to

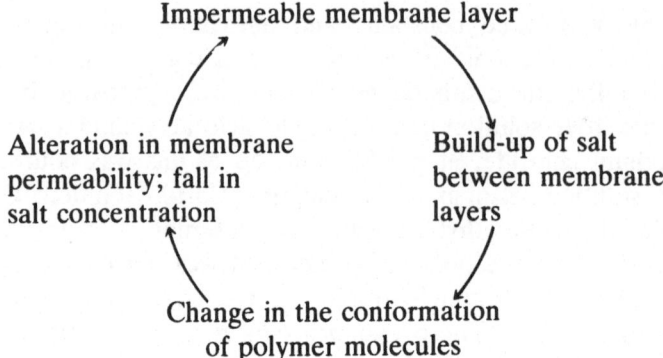

consist of two layers, an uncharged one in which the carboxylate groups of the polyacid were neutralized with Ca^{2+} ions, and an ionic one, in which the carboxylates were unneutralized. They were 500 to 1000 Å thick, and slightly porous.

These membranes showed a distinct region of hysteresis in the voltage–current curve, with a sharp transition from low membrane resistance to high membrane resistance occurring at a current density different from that of the inverse transition. Such electrical behaviour produced the familiar oscillating current in the region 0·5–1·3 V. One such oscillating current trace is shown in Fig. 7, having an amplitude of 5 mA/cm² and a period of 30 ms. The spikes were asymmetric, and had a rising phase of about 10 ms and a falling phase of about 20 ms. In general the generation of spike potentials could be sustained for only a few minutes, after which the membrane properties were altered irreversibly, presumably because of some permanent change in the structure of the membrane.

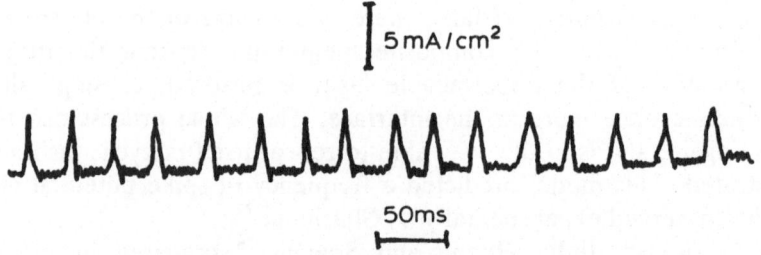

FIG. 7. Oscillatory activity in a poly(glutamic acid)–Ca^{2+} membrane. From ref. 78 reproduced by courtesy of the author.

The frequency and amplitude of the oscillatory waveform was found to vary from membrane to membrane, through frequently the waveform assumed the form of repetitive short 'spikes', as in Fig. 7. The frequencies observed ranged from 20 to 200 Hz, while the amplitudes varied between 3 and 10 mA/cm². These variations may arise from differences in membrane thickness, matrix porosity and other uncontrolled variations in the preparative procedures.

4.2.2 Persistent Electrical Polarization

The other important physical property possessed by excitable biological membranes is persistent electrical polarization, and this, too, has been modelled in synthetic membranes.[73,74] As a phenomenon, persistent electrical polarization has been of interest for many years, and led Heaviside in 1885 to the concept of an electret,[73] i.e. an electrical analogue of the magnet in which there are permanent electrical poles, one positive, the other negative. Electret formation was demonstrated experimentally for the first time in 1922 by Eguchi,[74] using carnauba wax.

In one particular study by Bornzin and Miller,[76] synthetic membranes containing electrets were prepared in an attempt to model neural excitation. The following procedure was used. Casting solutions containing the materials listed in Table 9 were made up using a mixture of methanol and water as solvent. These solutions were cast to give dry films 0·05–0·07 mm thick, which were then cured at 70°C. The cured films were clear and flexible, and contained 8 mol % of sulphonate groups. These films were charged to form electrets by heating in an electric field under vacuum within the range 35–98°C. The authors assumed that this process caused mobile ions to be displaced down the electric field during heating, and that the resulting

TABLE 9
COMPONENTS OF MODEL EXCITABLE MEMBRANE[76]

Substance	Proportion (% wt)
Sodium poly(styrenesulphonate), MW (2–4) × 10⁶	15·2
Poly(vinyl alcohol)	30·9
Hexamethoxymelamine	51·7
p-Toluenesulphonic acid	2·2

dipoles were stabilized by hydrogen bonding. Such membranes were found to be able effectively to store charge of up to about 1 C/g, and furthermore, their capacity to do so increased when they were swollen with water.

A number of workers[73-76,84,85] established that persistent electrical polarization is compatible with electrical conductivity in these polymer membranes, consisting of sodium poly(styrenesulphonate) on various polar matrices such as poly(vinyl alcohol) and poly(vinylacrylamide) because the mechanisms of the two phenomena are different: conduction is electronic, while the persistent polarization probably arises because of the slight displacement of the sodium ions from the fixed sulphonate sites.

4.3 Natural Membrane Systems

With the exception of Huang and Spangler's study,[78] which involved the use of poly(glutamic acid), none of these model membranes closely reflects the chemical nature of actual biological membranes. Glycoproteins, which are implicated in a variety of membrane functions, have been found to have a polyanionic structure, and this is clearly significant in view of their biological role.[86] Glycoproteins are macromolecules consisting of oligosaccharide chains with terminal acidic residues, these acid moieties being of the sialic acid group. Sialic acids are N- or O-acyl derivatives of neuraminic acid, which is illustrated in Fig. 8.

The anionic character of the glycoproteins appears to originate in the ionization of these acid groups, and enables glycine proteins to carry a substantial negative charge. In this characteristic, they are physically similar to the polymers used in the synthetic systems which model neural excitability. In fact, there is biochemical evidence[86] that

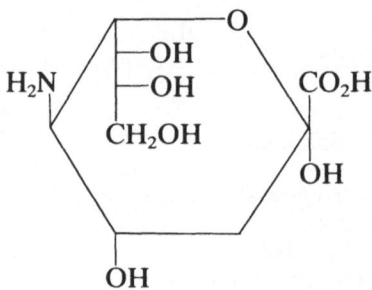

FIG. 8. Neuraminic acid (α-D-pyranose form).

naturally occurring ionic polymers are present in membranes, and are responsible for the biologically essential property of variable and selective permeability which these membranes exhibit.

5 CONCLUSION

The studies described in this chapter clearly demonstrate that ionic polymers are extremely useful as membranes. Their ion-clustered morphology enables them to exclude solutes selectively and, as a result, they have found many diverse applications, ranging from electrolytic chlor-alkali production to desalination and from artificial kidney dialysis to ion-selective electrodes. Ionic polymer membranes have also been used as models to improve our understanding of the behaviour of excitable membranes in biology. As a class, they are clearly versatile, and the expectation is that they will figure prominently as membrane processes continue to grow in importance.

REFERENCES

1. GRAHAM, I. T., *Phil. Trans. Roy. Soc. London*, 1854, **144,** 177.
2. MUIR, W., U.S. Patent 3 616 927, 2 Nov. 1971.
3. BAKER, R. W. and LONSDALE, H. K. in *Controlled Release of Biologically Active Agents* (A. C. Tanquary and R. E. Lacey, Eds), 1974, Plenum Press, New York.
4. *Kirk–Othmer Encyclopedia of Chemical Technology*, 3rd edn, 1978, Wiley–Interscience, New York.
5. LEITNER, G. F., *Tech. Proc. Ann. Conf. Natl. Water Suppl. Imp. Assoc.*, 1979, No. 6, USA; *Chem. Abs.*, **95** 156248.
6. GREGOR, H. P. and GREGOR, C. D., *Sci. Am.*, 1978, **239,** 112–28.
7. BROWN, H. P., *Rubber Chem. Technol.*, 1957, **30,** 1347.
7a. LONGWORTH, R., in *Developments in Ionic Polymers—1* (A. D. Wilson and H. J. Prosser, Eds), 1983, Applied Science Publishers, London, Ch. 3.
8. MACKNIGHT, W. J. and EARNEST, T. R., *J. Polym. Sci., Macromolecular Rev.*, 1981, **16,** 42.
9. YEO, S. C. and EISENBERG, A., *J. Appl. Polym. Sci.*, 1977, **21,** 875.
10. URIHASHI, H., *Chemtech*, 1980, 118.
11. MAURITZ, K. A. and HOPFINGER, A. J., *Modern Aspects of Electrochem.*, 1982, **14,** 425.
12. EISENBERG, A., *Macromolecules*, 1971 **4,** 125.
13. EISENBERG, A. and KING, M., *Ion Containing Polymers*, 1977, Academic Press, New York.

300   PAULINE J. BROOKMAN AND JOHN W. NICHOLSON

14. PETTIT, L. D. and BRUCKENSTEIN, S., *J. Am. Chem. Soc.*, 1966, **88**, 4783.
15. EISENBERG, A., *Macromolecules*, 1970, **3**, 147.
16. TOBOLOSKY, A. V., LYONS, P. F. and HATA, N., *Macromolecules*, 1968, **1**, 515.
17. WARD, T. C. and TOBOLOSKY, A. V., *J. Appl. Polym. Sci.*, 1967, **11**, 2403.
18. YARUSSO, D. J. and COOPER, S. L., *Macromolecules*, 1983, **16**, 1871.
19. LONGWORTH, R. and VAUGHAN, D. J., *Nature*, 1968, **218**, 85.
20. BONOTTO, S. and BONNER, E., *Macromolecules*, 1968, **1**, 510.
21. RISEN, W. T., *Macro 82 Macromolecular Micro Symp. Abs. M*, 1982, 899.
22. LOWRY, S. R. and MAURITZ, K. A., *J. Am. Chem. Soc.*, 1980, **102**, 4665.
23. FALK, M., in ref. 39, Ch. 8.
24. PAINTER, P. C., BROZOSKI, B. A. and COLEMAN, M. M., *Macromol. Symp. 28th*, 1982, p. 53; *Chem. Abs.* **99** 105984.
25. BROZOSKI, B. A., COLEMAN, M. M. and PAINTER, P. C., *Macromolecules*, 1984 (D), **17**, 230.
26. NEPPEL, A., BUTLER, I. S. and EISENBERG, A., *Macromolecules*, 1979, **12**(5), 948–52.
27. RODMACQ, B. C., COEY, J. M. D. and PINERI, M., ref. 39, Ch. 9.
28. MEAGHER, A., RODMACQ, B., COEY, J. M. D. and PINERI, M., *Reactive Polymers*, 1984, **2**, 51.
29. KOMOROSKI, R. A. and MAURITZ, K. A., *J. Am. Chem. Soc.*, 1978, **100**, 7487.
30. DIEBLER, H. and EIGEN, M., *Zait. Phys. Chem.* (*Frankfurt*), 1959, **20**, 299.
31. EIGEN, M. and TAMM, K., *Electrochem.*, 1962, **66** (93), 107.
32. KOMOROSKI, R. A. and MAURITZ, K. M., in ref. 39, Ch. 7.
33. GIERKE, T. D., 'Ionic clustering in Nafion perfluorinated sulphonic acid membranes and its relationship to hydroxyl rejection and chlor-alkali efficiency', *Electrochem. Soc. 152nd Meeting, Oct. 1977.*
34. YEAGER, H. L., in ref. 39, Ch. 3.
35. GIERKE, T. D. and HSU, W. Y., in ref. 39, Ch. 13.
36. HSU, W. Y., BARKLEY, J. R. and MEAKIN, P., *Macromolecules*, 1980, **13**, 198.
37. HULL, N. F. and CONANT, J. B., *J. Am. Chem. Soc.*, 1927, **49**, 3047.
38. OLAH, G. A., PRAKASH, G. K. S. and SOMMER, J., *Science*, 1979, **206**, 4414.
39. EISENBERG, A. and YEAGER, H. L. (Eds), 'Perfluorinated ionomer membranes', *ACS Symposium Series*, Vol. 180, 1982, American Chemical Society, Washington.
40. PURCELL, R. W., *The Modern Inorganic Chemicals Industry*, (R. Thompson, Ed.), 1977, Chemical Society, London, p. 106.
41. YEAGER, H. L., KIPLING, B. and DOTSON, R. L., *J. Electrochem. Soc.*, 1980, **127**, 303.
42. DOTSON, R. L. and WOOD, K. E., in ref. 39, p. 311.
43. COVITCH, M. J., *Proc. Electrochem. Soc.*, 1983, **83**(3), 311 (Proceedings of Symposium on Membranes and Ionic and Conductive Polymers, Pittsburgh, PA, 1982).

44. JACKSON, C. and KELHAM, S. F., *Chem. Ind. (London)*, 1984, 397.
45. YEO, R. S. and McBREEN, J., *J. Electrochem. Soc.*, 1979, **126**, 1682.
46. YEAGER, H. L., O'DELL, B., TWARDOWSKI, Z. and CLARKE, L. M., *J. Electrochem. Soc.*, 1982, **129**, 85.
47. YEAGER, H. L., TWARDOWSKI, Z. and CLARKE, L. M., *J. Electrochem. Soc.*, 1982, **129**, 328.
48. CAIRNS, E. J. and WITHERSPOON, R. R., *Kirk–Othmer Encyclopaedia of Chemical Technology*, 3rd edn, Vol. 3, 1978, Wiley–Interscience, New York, p. 545.
49. SPILLMAN, R. W., SPOTNITZ, R. M. and LUNDQUIST, J., *J. Chem. Tech.*, 1984, **14**(3), 176.
50. KESTING, R., *Synthetic Polymeric Membranes*, 1971, McGraw–Hill, New York.
51. STECK, A. and YEAGER, H. L., *Anal. Chem.*, 1980, **52**, 1215.
52. MARTIN, C. R. and FRIESER, H., *Anal. Chem.* 1981, **53**, 903.
53. OESCH, U. and SIMON, W., *Anal. Chem.*, 1980, **52**, 692.
54. YEAGER, H. L. and KIPLING, B., *J. Phys. Chem.*, 1979, **83**, 1836.
55. BAILEY, P. L., *Analysis with Ion-selective Electrodes*, 1979, Heyden, London.
56. JOHNSON, J. S., MINTURN, R. E. and WADIA, P. H., *J. Electroanal. Chem. Interfacial Electrochem.*, 1972, **37**, 267.
57. HARAYA, K., NAKANE, T. and YOSHITOME, H., *Nippon Kaisui Gakkaishi*, 1978, **32**(4), 197; *Chem. Abstr.*, **90**, 175184.
58. UNITIKA Ltd, Jpn. Kokai Tokkyo Koho, 82 156 003, 27 Sept. 1982; *Chem. Abstr.*, **98**, 73576.
59. HEGAZY, E. S. A., ISKIGAKI, I., RABIE, A., DESSOUKI, A. M. and OKAMOTO, J., *J. Appl. Polym. Sci.*, 1983, **28**(4), 1465.
60. MUIR, W. M., GRAY, R. A., COURTNEY, J. M. and RITCHIE, P. D., *J. Bio. Med. Mater. Res.*, 1973, **7**(1), 3.
61. HABERT, A. C., HUANG, R. Y. M. and BURNS, C. M., *J. Appl. Polym. Sci.*, 1979, **24**, 489.
62. HABERT, A. C., HUANG, R. Y. M. and BURNS, C. M., *J. Appl. Polym. Sci.*, 1979, **24**, 801.
63. RUDMAN, A. R., VENGEROVA, N. A., KALYUZHNAYA, R. I., EL'TSEFON, B. S. and ZEZIN, A. B., *Khim.–Farm. Zh.*, 1979, **13**(3), 82; *Chem. Abstr.*, **90**, 192514.
64. SACHS, S. B. and LONSDALE, H. K., *Proc. Int. Symp. Fresh Water Sea, 3rd*, 1970, **2**, 561; *Chem. Abstr.*, **74**, 79410.
65. MUIR, W. M., COURTNEY, J. M. and GRAY, R. A., in *Developments in Biomedical Engineering* (M. M. Black, Ed.), 1972, Crane, Russak, New York, pp. 92–103.
66. HUANG, R. Y. M., GAO, C. J. and KIM, J. J., *J. Appl. Polym. Sci.*, 1983, **28**, 3063.
67. LYSAGHT, M. J., in *Ionic Polymers*, (L. Holliday, Ed.), 1975, Applied Science Publishers, London.
68. KISER, E. J. and LATTY, J. A., Ger. Offen. 2 816 088, 2 Nov. 1978; *Chem. Abstr.*, **90**, 122675.
69. SATA, T. and IZUO, R., *Colloid and Polymer Sci.*, 1978, **256**, 757.

70. SATA, T., YAMANE, R. and MIZUTANI, Y., *J. Polym. Sci., Poly Chem. Ed.*, 1979, **17**, 2071.
71. RAFIKOV, S. R., CHELNOKOVA, G. N., GOL'DING, I. R. and TOROTSEUA, T. N., *Izobret. Prom. Obraztsy. Tovarnye Znaki*, 1967, **44**(9), 119; *Chem. Abstr.*, **68**, 69731.
72. SCHROEDER, J. P. and SOPCHAK, W. P., *J. Polym. Sci.*, 1960, **47**, 417.
73. LINDER, C. and MILLER, I. F., *J. Phys. Chem.*, 1972, **76**, 3434.
74. LINDER, C. and MILLER, I. F., *J. Electrochem. Soc.*, 1973, **120**, 498.
75. MILLER, I. F. and MAYORAL, J., *J. Phys. Chem.*, 1976, **80**, 1387.
76. BORNZIN, G. A. and MILLER, I. M., *J. Electrochem. Soc.*, 1978, **125**, 409.
77. SHASHOUA, V., *Nature (London)*, 1967, **215**, 846.
78. HUANG, L. M. and SPANGLER, R. A., *J. Membrane Biol.*, 1977, **36**, 311.
79. HODGKIN, A. L. and HUXLEY, A. F., *J. Physiol.*, 1952, **117**, 500.
80. HODGKIN, A. L., *Biol. Rev.*, 1951, **26**(4), 339.
81. WOBSCHALL, D., *J. Theoret. Biol.*, 1968, **21**, 439.
82. KATACHALSKY, A., 'Membrane thermodynamics', in *The Neuroscience: A Study Program*, 1967 Rockefeller University Press, New York, as quoted in ref. 78.
83. KATACHALSKY, A. and SPANGLER, R. A., *Q. Rev. Biophys.*, 1968, **1**, 127, as quoted in ref. 78.
84. RATNER, B. D. and MILLER, I. F., *J. Polym. Sci., Part A-1*, 1972, **10**, 2425.
85. LINDER, C. and MILLER, I. F., *J. Polym. Sci. Part A-1*, 1973, **11**, 1119.
86. BONFILS, C., NATO, F., BOURRILLON, R. and BALNY, C., *FEBS Lett.* 1981, **123**(2), 222.

Chapter 7

APPLICATION OF IONIC POLYMERS IN MEDICINE

F. G. HUTCHINSON

Imperial Chemical Industries PLC, Pharmaceuticals Division, Macclesfield, UK

1 INTRODUCTION

The major constituent of soft mammalian tissue is water, and almost all organic matter in living cells consists of macromolecules of many different types such as proteins and enzymes, nucleic acids and polysaccharides. During the course of evolution, these high molar mass components in living organisms evolved from primordial biomolecules to highly specific macromolecules, each of which is designed to serve a particular function. Thus, lipid–protein membranes evolved for compartmentalisation, polynucleic acids for memory and replication, polypeptides and proteins as enzymes and hormones, and polysaccharides for energy storage.

All of these complex molecules are necessary for the integrity and effective functioning of living cells. In the physiological environment, changes induced in cells or tissue by some external influence on the modes of behaviour of a particular macromolecular component of a cell will in turn often modify the behaviour of other different macromolecular assemblages in the organism. For example, the process of endocytosis, by which pharmacologically active agents can be taken up by cells, occurs as a series of events. Initially, the drug interacts or binds to a proteinaceous receptor (a 'lock–key' effect) at the cell surface. The formation of this ligand leads to modification of the lipid–protein membrane such that receptor–agent complexes tend to cluster, and ultimately leads to internalisation of the agent within

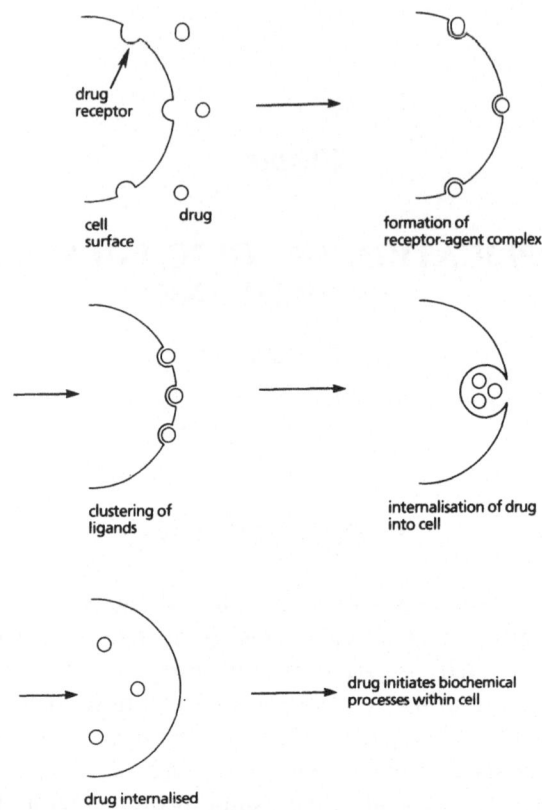

FIG. 1. Cellular uptake of drugs–endocytosis.

the cell (Fig. 1). Once the agent has entered the cell it initiates the biochemical processes leading to a demonstrable and characteristic pharmacological effect.

A particular objective of biochemists has been to elucidate the structure and behaviour of these various biological macromolecules in living tissue and how cell function is modulated by external agencies. This understanding has led to important advances in biochemistry and biology and is exemplified particularly well by the use of microbiological and genetic engineering techniques leading to the synthesis of biological macromolecules such as monoclonal antibodies and pharmacologically active, high molar mass polypeptide hormones. In the broadest sense these can be considered as polyelectrolytes as they are based on α-amino acids which include not only neutral types but both acidic (L-aspartic and L-glutamic acids) and basic (L-lysine and L-arginine) ones.

Concurrently with these notable advances in biochemistry and biology polymer scientists have become increasingly involved in the development of separation, purification and analytical techniques to identify the macromolecular components of living tissue. This merging of bioscience and polymer science has inevitably provided the stimulus for the chemical synthesis of new macromolecular entities which mimic physiologically active natural materials and which are able to modify cell behaviour or viability.

Polyelectrolytes are a particular class of water-soluble polymers whose applications in medicine, pharmacy and biology have been the subject of detailed investigation over the last two decades. These studies have encompassed macromolecules which possess intrinsic biological activity due to a direct interaction of the high molar mass species, or segments of it, with living cells or tissue[1-4] and polymers which are pharmacologically inert and which are used pharmaceutically in drug formulations to give the most practical and effective dosage forms.[5-7]

2 POLYELECTROLYTES HAVING BIOLOGICAL ACTIVITY

A distinction can be made between polyelectrolytes that are intrinsically biologically active because their inherent structure results directly in some interaction at the cellular level and those which, although pharmacologically inert, are rendered active by chemical binding or linking of a drug or hormone to the inert polymer.

2.1 Interaction of Polyelectrolytes with Plasma Membranes

The binding of macromolecules to cell membranes occurs by both specific and non-specific interactions. Highly specific interactions resemble those between receptor and hormone and may indeed be of a 'lock and key' type (Fig. 1). Such interactions occur with polypeptide hormones or with polyelectrolytes containing antibody, drug or hormone ligands. This type of interaction can be used for targeting of drugs to specific organs or cells and is discussed later (Section 4.3.3).

Non-specific interactions can arise between electric charges or chemical groups which are more or less uniformly distributed on the polyelectrolyte and on the cell membrane. With polycations or polyanions, these non-specific interactions are governed by Coulombic forces which are so strong that other non-specific types, which are hydrophobic or hydrophilic in character, can be disregarded.

The strength of the interaction of a polyelectrolyte with the cell membrane increases with the number of interacting groups within the polymer chain and so binding generally increases with increasing molar mass.[8,9] As the greatest biological response is elicited by the strongest binding, it is implied that the most appropriate polymers are those of very high molar mass. However, the use of such polymers is contra-indicated, in practice, as the kidney threshold for excretion of water-soluble polymers is 50 000 or so and also by their toxicity.[10,11] Consequently it has been found necessary to use polyelectrolytes of low molar mass (<50 000) and which additionally have been well fractionated to give a narrow molar mass distribution. Fractionation is often necessary as the presence of very large molecules can lead to retention of these molecules in the body and even to adverse effects such as erythrocyte aggregation.

2.1.1 Polycations

Cell surfaces have an overall negative charge which leads to strong Coulombic interactions between cells and polycations[12,13] (Fig. 2(a)) and cells can even acquire a net positive charge (Fig. 2(b)). The adsorption of polycations onto the plasma membrane is a function of both molar mass and the charge density on the polyelectrolyte. Thus, in those cases where it is possible for cells to acquire a net positive charge following interaction, aggregation of cells can result (Fig. 2(c)). The effect of charge density has been demonstrated in the haemo-agglutinating activity of cationic derivatives of polysaccharides[14] and agglutination of other cells.[13]

The interaction of a polycation with the electronegatively charged membrane can perturb the cell surface, thereby modifying the permeability. Thus in the presence of polycations the permeability of the membrane is modified and entry of anions into the cell is facilitated.[15] Conversely, disruption of the cell membrane can allow intracellular fluids or contents to escape into the extracellular medium, so affecting the viability of the cell and in the extreme condition even leading to its death.

In view of these various effects it is not surprising that a number of polycations show potent antimicrobial and antifungal activity. Examples of these antimicrobial polymers are the polybiguanides[16]

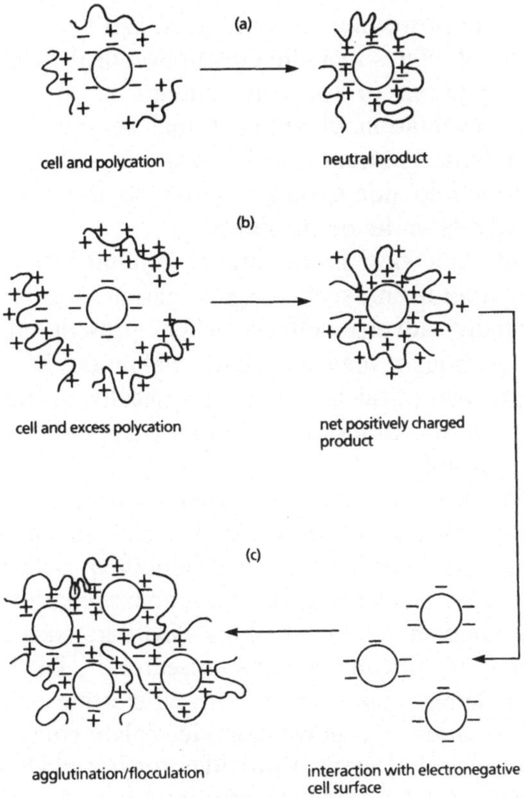

FIG. 2. Interaction of polycations with plasma membranes.

having the structure

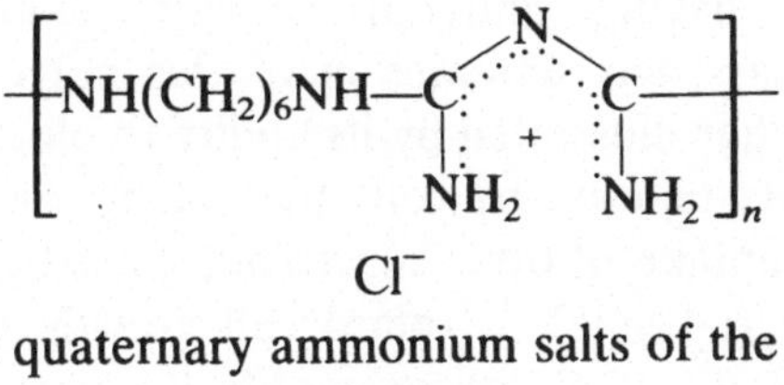

$$\left[NH(CH_2)_6 NH-\underset{NH_2}{\overset{N}{C}}\cdots\underset{NH_2}{\overset{+}{C}} \right]_n$$

Cl⁻

and polymeric quaternary ammonium salts of the type[17]

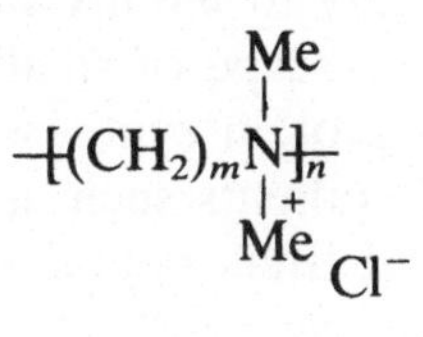

$$\left[(CH_2)_m \underset{Me}{\overset{Me}{\underset{|}{N}}} \right]_n$$

Cl⁻

These materials owe their antimicrobial and antifungal properties to a combination of effects including disruption of the cell membrane and possibly precipitation of the cell contents by the polycations. The synthesis and evaluation of antimicrobial polymers[2] and polycations incorporating benzalkonium salts[18] is possible and the activity of these is again ascribed to non-specific binding to the cell membrane and consequent effects on its permeability.

In view of their potent binding to cell membranes and adverse effects arising therefrom, such as agglutination of cells, polycations can elicit potentially adverse effects when present in the systemtic circulation. Although their use and evaluation has generally been limited to extra-corporal use, e.g. as topically administered antibacterial agents, there is other evidence that polycations may offer a more general medical utility.

Antitumourogenic effects have been observed with a number of polycationic polymers including those based on diethylaminostyrene and amide/amine copolymers[19] and modified polyethyleneimines.[20] Polyvinylpyridines likewise have antitumourogenic effects[19] and have also been studied in relation to lung disorders such as experimental silicosis[21] and in pneumoconiosis research.[22] Partially N-oxidised polyallyldiethylamine antagonises the anticlotting activity of heparin due to the formation of a polycation electrolyte complex.[23]

Evaluation of poly(L-lysine) provides further evidence of the complexity of the cellular interaction of polycations. In the last decade or so it has been shown that this poly(α-amino acid) interacts with various cell types in very different but specific ways. Thus, it has been demonstrated that poly(L-lysine) alone has unusual antineoplastic activity,[10] and an affinity for malignant tissue,[24] and it is reported to have some activity against murine tumours.[25] Indeed, poly(L-lysine) was recognised as having antiviral activity[26] and antibacterial activity[27] over 30 years ago. Interaction of this polycation with the cell membrane is further illustrated by its ability to block development of bacteriophage,[28] its take-up by cells by endocytosis as well as its enhancement of the uptake of other macromolecules by cells.[29]

The ability of poly(L-lysine) to retard cell growth and kill cells is probably associated with its ability to modify the permeability of the plasma membrane, allowing the escape of small molecules which are required for DNA, RNA and protein synthesis. Thus poly(L-lysine) results in the discharge of cell contents such as potassium, inorganic phosphate, free amino acids and small peptides[30] and, depending on the level of depletion, cell death may result.

These properties are not unique to poly(L-lysine), as related polycations can induce similar effects. For example, poly((L-ornithine) promotes loss of potassium from certain mouse cells and can retard cell growth,[31] and poly(L-arginine) can activate membrane phospholipase. This latter observation suggests that this polycation modifies membrane permeability by promoting hydrolysis of membrane phospholipids.

2.1.2 Polyanions

In comparison with polycations, polyanions are adsorbed onto cell surfaces to a far lesser degree and this low degree of adsorption is readily saturated. Whereas 1 pg of diethylaminoethyldextran is bound to one human fibroblast in serum-free medium,[32,33] only about 0·1 pg of dextran sulphate is adsorbed per cell.[34] Dextran sulphate and other polyanions increase the electronegative charge when adsorbed on the cell membrane and it is probably this increase, leading to enhanced Coulombic repulsion, that gives decreased adhesion between cells in the presence of polyanions.[35,36] Polyanions are, however, capable of modifying the permeability of cell membranes by creating holes which may then be resealed by proteins present in the plasma or in ascitic fluid,[34] but the presence of calcium ions is necessary for restoration of the intact membrane to occur.[37] Although polyanions disrupt the cell membrane, this disruption is not accompanied by an increased uptake of proteins into cells.[38,39] In view of this reversibility of effect and the non-agglutination of cells by polyanions, possible biological and medical applications of these electronegatively charged macromolecules have been pursued much more vigorously than has been the case with polycations.[3,4]

Knowledge of the effect of polyelectrolytes on cell division and proliferation is essential to the understanding of the pharmacology of these polymers. Following very early observations that in vitro exposure of tumour cells to heparin prior to implantation in the chorioallantoic membrane of the chick embryo resulted in tumour growth, many polyanions have been evaluated as cell-growth regulators or inhibitors. The antimitotic behaviour of these high molar mass species is complex and is related to the multifarious effects they exercise on the endogenous polysulphates, polyphosphates and polycarboxylic polyanions which are present in living organisms and which control, or are involved in, such biological processes as enzyme activity and cell division. The growth-regulating properties of many polymers including dextran sulphate, heparin, sulphated chitosan and cellulose esters, and polyethylene sulphonate have been reviewed[1-4,40-42] and discussed with respect to their place in the intracellular environment and their role in cell physiology.

The biological and polymer-related activity of polyanions are summarised in Table 1.

Polyanions have been shown to possess a variety of biological activities and are particularly active against bacteria, fungi, viruses and both virally induced and non-viral related malignant diseases. They also elicit a number of pharmacological effects in the reticuloendothelial system. Being of high molar mass, they are not absorbed from the alimentary tract and so are ineffective when administered orally. However, as they are, in general, water-soluble they may be given by injection (intravenously or intraperitoneally), and following injection they are distributed throughout the body by the systemic circulation and cellular and lymphatic transport. After administration, and depending on molar mass (particularly when >50 000) these

TABLE 1
BIOLOGICAL ACTIVITY OF POLYANIONS

Activity	Polymer
Antimitotic	Many.[40,41]
Antitumour	Poly(methacrylic acid);[43,44] acrylic acid–maleic anhydride,[45] ethylene–maleic acid,[44] and divinyl ether–maleic anhydride[45,48-58] copolymers, poly(ethylene sulphonate);[46,47] poly(xenyl phosphate).[59,60]
Antiviral	Poly(methacrylic acid);[61-68] acrylic acid–maleic anhydride copolymer;[69] poly(acetalcarboxylic acid);[70] maleic anhydride copolymers;[63] divinyl ether–maleic anhydride copolymer;[64,71-75] poly(vinyl sulphate).[76]
Antimicrobial	Various;[2] poly(acrylic acid);[77] acrylic acid–ethylene[18] and divinyl ether–maleic anhydride[77-79] copolymers.
Immunological	Divinyl ether–maleic anhydride copolymer.[55,80-88]
Anticoagulant	Heparin; dextran sulphate; sulphated polysaccharides;[3,4] poly(ethylene sulphonate);[1,89-92] poly(sulphoxyethyl methacrylate);[93] propylene copolymers;[89,92] styrene–isoprene[94] and divinyl ether–maleic anhydride copolymers.[95,96]

macromolecules can accumulate in the reticuloendothelial system (lungs, liver and spleen).

Polyanions exhibit biological behaviour that is similar to glycoproteins and nucleotides, which are involved in the modulation of pharmacological activities, particularly with respect to their antifungal/antimicrobial behaviour and immunomodulation. These diverse effects and the activities mentioned above have been discussed by Regelson,[40,41] who related the polyanion function to their role in cell physiology, with particular reference to such features as embryogenesis, growth control, cell division and proliferation, and the effects of polyanions on calcium (intracellular and extracellular), plasma membranes, colloidal effects and inhibition/activation of enzymes. Regelson has also shown that, depending on the molar mass of the polyanion, depression or stimulation of activity in the reticuloendothelial system can occur.[100]

Perhaps, the most important biological activities that polyanions possess is their mitotic inhibitory effects and their potential utility as antineoplastic agents. Cells in connective tissue in the intracellular matrix are bound together by natural anionic macromolecules and these provide the internal cohesion necessary for cell or tissue viability.[101] Adsorption of a synthetic polyanion onto a cell can disrupt this natural cohesion in a variety of ways. The first polyanion to be studied in detail was the naturally occurring sulphonated polysaccharide, heparin, because of its anticoagulant effects but it was also shown to impede the growth of tumours by interfering with calcium binding.[102] This antimitotic activity of heparin (and related sulphomucopolysaccharides) implies that polyanions exerted some control over the physiological growth controlling function via a number of effects, and subsequent studies have shown that polyanions are potent interferon inducers,[103] are enzyme inhibitors,[104] exert diuretic effects,[40,41] block adrenocortical function,[40,41] affect the isoelectric points of proteins,[40,41] affect blood viscosity and oxygen diffusion[105] and affect intracellular calcium and cyclic nucleotide function.[106]

Although polyanions possess antiviral, antitumour, anticoagulant and immunomodulation properties and their potential medical utility has been extensively demonstrated, their use is severely constrained by the adverse toxicological effects they induce. These can include anaemia, induction of leukocytosis and sensitisation to biological endotoxins[107] and interference with drug metabolism by the liver; in human patients, they have caused pyrexia, thrombocytopaenia and temporary blindness.[3] However, preparation of lower molar mass types with much narrower distributions can markedly reduce these toxic side effects without significantly altering the efficacy.

3 METALLOBIOPOLYMERS

Metallobiopolymers are a further class of polyelectrolytes whose utility in medical applications has been re-evaluated. The resurgence of interest in the use of polymers containing biologically active metal ions is based on the potential reduction of the toxicity of heavy-metal salts when these are complexed or associated with polymers. In some cases this advantage has been realised and it has been demonstrated that when combined with polymeric materials the metal ions retain their biological activity but toxicity is much reduced. Mycotic infections have been treated with acrylic–divinylbenzene copolymers containing complexed copper and cobalt salts.[97] Water-soluble non-ionic polymers such as dextran can form complexes with metal ions such as iron and antimony and the dextran–antimony complex has been evaluated in the treatment of parasitic infections.[98,99] Tin and organoarsenic polymers possess antifungal and antibacterial properties whilst hetero-polymolybdates are claimed to have anticoagulant and lipolytic properties.

4 PHARMACEUTICAL APPLICATIONS

The medical use of most of the polyelectrolytes which have intrinsic biological activity when present in the systemic circulation is compromised in many cases by the toxic side effects elicited by the water-soluble polymer. In contrast, pharmaceutical applications of polyelectrolytes are extensive, particularly with respect to oral dosage forms. The polymers of use are either inert or biologically inactive in the local physiological environment. Additionally, the high molar mass polyelectrolytes are not absorbed from the gastrointestinal tract and cannot enter into the systemic circulation, and so do not present systemic toxicity problems.

Polyelectrolytes are widely used as processing and formulation aids, as they offer the formulator the opportunity to design and manufacture the most effective and practical dosage form for drug delivery. Drug formulations may be solids, suspensions, creams or liquids and their use as drug delivery systems has to meet three important needs. The formulation has firstly to ensure that the drug arrives at the target tissue and secondly to be maintained there at an adequately high level for sufficient time for it to exercise optimal therapeutic effect. The third requirement is to fabricate a delivery system that meets these two needs. The most common routes of administration are oral and parenteral (intramuscular, intravenous and subcutaneous injections) but other routes of administration, such as topical, are becoming increasingly important. As the route of administration is often far removed from the site of drug action the dosage form has to release the drug into some aqueous physiological absorption pool from which it may be absorbed into the systemic circulation. This is summarised in Fig. 3, from which it can be seen that the blood concentration of drug and its persistence in the systemic circulation is in major part determined by its rate of transport out of the dosage form into the absorption pool.

Many water-soluble polymers, including polyelectrolytes, have been used in drug delivery systems of varying complexity and sophistication. Their uses are summarised in Table 2.

4.1 Polyelectrolytes as Formulation/Processing Aids
The oral route of administration is the most common, and the compressed tablet is the most commonly used dosage form as this offers both precision of dosage and convenience. In general, manufacture of compressed tablets involves the use of non-ionic polymers as binders or granulating agents in order to generate materials suitable for compression but polyelectrolytes such as alginic acid, gelatin and modified celluloses have been used.[108] From Fig. 3 it can be seen that

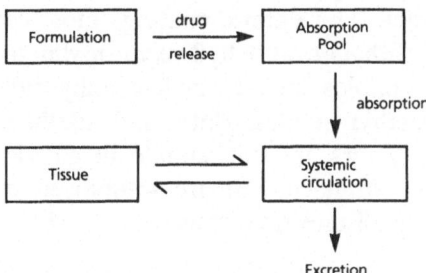

FIG. 3. Absorption of drug into systemic circulation following release from dosage form.

TABLE 2

PHARMACEUTICAL APPLICATIONS OF POLYELECTROLYTES

Formulation processing aids	Granulating and binding agents, tablet coatings, tablet disintegration.
Controlled/sustained release	Enteric coatings, to extend duration of release and give sustained pharmacological effect; microcapsules for sustained release or repeat-action dosage forms.
Dispersing/flocculating agents	To maintain stability of suspensions or to give easily dispersed sediments.
Drug carriers	To control administration and/or metabolism of a drug substance; drug targeting.

high drug concentrations in the aqueous physiological absorption pool in general give optimal bioavailability of drugs as the drug absorption rate is proportional to drug concentration in the absorption pool. Rapid disintegration of the tablet generates such high concentrations. Consequently, hydrophilic agents are often incorporated into tablets, which swell or expand in gastric fluids so overcoming the binding forces that have been generated during compression. Both synthetic and natural polymers have been used as disintegrants and these include alginates and cellulose derivatives.[108]

Tablet coating is an important formulation aid, as polymer films give only a small increase in tablet weight while giving a durable protective coating which is not only resistant to cracking but can also act as a protective barrier to light or air. Enteric coatings are particularly useful as the solubility of these (to give polyelectrolytes) is pH-dependent. Consequently they can be used for targeting release in the gastrointestinal tract. The enteric coatings most commonly used are cellulose acetate phthalate and hydroxypropylmethylcellulose phthalate, but other examples include maleic anhydride copolymers and copolymers of methyl methacrylate and methacrylic acid. These enteric coating polymers are not soluble in gastric fluids at pH < 5, such as that in the stomach, but are solubilised at pH > 6–7 in the upper part of the small intestine; thus they provide:

(i) protection of the stomach from drugs which give local irritation or induce nausea or distress;

(ii) protection of acid-sensitive drugs from gastric fluids;

(iii) targeted delivery of drug for local action, e.g. intestinal antiseptics;

(iv) sustained release formulations.

4.2 Polyelectrolytes as Dispersing/Flocculating Aids

Water-soluble polyelectrolytes are widely used in liquid pharmaceutical preparations as stabilisers (to give a stable dispersed phase) or to induce controlled flocculation. The effectiveness of a polymer as a suspending agent for solids is determined by the bulk viscosity of the polymer solution. High viscosities decrease the sedimentation rates of suspended particles and also inhibit the coalescence of particles in liquid dispersions. In addition to effects on sedimentation rate, adsorption of polyelectrolytes onto the surface of the dispersed phase can result in modification of surface properties, giving rise to further stabilisation due to entropic and enthalpic effects (steric stabilisation)[109] and Coulombic repulsion of surface-absorbed polyelectrolyte (Fig. 4(a)).

Polyelectrolytes, which can provide stabilisation when effectively coating the particle, can, conversely, promote flocculation when absorbed to give low surface coverage—particularly when the surface of the dispersed phase carries the opposite charge to that on the polyelectrolyte chain. This flocculation is believed to occur due to bridging of polymer between particles (Fig. 4(b)). In general, pharmaceutical suspensions involve particles $>1\,\mu m$ and so these will inevitably separate on standing to give a sedimented product. Consequently, the formulator is more concerned about decreasing sedimentation rate and ensuring that the caked product can be resuspended rather than obtaining a totally stable dispersion. Completely deflocculated systems are extremely difficult to resuspend as a dense compact cake is generated. However, loose dispersible sediments can be generated using polyelectrolytes at low surface coverage. The high molar mass polyelectrolyte reduces the co-ordination number of a particle with respect to its nearest neighbours because of both steric and charge repulsion effects to give a sediment that is easily redispersed.[110,111]

Polyanionic polymers which have been used as dispersing or flocculating agents include sodium carboxymethylcellulose, and sodium salts of polyvinyl and alginic acids. By controlling polymer structure, polyelectrolytes having defined rheological properties can be prepared. For example, a high molar mass polymer of acrylic acid

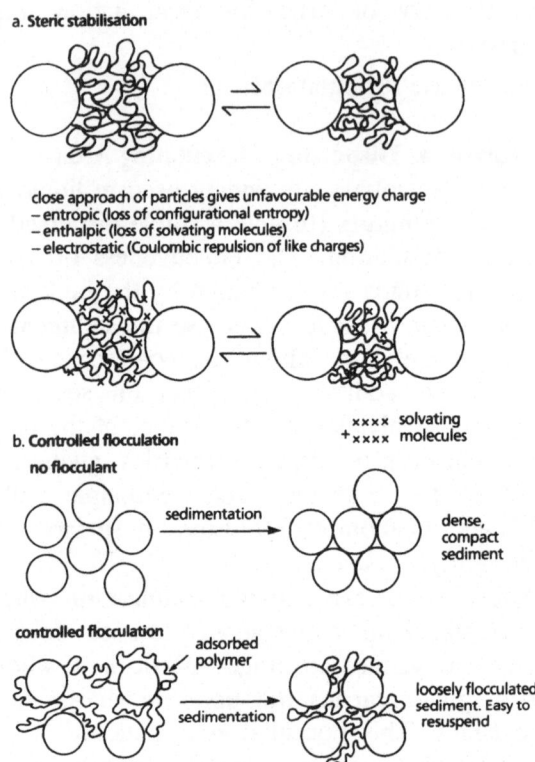

FIG. 4. Use of polyelectrolytes as dispersing/flocculating agents.

partially crosslinked with allyl sucrose gives a product for which viscosity is a function of pH.[112]

4.3 Polyelectrolytes for Use in Controlled Drug Delivery

The stimulus for controlled drug delivery technology, which encompasses both sustained and targeted drug delivery, is the need to provide safer and more effective means of delivery of drugs, and its rationale and scope is discussed at length elsewhere.[113,114] Classical dosage forms are often inefficient with respect to the supply of drug to the systemic circulation and in particular to the supply of drug at the target site of action. Sustained release dosage forms ensure that entry of drug into the systemic circulation occurs at an effective level for biological effect over a prolonged interval of time, thereby making more effective use of the payload in the delivery system. This offers some improvement on classical dosage forms but is still relatively non-selective as it offers only a changed temporal relationship of drug in the whole of the systemic circulation (Fig. 5). A totally different concept which presents the most ambitious objective in formulation science is to programme the delivery system to deliver drug at a specific site of action by incorporating into the dosage form a built-in recognition system for a specific cell type or tissue. These considerations have led to the use of polyelectrolytes not only as drug release materials *per se,* but as macromolecular carriers for drugs in drug targeting.

4.3.1 Polyelectrolytes for Controlled Drug Release

Water-soluble polyelectrolytes have utility as membranes or matrices in simple controlled or sustained delivery dosage forms and, in particular, enteric products such as cellulose acetate phthalate have been used to control the release of drug from tablets or microcapsules (see Section 4.1). Microcapsules containing drug can be prepared by a variety of methods including spray-drying, air suspensions and coacervation phase separation[115-117] and since the advent of the Spansule capsule[118] this technology has had an important role in modern pharmaceutical and formulation science.

4.3.2 Ionic Polymers as Drug Carriers for Sustained Release

Controlled drug release from sustained action dosage forms using polyelectrolytes as the carrier is based essentially on two principles.

(i) *Physical*: transport of drug from the dosage form is governed by physical characteristics. For instance, the drug or dosage form may be given an enteric coating (as discussed previously) or the

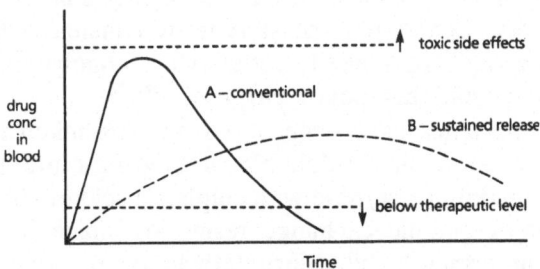

FIG. 5. Comparison of conventional and sustained release formulations.

drug may be complexed with the carrier, effectively lowering the solubility of the drug in physiological fluids.

(ii) *Chemical*: drug availability is controlled by ionic or covalent combination with polyelectrolytes. When the drug is covalently linked to the polymer it may occur as a pendant side group or as an in-chain comonomer unit. Both water-soluble and -insoluble (but swellable) polyelectrolytes have been used to achieve prolonged action dosage forms.[7]

Formulations using Ion-exchange Resins. Ion-exchange resins are water-insoluble but water-swellable crosslinked polyelectrolytes, and with such carriers only drugs having acidic or basic groups in their chemical structure can be used. Sustained release of the agent rests on the principle that insoluble drug–resinates are generated when positively or negatively charged agents combine with ion-exchange resins. Thus for a basic drug a cationic ion-exchange resin is required:

$$R{-}SO_3^- \, H^+ + D \rightleftharpoons R{-}SO_3^- \, DH^+$$

(where $R{-}SO_3^- \, H^+$ is a cationic ion-exchange resin and D is a basic drug) and for an acidic drug an anionic ion-exchange resin is needed:

$$R{-}NH_3^+ \, OH^- + DH \rightleftharpoons R{-}NH_3^+ \, D^- + H_2O$$

(where $R{-}NH_3^+ \, OH^-$ is an anionic ion-exchange resin and DH is an acidic drug).

This concept was utilised over 35 years ago to complex quinine,[119] and since then a variety of ion-exchange resins have been evaluated. The use of these resins for sustained drug release is limited to oral presentations and in general emphasis has focused on organic synthetic resins.[120] Carboxylic acid type exchangers are prepared by the polymerisation of organic acids such as acrylic or methacrylic acids in the presence of a dimethacrylate or divinylbenzene to yield crosslinked networks. Copolymers of styrene and maleic anhydride with divinylbenzene and crosslinked methacrylate copolymers have been advocated for specific therapeutic purposes.[121,122]

Most drug–resinates have been based on crosslinked polystyrenes. Cationic types containing sulphonic acid groups are prepared by reacting crosslinked polystyrene with sulphuric acid or chlorosulphonic acid. The major anionic exchange resins are made in a variety of ways,[123,124] but primarily by chloromethylation of polystyrene beads with subsequent treatment with ammonia or amines. Alternative ion-exchangers, such as those based on polysaccharides, have found only limited use in the application.[7]

Ion-exchange resins based on crosslinked polyelectrolytes have been shown to have utility for the preparation of sustained release formulations of a wide range of pharmaceutical agents[7] including sympathomimetics, cough suppressants, antihistamines, anticholinergics, anthelmintics and antibacterials. The release of these agents from the drug–resinate combination is a function of acid–base strengths of the resin and drug, particle size of the resin, its porosity and swellability.

Formulations based on soluble polyelectrolytes. Soluble polyelectrolytes, including poly(methacrylic acids), polysaccharides and poly(uronic acids), are frequently used in drug formulations as formulation aids and disintegrating agents. Many of these polyelectrolytes form ionic complexes to give drugs having a much diminished solubility in gastric fluids and in which the complex of drug and polymer swells or dissolves slowly. As drug is reversibly bound to the polyelectrolyte, these mixtures can release drug in a sustained fashion:

$$P—CO_2^- \, DH^+ + H_2O \rightleftharpoons P—CO_2^- \, H^+ + D$$

(in solution) where D is a basic drug and $P—CO_2^-$ is a soluble polyanion, or;

$$R—NH_3^+ \, D^- + H_2O \rightleftharpoons P—NH_3^+ \, OH + DH$$

(in solution) where DH is an acidic drug and $P—NH_3^+$ is a polycation.

The release from the polymeric salt following oral administration can be divided into various stages. In the absence of aqueous fluids, the polysalt is incapable of releasing the agent and, initially, gastric fluids have to penetrate the dosage form, so allowing free dissociated drug to be transported into the absorption pool (Fig. 3). Further ingress of fluids gives swelling of the polysalt to form a transient and slowly eroding gel barrier through which dissociated drug has to diffuse in order to be available for absorption. The gel barrier generated can have a profound effect on release of drug and is a function of molar mass and its capacity to undergo exchange with the penetrating ions present in the gastric fluids.[125] Eventually, dissolution of the matrix will give release of the total payload into the gastrointestinal fluids.

Using polyacids such as poly(acrylic acid), modified celluloses, dextran sulphate, alginic acid and polygalactouronates, dosage forms have been prepared which give sustained release of antibiotics, therapeutic amines and pilocarpine.[7]

4.3.3 Drugs Covalently Bound to Polyelectrolytes

Polyelectrolytes containing covalently bound drug can be used as carriers for two specific purposes:

(a) sustained action;
(b) targeted drug delivery.

Polymeric drugs for sustained action. When drugs are incorporated into the macromolecule, either as pendant side chains or as in-chain comonomer units, sustained biological activity can be effected in two distinct ways.

Firstly, polyelectrolytes are excreted, via the kidney, at a rate determined by the molar mass of the polymer. Consequently, covalent attachment by a non-degradable group to the polyelectrolyte can generate a pharmacologically active macromolecule, whose half-life in the systemic circulation is considerably longer than for the free, low molar mass drug. That is, biologically unstable drugs of short half-life can be converted to 'long-acting drugs'.

Secondly—and in contrast to the above case, where drug metabolism and clearance from the systemic circulation is controlled by the molar mass—the drug may be linked to the polymer through an enzymatically unstable or hydrolytically unstable 'spacer' group. Scission of this group generates the drug in a free and biologically active form where X—X is the 'spacer' group and D is drug.

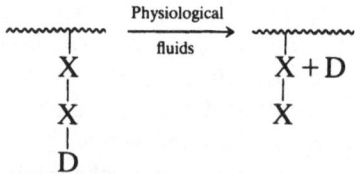

Many different sorts of polyelectrolyte have been used to achieve long acting preparations by these two approaches.

Polysaccharides such as peptic acid, alginic acid, cellulose and urogenic acids have been used as carriers for antigens, enzymes and hormones.[126,127] Other polyanionic carriers include poly(acrylic acid)[128] and poly(methacrylic acid)[129] and polycationic polymers include polyethyleneimine.[20]

Covalent combinations of drug and polyelectrolytes are detailed in a number of reviews.[130–134]

Polyelectrolytes as drug carriers in drug targeting. To be effective, a drug must reach the proper group of cells in the organism or tissue and bind to or associate with the appropriate receptors. Dosage forms in general do not possess precise selectivity for delivery of drugs at specific organs or tissue, and as early as 1906 Ehrlich[135] proposed the possibility of drug targeting. Despite its early appearance it is only latterly that this concept has been tested in depth, following recent advances in synthetic, biological and immunological techniques. The concepts and techniques whereby polyelectrolytes—including natural ionic polymers such as polypeptides, glycoproteins and polysaccharides, and synthetic polymers such as poly(L-lysine)—can be used in targetable delivery systems have been reviewed.[136,137] Here, biologically reactive macromolecules and polymers containing antibody and/or hormone or drug ligands are described. The incorporation of these ligands into the macromolecular structure confers on the macromolecular complex the ability to interact with specific cells or tissue. If the complex also contains a drug, e.g. an antineoplastic agent, then it is possible to direct it specifically to the cells or tissue requiring treatment. The take-up of drug into the cell may occur by endocytosis or pinocytosis of the total complex, following which the drug exerts its biological effect. Alternatively, the macromolecular complex may be directed to, and bound at, a specific cell surface and the active agent is taken up by the cell following scission (for example enzymatic degradation) of a 'spacer' group linking it to the carrier complex. Targeting of drugs has in general concentrated on the use of antineoplastic agents and toxins in anticancer treatments but its more general applicability to other disease states is an exciting prospect.

In view of the complexity and ramifications—synthetic, chemical, biological and immunological—of events giving a specific cell recognition property and post-cellular interaction events—endocytosis, pinocytosis and biological effects elicited—the reader should consult refs 136 and 137, and the references therein.

5 BIOMEDICAL/PROSTHETIC USE OF POLYELECTROLYTES

In the field of biomaterials, considerable interest has focused on the utilisation of the bulk or surface properties of hydrogels for biomedical applications for replacement or augmentation of soft tissue. The physical characteristics of hydrogels more than any other synthetic material, resemble those of living tissue. These hydrogels comprise three-dimensional crosslinked networks which swell but do not dissolve in water; the crosslinking may be covalent, ionic or a combination of the two.

Anionic, cationic hydrogels and anionic–cationic polyelectrolyte complexes are based on copolymers of acrylic and methacrylic acids, dimethylaminoethyl methacrylate and cationic and anionic derivatives of polystyrene.[138,139] In comparison with non-polyelectrolyte hydrogels, e.g. crosslinked poly(hydroxyethyl methacrylate), polyelectrolyte forms have the major disadvantage that their degree of hydration is markedly dependent on the pH, ionic strength and temperature of the hydrating physiological medium. Nonetheless, they have found utility in some biomedical applications,[138,139] particularly where such changes in hydration do not prejudice their use. For example, chitosan (a basic cellulose); in association with collagen, has been used as a wound dressing.[140]

Polyelectrolytes have also been used to replace or augment hard tissue such as bone. Thus, polyacrylic acids have been used in the development of dental cements for the permanent restoration of teeth. The poly(acrylic acid) is crosslinked *in situ* using certain types of ion-leachable glasses. Mixing of these glasses with a solution of the poly(acrylic acid) gives a hard insoluble gel as polyvalent cations are leached from the glass and these then crosslink the polyanionic chains.[141]

6 SUMMARY

The purpose of this review has been to give to researchers a comprehensive, but not necessarily detailed, overview of current knowledge, developments and trends regarding the utility of polyelectrolytes in biology, pharmacy and medicine. The use of synthetic polymers[142] encompasses a very diverse range of therapeutic, pharmaceutical and biomedical effects, or even elements of all three. Polymer and material science and the various biosciences (biology and biochemistry) are the provinces of specialists. Thus, although polyelectrolytes (and other polymers) have enormous potential in these rapidly expanding fields, it can only be realised by a multidisciplinary and cooperative approach involving, on one hand, the mechanisms of interaction between biopolymers and living systems and, on the other, exhaustive evaluation of the bulk, solution and configurational properties of the polymers themselves.

REFERENCES

1. REGELSON, W., *J. Polym. Sci. Polym. Symp.* (1979), No. 66, 483.
2. DONARUMA, L. G., EDSWALD, J. K., KITOH, S. and WARNER, R. J. in *Polymeric Antimicrobial Drugs, Biomedical Polymers: Polymeric Materials and Pharmaceuticals for Biomedical Use*, E. P. Goldberg and A. Nakajima (Eds.), 1980, Academic Press, New York, p. 401.
3. DONARUMA, L. G. and VOGL, O. (Eds), *Polymeric Drugs*, 1978, Academic Press, New York.
4. DONARUMA, L. G., OTTENBRITE, R. M. and VOGL, O. (Eds), *Anionic Polymeric Drugs*, 1980, Wiley-Interscience, New York.
5. HUTCHINSON, F. G., *Medical and Pharmaceutical Applications of Water-soluble Polymers in Chemistry and Technology of Water-soluble Polymers* C. A. Finch (Ed.), 1983, Plenum Press, New York, p. 267.
6. LACHMAN, L., LIEBERMAN, H. A. and KANIG, H. L. (Eds), *The Theory and Practice of Industrial Pharmacy*, 1976, Lea and Febiger, Philadelphia.
7. SCHACHT, E. H., 'Ionic polymers as drug carriers' in *Controlled Drug Delivery*, S. D. Bruck (Ed.), Vol. 1, *Basic Concepts*, 1983, CRC Press, Boca Raton, Florida, Ch. 6.
8. BARFOD, N. M. and LARSEN, B., *Biochim. Biophys. Acta*, (1976), **427**, 197.
9. BARFOD, N. M. and LARSEN, B., *Eur. J. Cancer*, (1974), **10**, 765.
10. ARNOLD, L. J., DAGAN, A., GUTHEIL, J. and KAPLAN, N. O., *Proc. Natl. Acad. Sci. USA*, (1979), **76**, 3246.
11. BRESLOW, D. S., EDWARDS, E. I. and NEWBURG, N. R., *Nature* (1973), **246**, 160.
12. WALLACH, D. F. H., *The Plasma Membrane*, 1973, Springer–Verlag, New York.
13. KATCHALSKY, A., *Biophys. J.* (1964), **4**, 9.
14. WOOD, J. W. and MORAN, P. T., *J. Polym. Sci.* (1963), Part A, **1(11)**, 3511.
15. SHEN, W.-C. and RYSER, H. J.-P., *Mol. Pharmacol.* (1979), **16**, 614.
16. *Martindale—'The extra pharmacopoeia'*, 28th edn, (1982). The Pharmaceutical Press, London, 1745.
17. REMBAUM, A., US Patent 3655814 (1972).
18. ACKART, W. B., CAMP, R. L., WHEELWRIGHT, W. L. and BYCK, J. S., *J. Biomed. Mat. Res.* (1975), **9**, 55.
19. FERRUTI, P., DANUSSO, F., FRANCHI, G., POLENTARUTTI, N. and GARRATTINI, S., *J. Med. Chem.* (1973), **16**, 496.

20. PRZYBYLSKI, M., FELL, E., RINGSDORF, H. and ZAHARKO, D. S., *Makromol. Chem.* (1978), **179,** 1719.
21. SCHLIPKOTER, H. W. and BROCKHAUS, A., *Germ. Med. Monthly* (1980), **5,** 270.
22. HOLT, P. F., *Brit. J. Industr. Med.* (1971), **28,** 72.
23. MARCHISIO, M. A., SBERTOLI, C., FARINA, G. and FERRUTTI, P., *Eur. J. Pharmacol.* (1970), **12,** 236.
24. ANGHILERI, L. J., HEIDBREDER, M. and MATHES, R., *J. Nucl. Biol. Med.* (1976), **20,** 79.
25. RICHARDSON, T., HODGETT, J., LINDNER, A. and STAHMANN, M. A., *Proc. Soc. Exp. Biol. Med.* (1959), **101,** 382.
26. STAHMANN, M. A., GRAF, L. H., PATTERSON, E. L., WALKER, J. C. and WATSON, D. W., *J. Biol. Chem.* (1951), **189,** 45.
27. BICHOWSKI-SLOMNICJI, L., BERGER, A., KURTZ, J. and KATCHALSKI, E., *Arch. Biochem. Biophys.* (1956), **65,** 400.
28. SHAHTIN, C. and KATCHALSKI, E., *Arch. Biochem. Biophys.* (1962), **99,** 508.
29. RYSER, H. J.-P., 'Poly(amino acids) as enhancers in the cellular uptake of macromolecules' in *Rehovot Symposium of Peptides,* E. R. Blout, F. A. Bovey, M. Goodman and N. Lotan (Eds), 1974, Wiley, New York, p. 617.
30. KORNGUTH, S. E. and STAHMANN, M. A., *Cancer Res.* (1961), **21,** 907.
31. OGAWA, K. and ICHIHARA, A., *J. Biochem.* (1978), **83,** 519.
32. PITHA, P. M., HARPER, H. D. and PITHA, J., *Virology* (1974), **59,** 40.
33. PRESS, G. D. and PITHA, J., *Mech. Ageing Dev.* (1974), **3,** 323.
34. McCOY, G. D., RESCH, R. C. and RACHER, E., *Cancer Res.* (1976), **36,** 3339.
35. BREMERSKOV, V., *Nature* (1973), **246,** 174.
36. TAKEMOTO, K. K. and SPICER, S. S., *Ann. NY Acad. Sci.* (1965), **130,** 365.
37. KASAHARA, M., *Arch. Biochem. Biophys.* (1977), **184,** 400.
38. RYSER, H. J.-P., *Science* (1968), **159,** 390.
39. RYSER, H. J.-P., TERMINI, T. E. and BARNES, P. R., *J. Cell. Physiol.* (1975), **87,** 221.
40. REGELSON, W., *Adv. Chemother.* (1968), **3,** 303.
41. REGELSON, W., *Adv. Cancer Res.* (1968), **1,** 223.
42. REGELSON, W., *Polym. Sci. Technol.* (1973), **2,** 161.
43. ILIEV, I., GEORGIEVA, M. and KABIANOV, V., *Russ. Chem. Rev.* (1974), **43,** 69.
44. REGELSON, W., KUHAR, M., TUNIS, M., FIELDS, J. E., JOHNSON, J. H. and GLUESENKAMP, E. W., *Nature* (1960), **186,** (4727), 778.
45. OTTENBRITE, R., GOODELL, E. M. and MUNSON, A. E., *Polymer* (1977), **18,** 461.
46. REGELSON, W. and HOLLAND, J. F., *Nature* (1965), **181** (4 Jan.), 56.
47. REGELSON, W. and HOLLAND, J. F., *Clin. Pharmacol. Ther.* (1962), **3,** 730.
48. KAPILA, K., SMITH, C. and RUBIN, A. A., *J. Reticuloendothelial Soc.* (1971), **9,** 447.

49. KAPLAN, A. M., MORAHAN, P. S. and REGELSON, W., *J. Natl. Cancer Inst.* (1974), **53**, 1409.
50. MORAHAN, P. S., MUNSON, J. A., BAIRD, L. G., KAPLAN, A. M. and REGELSON, W., *Cancer Res.* (1974), **34**, 506.
51. KAPLAN, A. M., *Fogarty International Center Proc.* (1975), 28.
52. MOHR, S. J., CHIRIGOS, M. A., FUHRMAN, F. S. and PRYOR, J. W., *Cancer Res.* (1975), **35**, 3750.
53. SNODGRASS, M. J., MORAHAN, P. S. and KAPLAN, A. M., *J. Natl. Cancer Inst.* (1975), **55**, 455.
54. KAPLAN, A. M. and MORAHAN, P. S., *Ann. NY Acad. Sci.* (1976), **276**, 134.
55. MORAHAN, P. S., SCHULLER, G. B., SNODGRASS, M. J. and KAPLAN, A. M., *J. Infectious Diseases* (1976), **133**, Suppl., A249.
56. MORAHAN, P. S. and KAPLAN, A. M., *Int. J. Cancer* (1976), **17**, 82.
57. HIRANO, T., KLESSE, W. and RINGSDORF, H., *Makromol. Chem.* (1980), **180**, 1125.
58. MORAHAN, P. S., EDELSON, P. J. and GASS, K., *J. Immunol.* (1980), **125**, 1312.
59. MUEHLBAECHER, C., STRAUMFJORD, J. V., HUMMEL, J. P. and REGELSON, W., *Cancer Res.* (1959), **19**, 907.
60. STRAUMFJORD, J. V. and HUMMEL, P. M., *Cancer Res.* (1959), **19**, 913.
61. DeCLERCQ, E. and DeSOMER, P., *Appl. Microbiol.* (1968), **16**, 1314.
62. DeCLERCQ, E., *J. Virol.* (1968), **2**, 878, 886.
63. MERIGAN, T. C. and FINKELSTEIN, M. S., *Virology* (1968), **35**, 363.
64. DeCLERCQ, E. and DeSOMER, P., *Proc. Soc. Exp. Biol. Med.* (1969), **132**, 669.
65. NIBLACK, J. F., *Ann. NY Acad. Sci.* (1970), **173**, 536.
66. BILLIAU, A., MUYEMBE, J. J. and DeSOMER, P., *Infect. Immun.* (1972), **5**, 854.
67. LEVY, H. B., ref. 3, p. 305.
68. CHIRIGOS, M. A., TURNER, W., PEARSON, J. and GRIFFIN, W., *Int. J. Cancer* (1969), **4**, 267.
69. OTTENBRITE, R. M., GOODELL, E. M. and MUNSON, A. E., *ACS Polymer Preprints* (1977), **18**(1), 581.
70. CLAES, P., BILLIAU, A., DeCLERCQ, E., DESMUYER, J., SCHONNE, E., VARDERHAEGE, H. and DeSOMER, P., *J. Virol.* (1970), **5**, 313.
71. FELZ, E. T. and REGELSON, W., *Nature* (1962), **196** (17 Nov.), 642.
72. MERIGAN, T. C. and REGELSON, W., *N. Engl. J. Med.* (1967), **277**, 1283.
73. REGELSON, W., *Adv. Expl. Med. Biol.* (1967), **1**, 315.
74. DeCLERCQ, E. and MERIGAN, T. C., *J. Gen. Virol.* (1969), **5**, 359.
75. GAZDAR, A. F., *Proc. Soc. Expt. Biol. Med.* (1972), **139**, 1132.
76. CAME, P. E., LIBERMAN, M., PASCALE, A. and SHIMONASKI, G., *Proc. Soc. Expt. Biol. Med.* (1969), **131**, 443.
77. REMINGTON, J. S. and MERIGAN, T. C., *Nature* (1970), **226** (25 Apr.), 361.
78. PINDAK, F. F., *Infect. Immun.* (1970), **1**, 271.

79. GIRON, D. J., SCHMIDT, J. P., BALL, R. J. and PINDAK, F. F., *Antimicrob. Agents Chemother.* (1972), **1**, 80.
80. KOPUSTON, M. A. and MENDELSON, J., *Arthritis Rheum* (1969), **12**, 463.
81. BRAUN, W., REGELSON, W., YAJIMA, Y. and ISHIZUKA, M., *Proc. Soc. Expt. Biol. Med.* (1970), **133**, 171.
82. HIRSCH, M. S., BLACK. P. H., WOOD, M. L. and MONACO, A. P., *Proc. Soc. Expt. Biol. Med.* (1970), **134**, 309.
83. PEARSON, J. W., CHIRIGOS, M. A., CHAPARAS, S. D. and SHER, N. A., *J. Nat. Cancer Inst.* (1974), **52**, 463.
84. BAIRD, L. G. and KAPLAN, A. M., *Cell. Immunol.* (1975), **20**, 167.
85. LEVINE, H. I., MARK, E. H. and FIEL, R. J., *ACS Polymer Preprints* (1978), **19**(2), 570.
86. PUCCETTI, P., SANTONI, A., RICCARDO, C., HOLDEN, H. T. and HERBERMAN, R. B., *Int. J. Cancer* (1979), **24**, 819.
87. BAIRD, L. G. and KAPLAN, A. M., ref. 4, p. 185.
88. BARTOCCI, A., CHIRIGOS, M. and READ, E., *J. Immunopharmacol.* (1980), **2**, 149.
89. PLATE, N. A., ref. 3, p. 63.
90. DUNCAN, C. H., BEST, M. M. and McGAFF, C. J., *J. Lab. Clin. Med.* (1958), **52**, 809.
91. KUO, P. T., HOPKINS, F. J. and WURZEL, H., *Circulation Res.* (1958), **6**, 178.
92. FALB, R. D., *Polym. Sci. Technol.* (1974), **8**, 77.
93. SORM, M., NESPUREK, S., MRKVICKOVA, L., KALAL, J. and VORLOVA, Z., *J. Polym. Sci. Polym. Symp.* (1979), **66**, 349.
94. VAN DER DOES, L., BEUGELING, T., FROEHLING, P. and BANTJES, A., *J. Polym. Sci. Polym. Symp.* (1979), **66**, 337.
95. SHAMASH, Y. and ALEXANDER, B., *Biochem. Biophys. Acta* (1969), **194**, 449.
96. ROBERTS, P. S., REGELSON, W. and KINGSBURY, B., *J. Lab. Clin. Med.* (1973), **82**, 882.
97. SAMOUR, C. M., *Chem. Tech.* (1978)(Aug.), 494.
98. CASALS, J. B., *Br. J. Pharmacol.* (1972), **46**, 281.
99. MIKHAIL, J. W., MANSOUR, N. S. and KHAYYAL, M. T. *Expt. Parasitol.*, (1975), **37**, 348.
100. REGELSON, W., *L'Interferon* (1970), **6**, 353.
101. REGELSON, W. and OTTENBRITE, R. M., 'Biological activity of poly-anions: control of tumour growth and immunoadjuvant actions', *17th Microsymposium on Macromolecules, Aug. 1977, Prague, Czechoslovakia.*
102. REGELSON, W., *J. Med.* (1974), **5**, 50.
103. LEVY, H. B., ref. 3, p. 305.
104. REGELSON, W., *Med. Chem.* (1973), **8**(17), 160.
105. EVANS, J. C. and GABLE, P. V., *Radiology* (1967), **89**, 140.
106. OTTENBRITE, R. M., ref. 4, p. 22.
107. MUNSON, A. E. and REGELSON, W., *Proc. Soc. Biol. Med.* (1971), **137**(2), 553.
108. SHOTTON, E. and HERSEY, J. A., ref. 6, Ch. 10 and 11.

109. NAPPER, D. H., 'The role of polymers in the stabilisation of disperse systems' in *Chemistry and Technology of Water-soluble Polymers*, C. A. Finch (Ed.), 1983, Plenum Press, New York.
110. HAINES, B. A. and MARTIN, A. N., *J. Pharm. Sci.* (1961), **50**, 228, 753, 759.
111. HIESTAND, E. H., *J. Pharm. Sci.* (1964), **53**, 1.
112. ref. 16, p. 950.
113. ROBINSON, J. E. (Ed.), *Sustained and Controlled Release Drug Delivery Systems*, 1978, Marcel Dekker Inc., New York.
114. ZAFFARONI, A., *Chemtech* (1980) (Feb.), 82.
115. LUZZI, L. A., *J. Pharm. Sci.* (1970), **59**, 1367.
116. VANDEQAER, J. E. (Ed.), *Microencapsulation: Processes and Applications*, 1974, Plenum Press, New York.
117. KONDO, A., *Microcapsule Processes and Technology*, 1979, Marcel Dekker, New York.
118. US Patent 2738303 (1956).
119. SAUNDERS, L. and SRIVASTAVA, R., *J. Chem. Soc.* (1950), 2915.
120. HELFFERICH, F., *Ionic Exchange*, 1962, McGraw–Hill, New York, Ch. 2.
121. US Patent 2857311 (1958).
122. Ger. Offen. 1070381 (1959).
123. MATHER, N., NARANG, C. and WILLIAMS, R., *Polymers as Aids in Organic Synthesis*, 1980, Academic Press, New York.
124. HODGE, P. and SHERRINGTON, D., *Polymer Supported Reactions in Organic Synthesis*, 1980, Wiley–Interscience, New York.
125. FLORENCE, A. T., *Pharm. J.* (1974), **213**, 36.
126. Ger. Offen. 1915970.
127. British Patent 1174854 (1969).
128. BAUDIN, G., CASADOP, S., PIETRASANTA, Y. and PUCCI, B., *Chimica e Ind.* (1980), **62**, 421.
129. WEINER, B. Z., TAHAN, M. and ZILKHA, A., *J. Med. Chem.* (1972), **15**, 410.
130. RINGSDORF, H., *J. Polym. Sci. Polym. Symp.* (1975), **51**, 135.
131. FERRUTTI, P., *Il Farmaco Ed. Sci.* (1977), **32**(3), 220.
132. KIM, S., PETERSON, R. and FEIJEN, J., Polymeric drug delivery systems in *Drug Design*, Vol. X, 1980, Academic Press, New York, Ch. 5.
133. LEWIS, D., *Controlled Release of Pesticides and Pharmaceuticals*, 1981, Plenum Press, New York.
134. ZAFFARONI, A. and BONSEN, P., 'Controlled chemotherapy through macromolecules', in ref. 3.
135. EHRLICH, P., 'On the relationship between chemical constitution, distribution and pharmacological action', 1906, in *Collected Studies in Immunity*, Wiley, London, Chap. 34.
136. GOLDBERG, E. P. (Ed.), *Polymers in Biology and Medicine*, Vol. 2, Targeted drugs, 1983, John Wiley and Sons, New York.
137. GREGORIADIS, G. (Ed.), *Drug Carriers in Biology and Medicine*, 1979, Academic Press, New York.
138. ANDRADE, J. D. (Ed.), *Hydrogels for Medical and Related Applications*,

American Chemical Society Symp. Series No. 31, 1976, American Chemical Society, Washington, USA.
139. PEDLEY, D. G., SKELLY, P. J. and TIGHE, B. J., *Br. Polym. J.*, (1980), **1**.
140. *European Patent Application* 089152 (1983).
141. KENT, B. E. and WILSON, A. D., *Br. Dental J.* (1973), **135,** 322.
142. JUNIPER, R. (Ed.), *Pharmaceutical Applications of Polymers—a Selected Bibliography,* 1981, Rubber and Plastics Research Association of Great Britain, Shrewsbury, Shropshire, SY4 4NR, UK.

Chapter 8

ELECTRICAL AND CHEMICAL ASPECTS OF ELECTRODEPOSITION OF PAINT

Fritz Beck

FB6—Elektrochemie, Universität Duisburg, Federal Republic of Germany

1 INTRODUCTION

The technique of electrodeposition of paint (EDP, 'electrophoretic painting') was introduced in the early 1960s. Since that time, its importance for the automatic deposition of the first layer (primer) in industrial painting of metal massware such as car bodies has increased distinctly through the years. The process has now been widely adopted in the corresponding manufacturing lines.[1,2] Anodic systems were dominant in the first period of development; however, in the course of the last five years they have almost totally been replaced by cathodic paints.

It is the objective of this review to provide some general views concerning the electrical and chemical aspects of electrodeposition of paint.

2 ELECTROCHEMISTRY OF IONS

EDP systems contain water as a solvent, synthetic binders and pigments. The macromolecules constituting the binder carry ionogenic groups distributed along their backbone. The most important are carboxylic groups for anodic paints and amine groups for cathodic ones. These groups establish an acid–base equilibrium with bases or acids added to the bath to solubilize the macromolecule; see Fig. 2.

In this way, macro-ions are generated. From an electrochemical point of view, these systems can be regarded as diluted poly-electrolytes. They are also related to surfactants.

Common to both macro-ions is the presence of relatively long sections of polymer backbone carrying no charged groups. This leads to the formation of micelles due to the strong intermolecular interaction of these hydrophobic parts of the macro-ions.[3,4] On changing the pH and generating more charged groups, the micelle diameter decreases markedly due to an increase of electrostatic repulsion. In Fig. 1, the primary structures of macro-anions are compared with the secondary structures, which are formed in aqueous dispersions. Table 1 shows molar mass and densities of ionic groups for various water-based systems. It must be mentioned that strong association of macro-ions with counterions is typical for these systems because of the heavy accumulation of electric charges in a small volume.[5]

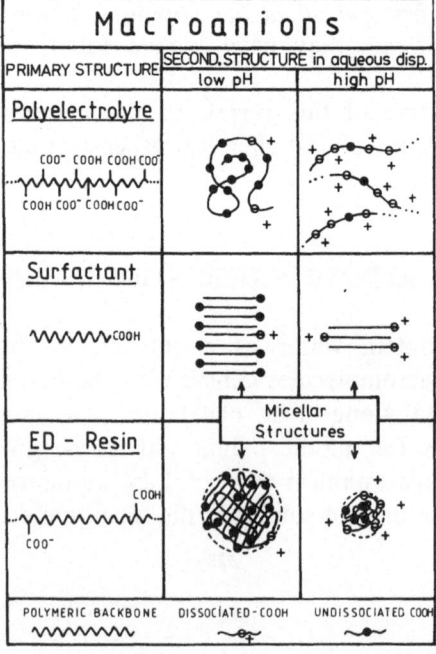

FIG. 1. Primary and secondary structures in acid and base solutions of macro-anion-forming molecules. Aggregation to micelles occurs in the case of structures with extended non-ionogenic sections.

TABLE 1
MOLAR MASS AND DENSITY OF IONIC GROUPS IN WATER-BASED
POLYMERS

System	Molar mass	Density of ionic groups (eq/kg)
Polyelectrolyte	1 000–100 000	10
Surfactant	100–1 000	2–4
Ion-exchange resin	Crosslinked	1
ED resin	1 000–30 000	0·3–2
Latex	10 000–100 000	0·001–0·3

As a consequence of the presence of micelles, the transport of the colloidally dispersed particles in an electric field $\vec{F}$ is due to electrophoresis. The migration velocity v is generally given by:

$$v = b \cdot \vec{F} \tag{1}$$

According to Stoke's law the mobility b, can be expressed by the equation:

$$b = \frac{ze_0}{6\pi\eta r} \tag{1a}$$

where z is the number of charges per particle, e_0 is the elementary charge, η is the specific viscosity and r is the radius of the particle, assumed to be globular. Hence, eqn (1) can be rewritten:

$$v = \frac{ze_0}{6\pi\eta r} \vec{F} \tag{2}$$

With the condenser model, introduced by Smoluchowski, eqn (2) can be rewritten as:

$$v = \frac{\varepsilon\varepsilon_0\zeta}{6\pi\eta} \vec{F} \tag{3}$$

where ζ is zeta potential, ε is electric permittivity and ε_0 is the electric field constant. As z in eqn (2) is large, the diffusion coefficient, given by the Planck–Einstein relation:

$$D = \frac{RT}{zF} b \tag{4}$$

is relatively small. Thus, diffusion fluxes are negligible in comparison with migration fluxes.[5,6]

3 ELECTRODE PROCESSES

The article to be painted is submerged in the gently stirred bath. It is usually made of (passive) iron and acts as one of the two electrodes necessary for the electrochemical cell. The situation resembles electroplating procedures from a superficial point of view. The principal difference is that the main electrode process is the electrolytic decomposition of water, the products entering only indirectly into the deposition process. At the anode, protons and oxygen are generated:

$$H_2O \rightarrow 2H^+ + \tfrac{1}{2}O_2 + 2e^- \tag{5}$$

The products of electrolysis at the cathode are hydroxide ions and hydrogen:

$$H_2O + e^- \rightarrow OH^- + \tfrac{1}{2}H_2 \tag{6}$$

At the anode, unwanted side reactions play some role, namely the anodic oxidation of organics and the anodic dissolution of the metal.[5,6] Neither reaction is possible at the cathode, which is one of the inherent advantages of cathodic electrodeposition. Typical electroorganic reactions[7] such as the anodic formation of Kolbe products,

$$2p{-}COO^- \xrightarrow{\;\;\times\;\;} p{-}p + 2CO_2 + 2e^- \tag{7}$$

or the cathodic cleavage of quaternary ammonium ions,

$$p{-}NR_3^+ + e^- \xrightarrow{\;\;\times\;\;} p{-}NR_2 + R^\cdot \tag{8}$$

sometimes stressed in the discussion of electrode processes, cannot proceed under the conditions of electrophoretic painting.

The production of electrolysis gases in the main reactions usually does not lead to gas bubbles, which would disturb the process. Instead, the gases dissolve in the electrolyte because of the low current densities involved. The formation of supersaturated solutions of electrolysis gases in the boundary layer rich in organic materials greatly facilitates the prevention of gas bubble formation. The very high viscosity in this zone further inhibits the nucleation of gas bubbles.[8]

The ionic electrolysis products accumulate in the diffusion layer and neutralize the ionic groups of the micelles; see Section 4. Thus, Faraday's law is approximately obeyed.[9] The mass m deposited is found to increase linearly with the charge Q passed through the cell:

$$m = m_e Q \tag{9}$$

As the equivalent weight of macro-ions is of the order of 1000 g, the electrochemical equivalent is usually of the order of 10 mg/C.

4 ELECTROCOAGULATION

After solubilization of the resinous material, a pH of 6–8 for anionic systems and 3–7 for cationic systems is established in the bath. Ions generated via water electrolysis accumulate in the diffusion layer, and thus the pH is strongly shifted from the equilibrium values in the bulk of the solution. Correction of free amine B or acid HA compensate for this effect to some extent and calculations lead to the following

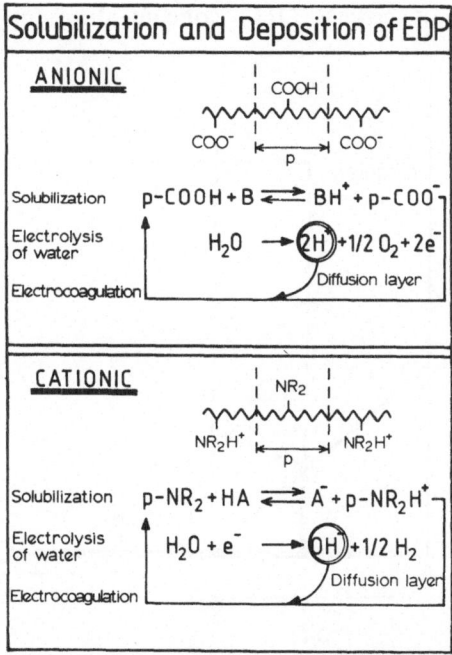

FIG. 2. Acid–base equilibria in ED systems. Interdependence of solubilization, water electrolysis and electrocoagulation for anionic and cationic binders.

figures:[5-7,10]

$$pH_0 = 3\text{--}4 \text{ at the anode}$$

$$pH_0 = 11\text{--}10 \text{ at the cathode}$$

Thus, insoluble material is generated where it is wanted, and an organic layer is deposited. This process is called electrocoagulation. Figure 2 shows the close interdependence of these main process stages in EDP.

Consequently, the diffusion layer model has been applied in the field of electrophoretic painting.[5-7,10] The idea of an acid surface layer causing anodic electrocoagulation was disclosed as early as 1933 by Beal.[11] At that time and up to the 1960s, the direct discharge of the macro-ions was erroneously considered to be the operating mechanism.

A closer examination of the deposition mechanism leads to the concentration profiles shown in Fig. 3. Thus, as the macro-cations approach the cathode, they penetrate into the diffusion layer of OH^- ions; this layer is much more expanded than those of the other species. When the zone of critical OH^- concentration, c_{OH^-}, is reached,

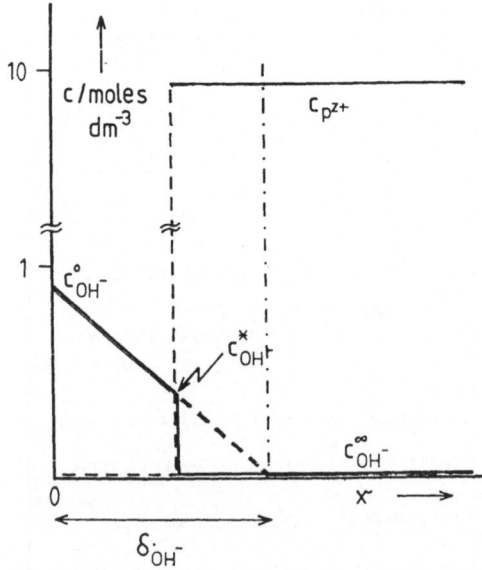

FIG. 3. Concentration profiles for OH^- ions and cationic polymers P^{z+} at the cathode in the region of the OH^- diffusion layer.

organic material precipitates. The schematic representation takes into account, approximately, the real ratio of *molar* concentrations of both reacting species.

This model of critical pH-value is apparently strongly supported by various experiments. Galvanostatic deposition in a non-stirred bath leads to an accumulation of OH^- ions at the cathode. The surface concentration of OH^- ions after time t is given by Sand's equation ($c_{OH^-}^0 \gg c_{OH^-}^\infty$):

$$c_{OH^-}^0 = jt^{1/2} \frac{2}{F(\pi D_{OH^-})^{1/2}} \tag{10}$$

After a transition time τ, when the critical OH^- concentration is reached at the interface, electrocoagulation should occur. If this is true, the product $j\tau$ must be a constant. This has been found to be so for anionic systems.[12] Experiments have been also performed using the rotating disc electrode with anionic[5,6,10] and cationic[13,24] systems. According to the Lewitsch equation, the thickness of the diffusion layer is given by

$$\delta_{OH^-} = 1 \cdot 75 v^{1/6} D^{1/3} \omega^{-1/2} \tag{11}$$

where v is kinematic viscosity and $\omega = 2\pi n$ (n is revolutions per second). From Fick's first diffusion law, the surface concentration of OH^- ions is expressed as:[5,6]

$$c_{OH^-}^0 = \frac{j\delta_{OH^-}}{2FD} \tag{12}$$

From this equation it follows immediately that at a given critical OH^- concentration, there must exist a critical current density j^* as well as a critical diffusion layer thickness δ^*. If the actual values drop below these critical limits, electrodeposition ceases to occur. From the last two equations, n^* should be proportional to j^2. This has been found over a wide range of conditions.[5,6,10,13]

From colloid science it is known that coagulation of a dispersed system can be promoted by the addition of electrolytes. Monovalent ions are the least effective ones (Schulze–Hardy rule). These ions 'neutralize' the charge in the electrochemical double layer which is a precondition for colloid stability. A critical concentration can be defined; for H^+ and OH^- ions it is appreciably lower than that for other ions like Na^+ or Cl^- due to the additional acid–base interaction.

The ions needed for coagulation are generated via electrolysis, as shown above. As already mentioned, the term 'electrocoagulation' has been applied to this process.

However, nucleation phenomena play an important role. This can be seen clearly in the galvanostatic experiments. The electrodeposition starts only after an induction period τ.[5,10] Generally, τ is much greater than the time constant

$$\tau = \frac{\delta^2}{4D} \tag{13}$$

required to build up a steady-state diffusion layer. When $\delta = 10^{-2}$ cm and $D = 10^{-4}$ cm^2 s^{-1}, then τ is of the order of a second. The induction period can reach minutes or even hours, if n approaches n^*. While the induction period τ, as defined by eqn (13), decreases rapidly with decreasing δ, quite the opposite is observed experimentally. These nucleation phenomona are not well understood at present.

ED resins with quaternary ammonium groups do not fit in the model of critical OH$^-$ ion concentration. Experiments with the rotating disc electrode show this clearly.[13] In this case, it is the strong accumulation of macro-ions at the cathode alone which leads to coagulation due to an increase of intermolecular forces. The viscosity in the neighbourhood of the electrode increases dramatically, and if a steady-state concentration profile of OH$^-$ ions is established, it is according to Fig. 4.

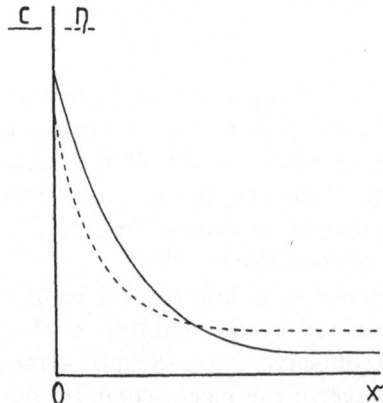

FIG. 4. Dependence of the concentration of polymers with quaternary ammonium groups (c) and of specific viscosity (η) of the bath (- - -) at the surface of a cathodically polarized substrate.

5   THICKNESS GROWTH: ELECTRIC ASPECTS

Deposition at constant current density j leads to a constant rate of thickness growth, if Faraday's law (eqn (9)) is obeyed:

$$\frac{dl}{dt} = \frac{m_e}{s} j \qquad (14)$$

where l = thickness, s = density of paint layer. The voltage–time relationship deviates from the linearity predicted by eqn (14) due to non-Ohmic behaviour of the film.

Under practical conditions, deposition with constant voltage is preferred. If Ohm's law is valid, the current density is correlated with the thickness l by the simple expression:

$$j = \frac{U \kappa_F}{l} \qquad (15)$$

where κ_F is the specific conductivity of the film. From eqns (14) and (15), a square-root law for thickness growth can be derived:

$$l = \sqrt{\left(\frac{2m_e}{s} U \kappa_F t \right)} \qquad (16)$$

This leads through eqns (15) and (16) to a current–time relationship $j \sim t^{-1/2}$. Once again, non-Ohmic behaviour of the film must be taken into consideration. A model based on an exponential growth of current with voltage[14] has been used by Pierce[4] to yield a film growth differential equation, which could only be solved by numerical integration.

Macroscopic potential distribution in the cell is represented schematically in Fig. 5. After the deposition of a film, most of the cell voltage (U_{cell}) drop occurs in the film (U_F) and the other components of U_{cell} are small in comparison:

$$U_{cell} = U_F + U_B + U_D + \varepsilon_a + \varepsilon_c \qquad (17)$$
$$U_{cell} \approx U_F$$

(U_B is the voltage drop in the bath; U_D is the voltage drop in the diffusion layers; ε_a and ε_c are potential drops at the electrodes). The situation is quite the opposite to that in electrostatic spraying of paint, where most of the voltage drop occurs in the air gap between the electrodes.

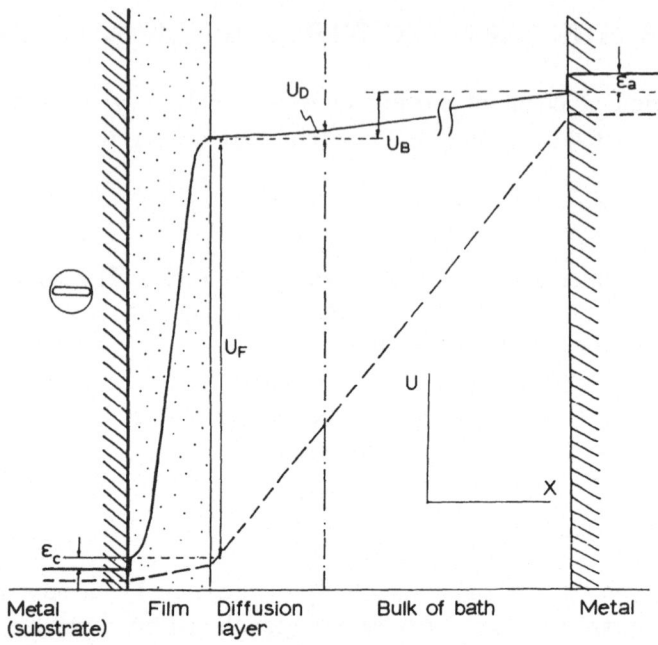

FIG. 5. Potential profiles for electric painting techniques: ————, cathodic electrodeposition of paint; - - - - -, electrostatic spray painting. Symbols are defined in the text.

One very important consequence is that throwpower must be superior in the EDP process. In practical situations, the electrical charge is able to penetrate into those regions such as slots in a substrate and crevices which are less accessible due to the shielding effect of those parts which have already been painted in the initial stages of the deposition process. A general theory of throwpower has been published by Furuno,[4,15] improving preceding treatments for special cases.[5,16] From this theory, the depth of penetration (L) is given by

$$L = \left(\frac{2A\kappa_B U}{PI_0} \right)^{1/2} \tag{18}$$

where A is the cross-section and P is the perimeter of the slot or pipe, κ_B is the specific conductivity of the bath and I_0 is the residual current of the film. It is clearly shown that throwpower is optimized by a high κ_B and a low I_0, implying a low κ_F.

As mentioned earlier, the film characteristics deviate appreciably from Ohm's law at higher voltages. The potential distribution becomes non-linear, with the current–voltage curve,[5,17,18] attaining an exponential character rather than a linear relationship. This is due to the low concentration of ions in the film (the dissociation constant of R—COOH in the film is as low as 10^{-12} mol dm^{-3}) in the presence of a high field,[17]

$$\vec{F} = \frac{U_F}{d} \approx 10^5 \text{ V cm}^{-1}$$

According to Poisson's law, a charge separation occurs:

$$-\frac{d^2U}{dx^2} = \frac{d\vec{F}}{dx} = \frac{1}{\varepsilon\varepsilon_0}\rho \qquad (19)$$

Assuming a homogeneous space charge density, $\rho = c_0 F$ where $c_0 =$ concentration of free charge carriers, integration of Poisson's law

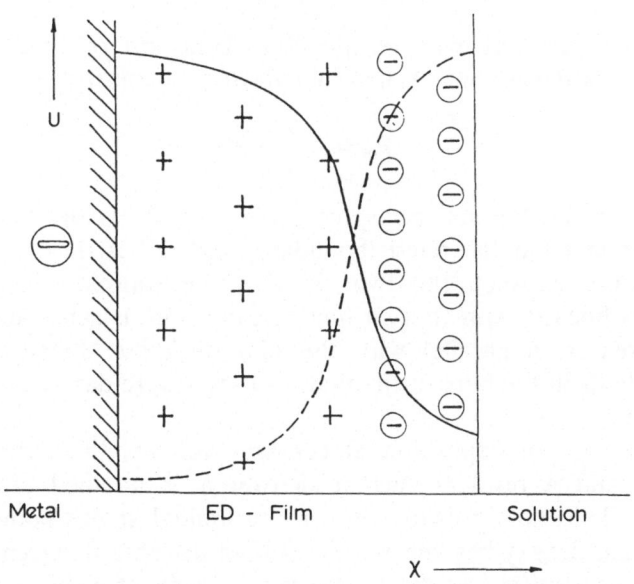

FIG. 6. Space charges of fixed ions (+) and OH$^-$ gegen-ions (−) in a cathodically deposited ED film: ————, potential due to internal field; - - - - - -, potential due to applied voltage.

yields

$$\frac{U_F}{d} = \frac{c_0 F}{2\varepsilon\varepsilon_0} d \qquad (20)$$

The field strength increases with depth of penetration into the space charge. The potential lines are strongly curved, as shown in Fig. 6 for a cathodic film. In order to drive a current through the system, the applied field has to overcome the internal field established by the space charges.

6 LIMITING THICKNESS OF THE LAYER

If the electrodeposition is performed under galvanostatic conditions, it should be possible to form very thick films in principle. However, a practical limit is the increasing generation of thermal energy with increasing film thickness. The electric power, P_e, consumed for this is given by:

$$P_e = A j^2 \rho d \qquad (21)$$

where ρ is the resistivity of the film. Transport of Joule's heat is achieved via the surface of the film, which is of constant area:

$$P_{th} = A \frac{\lambda}{\delta_T} \Delta T \qquad (22)$$

where λ is the thermal transport coefficient, δ_T is the thickness of transport layer at the phase boundary, and ΔT is the difference of temperature between the two sides of the transport layer. As d increases linearly with time, a quasi-steady state is achievable only if ΔT increases in parallel with the film thickness. Finally, residual solvent boils in the bulk of the film, causing severe damage of the film structure.

In the case of deposition at constant voltage, the current drops rapidly, and a residual current density j_0 is attained after a few minutes. Thickness growth seems to be limited at this point because the current density has become lower than the critical current density, j^*, for deposition. From an electrical point of view, there is a considerable drop in potential across the organic layer due to the space charges present. From eqn (20), a relationship $U \sim d_0^2$ follows for this situation.

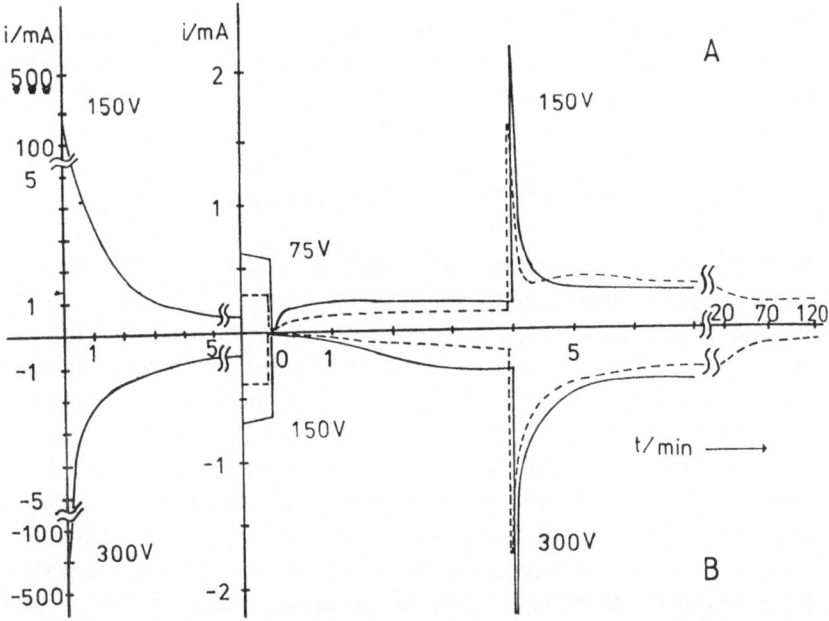

FIG. 7. Current–time curves for the electrodeposition of paint at constant voltage. The substrate was a rotating iron cone electrode, rotational speed $200\,min^{-1}$, $A = 11{\cdot}6\,cm^2$. In the region of residual current, the voltage applied originally was halved and later doubled again. A, Anionic ED resin (acrylate type); B, cationic ED resin; no pigmentation of the bath; - - - represents a second experiment performed subsequently.

On the other hand if we assume a simple Ohmic behaviour, the current density would be given by eqn (15), and the final thickness d_0 would be proportional to U. The data published until now are not suitable to distinguish between these two extremes. The low current density at the point where it has dropped to the residue value of j_0 seems to support the Ohmic picture. Experiments in which abrupt voltage changes have been initiated at a polarized film have been described by Cooke[19] and seem to indicate non-Ohmic behaviour in the non-steady state. We have repeated these experiments with two ED resins, and we find analogous effects (see Fig. 7).[20] Doubling the voltage leads to a strong increase of current, and it is only in the steady state, which is attained after some time, that the twofold current is established. Halving the applied voltage must be accompanied by partial redissolution of the film, whereas doubling causes additional deposition.

It can be seen from Fig. 7 that the residual current is nearly constant for a long period. The transport processes in the film occurring in that state have been discussed elsewhere.[5,17]

7 REDISSOLUTION KINETICS

The unpolarized film redissolves slowly in the electrolyte. For an anionic deposit, free amine molecules, which are present in an acid–base equilibrium in the bath (see Fig. 2), are responsible for this process. We have measured the rate of dissolution by following the rate of change of film resistance, R_0, with an alternating current bridge: R_0 versus t curves are reproduced in ref. 5. The steep portion in the middle region has been evaluated. The measurements have been performed at a rotating disc electrode to verify defined hydrodynamic conditions.[21] The rate of dissolution was found to be proportional to the square root of rotation speed (see Fig. 8). This indicates clearly a diffusion-controlled process. The redissolution seems to be proportional to the rate of diffusion of reacting amine molecules.

The question arises, whether a superimposed dissolution current density j_d is operative in the case of a polarized film. Pierce et al. have

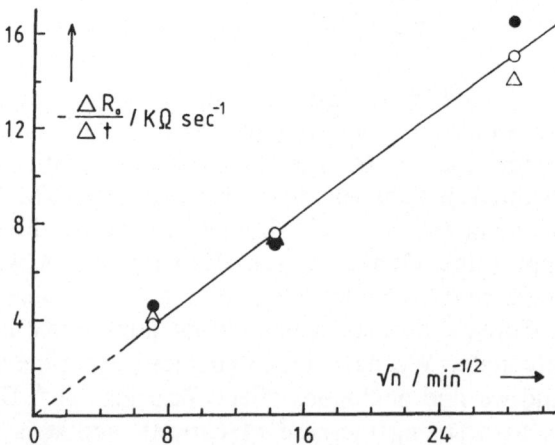

FIG. 8. Redissolution kinetics of an acrylate type anionic film at a rotating disc electrode. Rate of change of Ohmic resistance $\Delta R_0/\Delta t$ (as a measure of dissolution rate) vs square root of rotation speed.[21] The current densities for deposition are (mA cm^{-2}): $\bigcirc$, 4; $\bullet$, 3; $\triangle$, 2. R_0 at $t = 0$ was about 100 k-ohm.

made this assumption.[4] The end of thickness growth, once a residual current density j_0 is reached in a constant voltage deposition, can be interpreted in terms of kinetic equilibrium

$$j_0 = j_d \tag{23}$$

The rate of thickness growth at constant current density, described by eqn (14), has to be corrected by the term j_d:

$$\frac{dl}{dt} = \frac{m_e}{s}(j - j_d) \tag{24}$$

If we assume, that j_d increases with increasing convection rate of the bath, the current efficiency would decrease simultaneously. Corresponding experiments, preferably at the rotating disc electrode, have not yet been done. The existence of a limiting rotation speed n^* would then need to be reconsidered as the condition where $j = j_d$, rather than the approach to the critical pH-value.

8 POTENTIODYNAMIC ELECTRODEPOSITION OF PAINT

Voltammetric methods, normally applied widely in electrochemistry, have been almost totally neglected up to the present in EDP. As the depositing and growing film allows no steady state, (cyclic) potentiodynamic measurements must be used, where the attainment of a reproducible quasi-steady state seems feasible. The standard theory of voltage sweep methods involves the voltage scan rate, v_s:

$$v_s = \frac{dU}{dt}$$

as the most prominent parameter.[22-24] The complicated theory is based on a diffusion- and/or reaction-controlled Faraday conversion of the depolarizer. The possibility of application of this theory to the present problem is doubtful.

The only application of this method to EDP has been demonstrated by Bonora et al.[25-27] However, only a narrow variation of parameters have been investigated in this work, limited to aluminium as an electrode material, and only few attempts at quantitative treatment have been offered.

We have undertaken recently a more systematic investigation in this

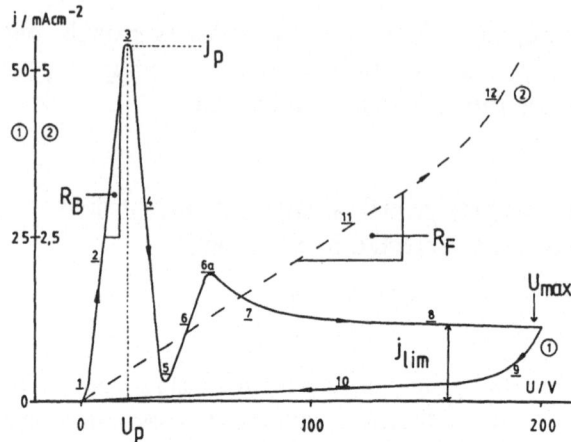

FIG. 9. Schematic representation of cyclic current voltage characteristic for potentiodynamic EDP. Numbered sections 1–12 relate to the chronological sequence. ①, ② are cycle numbers.

field.[28,29] Potentiodynamic electropainting at a rotating iron disc electrode has been investigated with three different EDP resins—two anodic of the acrylate type and one cathodic of the epoxide type—and a wide variation of conditions. Voltage scan rate ($v_s = 1$–$200\,V\,s^{-1}$) voltage range (40–200 V) and electrode rotation speed ($n = 60$ and 1000 rpm) were the most important parameters.

The (cyclic) voltammetric curves obtained generally exhibit three characteristic features (see Fig. 9):

(i) The current rises steeply at the start of the experiment. Bath resistance transforms the potentiodynamic curve simultaneously into a galvanodynamic curve. After a transition time τ, critical pH is attained at the phase boundary and electrocoagulation occurs. This leads to a rapidly decreasing current density. The sharp CD maximum thus established has a peak voltage, U_p, which increases with v_s according to the relation $\log U_p \sim \frac{1}{3} \log v_s$ in accordance with theory.

(ii) At high voltages, a limiting current density is observed, increasing with the square root of v_s. This could be quantitatively interpreted in terms of dynamic growth of film thickness governed by Ohmic ion transport in the film. The preceding part of the U versus j curve declines with $j \sim t^{-1/2}$, which indicates the prevalence of space charge effects.

(iii) Ohmic lines are measured in the course of the first reverse scan and in all quasi-steady state follow-up cycles. They are flatter by a factor of 1000 in regard to the initial Ohmic line and reflect low-voltage Ohmic behaviour of the EDP film. At high voltages positive current deviations occur due to Child's law.

The curves can be measured easily and reproducibly. Due to their salient features it is proposed to use them for characterization of ED paints.

9 CHEMICAL ASPECTS: ANIONIC AND CATIONIC SYSTEMS

A comprehensive survey of the preparative aspects of ED resins has been published by Schenck et al.[30] Two synthetic principles are used to introduce the ionogenic groups into the polymer:

(I) polymerization of monomers, which already carry these groups;
(II) reaction of a non-ionic polymer at reactive centres with reactants, which incorporate these ionogenic groups in the polymer backbone.

The ionogenic groups are nearly exclusively COOH groups in the case of anionic systems. Using principle I, acrylic acid copolymerization is a very well known route to obtain the required polymers. The comonomers may be acrylic esters:

$$n\,CH_2{=}CH + nm\,CH_2{=}CH \longrightarrow \left[-CH_2-CH{-}(CH_2-CH)_{\overline{m}} \right]$$
$$\underset{COOH}{|} \qquad \underset{COOR}{|} \qquad\qquad \underset{COOH}{|} \qquad \underset{COOR}{|} \Big]_n$$

$$(25)$$

Principle II is generally represented by 'maleinization' of polymers (or oils) containing C=C double bonds (see ref. 31):

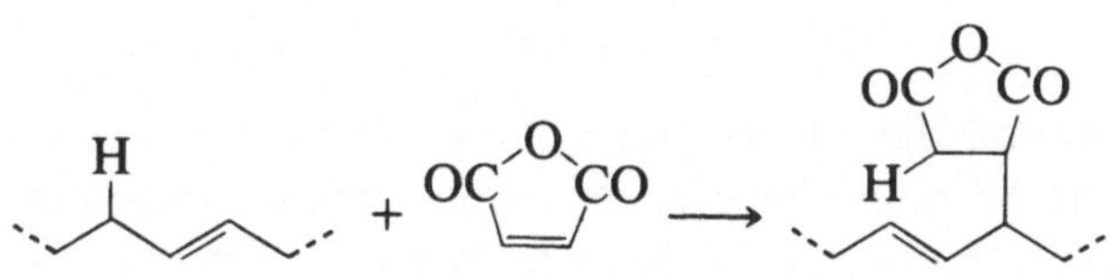

$$(26)$$

This is an ene reaction, in which maleic anhydride is added to an olefinic substrate possessing an allylic hydrogen atom. An allyl shift of the substrate double bond occurs. In this way, maleinized alkyds or epoxy esters have been synthesized. However, ester linkages along the polymer chain may lead to problems due to sensitivity to saponification.

Crosslinking reactions in the curing stage are readily performed because the film is acidic, and these reactions are generally acid-catalyzed.

The same preparative principles can be applied to the synthesis of cationic systems. The ionogenic groups are mainly amino groups. Quaternary ammonium groups have also been used. According to principle I, vinyl copolymerization of acrylic esters with dimethylaminoethyl methacrylate leads to an unsaponifiable backbone with tertiary amino groups (see ref. 32):

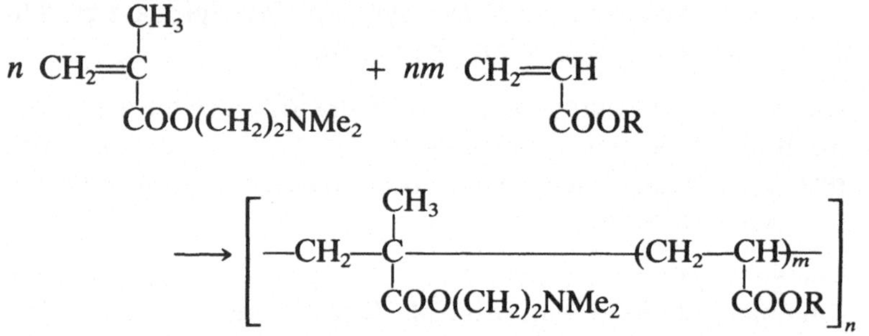

$$n \ CH_2{=}C \begin{matrix} CH_3 \\ | \\ | \\ COO(CH_2)_2NMe_2 \end{matrix} + nm \ CH_2{=}CH \begin{matrix} \\ | \\ COOR \end{matrix}$$

$$\longrightarrow \left[-CH_2-\overset{\overset{\displaystyle CH_3}{|}}{\underset{\underset{\displaystyle COO(CH_2)_2NMe_2}{|}}{C}}{\rule{2cm}{0pt}} (CH_2-CH)_{\overline{m}} \underset{COOR}{} \right]_n \quad (27)$$

Epoxy based resins are very important in cationic systems. Typically, an epichlorohydrin–bisphenol A epoxide is reacted with a tertiary amine to form quaternary ammonium salt groups according to principle II:[33]

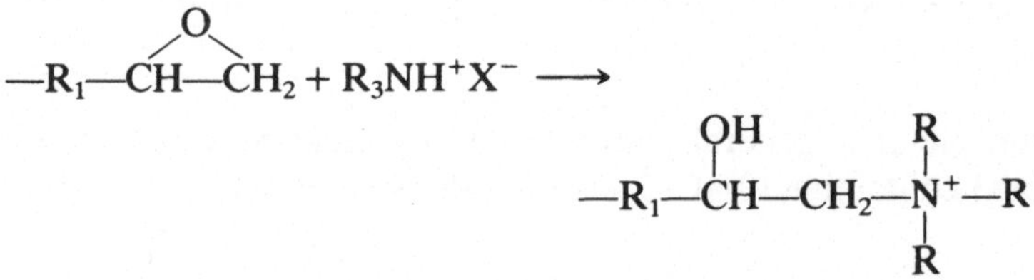

$$-R_1-\overset{\displaystyle O}{\overset{\diagup\diagdown}{CH-CH_2}} + R_3NH^+X^- \longrightarrow$$

$$\begin{matrix} OH & R \\ | & | \\ -R_1-CH-CH_2-N^+-R \\ | \\ R \end{matrix} \quad (28)$$

The cure of the highly basic films of cationic material is more problematic than in the case of acidic anionic materials. Chemical means have been developed to overcome this problem.

TABLE 2
CORROSION RESISTANCE (1) AND SUBSTRATE DISSOLUTION (2) FOR CATIONIC
AND ANIONIC AUTOMOTIVE PRIMERS

	Substrate			
	Steel		Zn-phosphated steel	
	Cationic	Anionic	Cationic	Anionic
(1) Salt spray corrosion,[a] $\frac{1}{64}$-inch units	4	12	1	2
(2) Substrate dissolution,[34] ppm Fe	55	2167	30	249
ppm Zn	—	—	83	13 300

[a] Measured as scribe creep after 336 h salt spray exposure.

Two important properties, corrosion resistance and metal compatibility, seem to be superior for cationic systems. This is shown in Table 2.[4,34]

Even with unphosphated steel, relatively good corrosion resistance can be achieved with cationic binders. The main reason for this is the corrosion-inhibiting action of amino groups together with aromatic groups present in the cationic systems. These distinct advantages outweigh some drawbacks such as the counter-electrode problem, doubled gas evolution at the substrate and slightly lower film thicknesses.[4,32]

10 CONCLUDING REMARKS

Electrodeposition of paint is a successful combination of paint application and electrochemical techniques. All the features for a fully automatic process are available. Whilst investment costs are relatively high due to unconventional equipment, manufacturing costs are only 50% of those of conventional spray painting.[1]

Reference to the technology of this process has only been mentioned briefly in this paper. However, appropriate design is of great importance. This holds also in achieving co-ordinated processes such as the continuous make-up of the bath via ultrafiltration, the preceding phosphating process and for the follow-up curing steps.

On theoretical aspects, much has been elucidated, but some problems still remain to be solved. Non-Ohmic behaviour and transport processes in the film are complex and require further investigation. Pigmented systems must also be included in basic studies. Colloid science must be regarded as the foundation not only for a proper characterization of the bath, but also to provide a better understanding of the deposition process.

ACKNOWLEDGEMENTS

Generous provision of EDP materials by BASF Farben & Fasern, Hiltrup (Dr Heilmann, Dr Streitberger), and by BASF Ludwigshafen, (Dr Sabelus, Dr Spoor) is gratefully acknowledged.

REFERENCES

1. BREWER, G. E. F., *American Electroplaters Society 68th Annual Technical Conference*, 1981, Paper J2.
2. BREWER, G. E. F., *J. Appl. Electrochem.*, 1983, **13**, 269.
3. PIERCE, P. E. and COWAN, C. E., *J. Paint Technology*, 1972, **44**, 61.
4. PIERCE, P. E., *J. Coat. Techn.*, 1981, **53**, 52.
5. BECK, F., *Progr. in Org. Coatings*, 1976, **4**, 1.
6. BECK, F., *Farbe und Lack*, 1966, **72**, 218.
7. BECK, F., *Electroorganische Chemie*, 1974, Verlag Chemie, Weinheim.
8. BECK, F., *Oberfläche Surface*, 1973, **14**, 79.
9. SAATWEBER, D. and VOLLMERT, B., *Angew. Makromol. Chem.*, 1969, **9**, 61.
10. BECK, F., *Chem.-Ing.-Techn.*, 1968, **40**, 575.
11. BEAL, C. L., *Ind. Eng. Chem.*, 1933, **25**, 609.
12. CATONNE, J. C., *J. Royon, Proc. Interfinish Basel, 1972*, 1973, Forster Verlag, Zürich, p. 258.
13. SPOOR, H. and SCHENCK, H.-U., *Farbe und Lack*, 1982, **88**, 94.
14. KOVAC-KALKO, Z. in *Electrodeposition of Coatings*, G. E. F. Brewer (Ed.), 1973, American Chemical Society, Washington, p. 149.
15. FURUNO, N. and OHYABU, Y., *Progr. Org. Coatings*, 1977, **5**, 201.
16. OLSEN, D. A., BOARDMAN, P. J. and PRAGER, ST., *J. Electrochem. Soc.*, 1967, **114**, 445.
17. BECK, F., *Ber. Bunsenges. Phys. Chem.*, 1968, **72**, 445.
18. COOKE, B. A., *Paint Technol.*, 1970, **34**, 12.
19. COOKE, B. A., NESS, N. M. and PALLUEL, A. L. L. in *Industrial Electrochemical Processes*, A. T. Kuhn (Ed.), 1971, Elsevier, Amsterdam, pp. 441–4.

20. GUDER, H. and BECK, F., unpublished work, 1983.
21. ELFF, K. and BECK, F., unpublished work, 1967 (BASF Aktiengesellschaft, Ludwigshafen).
22. NICHOLSON, R. S. and SHAIN, I., *Anal. Chem.*, 1964, **36,** 706; 1965, **37,** 178, 190.
23. BROWN, E. R. and LARGE, R. F. in *Physical Methods of Chemistry,* Part II A, A. Weissberger and B. W. Rossiter (Eds), 1971, Wiley Interscience, New York, Ch. 6.
24. BARD, A. J. and FAULKNER, L. R., *Electrochemical Methods,* 1980, J. Wiley & Sons, New York, Ch. 6.
25. BONORA, P. L. and CALVILLO, R., *Proc. Interfinish 1976.*
26. TROMBETTI, G., BIANCHINI, G. and BONORA, P. L., *Pitture e Vernici,* 1977, (2), 3.
27. BONORA, P. L., CALVILLO, R., TROMBETTI, G. and BIANCHINI, G., *Proc. Fatipec Congress,* 1978, 171.
28. BECK, F. and GUDER, H., *J. Appl. Electrochem.*, 1985, **15,** 825.
29. GUDER, H. and BECK, F., *Farbe und Lack,* 1985, **91,** 388.
30. SCHENCK, H. U., SPOOR, H. and MARX, M., *Progress Org. Coatings,* 1979, **7,** 1.
31. FORNEY, LE ROY S. and SHERRIN, T. J. in ref. 14, p. 88.
32. SCHENCK, H. U. and STOELTING, J., *J. Oil Col. Chem. Assoc.,* 1980, **63,** 482.
33. WISMER, M., PIERCE, P. E., BOSSO, J. F., CHRISTENSON, R. M., JERABEK, R. D. and ZWACK, R. R., *J. Coat. Technol.,* 1982, **54,** 35.
34. ANDERSON, D. G., MURPHY, E. J. and TUCCI, J., *J. Coat. Techn.,* 1978, **50,** 38.

INDEX